Environmental Sex Differentiation in Fish

Cover illustration

In Genetic Sex Differentiation (GSD) pathway of fishes, sex is determined before hatching by gene(s) harbored on sex chromosomes. In GSD + A, autosomal genes may alter GSD after hatching in some individuals. In Environmental Sex Differentiation (ESD) pathway, genes harbored on autosomes determine sex after hatching and genetic cascade of differentiation is regulated mostly by environmental factors.

Series on
Sex Differentiation in Fish

Environmental Sex Differentiation in Fish

T.J. Pandian
Visiting Professor
CAS, Marine Biology
Annamalai University
Tamil Nadu
India

CRC Press
Taylor & Francis Group
Boca Raton London New York

CRC Press is an imprint of the
Taylor & Francis Group, an **informa** business
A SCIENCE PUBLISHERS BOOK

CRC Press
Taylor & Francis Group
6000 Broken Sound Parkway NW, Suite 300
Boca Raton, FL 33487-2742

First issued in paperback 2019

ISBN-13: 978-1-4822-5437-2 (hbk)
ISBN-13: 978-0-367-37786-1 (pbk)

Library of Congress Cataloging-in-Publication Data

Pandian, T. J.
Environmental sex differentiation in fish / T.J. Pandian.
pages cm
"A CRC title."
Includes bibliographical references and indexes
ISBN 978-1-4822-5437-2 (hardcover : alk. paper) 1. Fishes--Reproduction. 2. Fishes--Effect of human beings on. 3. Fishes--Climatic factors. 4. Fishes--Sex ratio. I. Title.

QL639.2.P35 2015
597.156--dc23 2014035506

Visit the Taylor & Francis Web site at
http://www.taylorandfrancis.com

and the CRC Press Web site at
http://www.crcpress.com

Preface to the Series

Fishes constitute more than 50% of the living vertebrate species. The level of sexual plasticity and flexibility displayed by them has no parallel among other vertebrates. With most divergent expression of sex including many morphotypes within a sex, sexuality in fishes ranges from gonochorism to unisexualism and to an array of unbelievable patterns of functional hermaphroditism; their mode of reproduction also ranges from sexual to gynogenesis and hybridogenesis, each of these modes involves external fertilization and oviparity or internal fertilization and viviparity, which itself ranges from zygoparity to embryoparity. In fact, no other vertebrate group offers as much diversity as fishes to study the processes of sex determination and sex differentiation.

This book series is a continuum of my books on 'Sexuality in Fishes' and 'Sex Determination in Fish'. Information available on Sex Differentiation in Fish is rich and diverse, academically and economically so important that the original plan of bringing out a single volume has to be changed to three parts: 1. Genetic Differentiation, 2. Endocrine Differentiation and 3. Environment Induced Differentiation. An overview of a large number of publications concerning 'Sex Differentiation in Fish' suggests a transition from an experimental to analytical approach during the last decade. With the advent of molecular biology and arrival of more sophisticated tools and techniques, publications in this area are covered by increasing numbers and kinds of journals. Consequently there is a wide gap between molecular biologists including molecular endocrinologists publishing in high profile journals and fisheries scientists, who continue to publish in conventional fishery biology journals. A major objective of this book series is to bridge the widening gap between them. These days single author book series have become a rarity but an attempt has been made to comprehensively synthesize the available and relevant publications in this area.

October, 2011 T.J. Pandian
Madurai-625014

Preface

This book is the third in the series on Sex Differentiation in Fish and the last of the Treatise on Sexuality, Sex Determination and Sex Differentiation in Fish. The objective theme of the book is to comprehensively elucidate the negative effects of anthropogenic activities on skewing sex ratio of fishes by disrupting sex differentiation process and impaired reproductive output. Fisheries constitute one of the most valuable natural renewable resources for humanity and any reduction in their ability to reproduce as a result of human activity may have far-reaching socio-economic consequences. There are books on Toxicology of Fishes and Climate Change. But they are limited to their respective titles. In fishes, synthetic estrogens and their mimics feminize genetic males, whereas elevated temperature masculinizes genetic females. Human activity ranges very widely and its negative effects are so diverse on sex differentiation and reproductive output that it has been difficult to comprehend them into a single book. This book represents perhaps the first attempt to comprehensively address issues of negative effects of human activities on sex differentiation and reproductive output of fishes.

Sex differentiation is realized by a host of genes in the differentiation cascade. Unlike in mammals, the expression of these genes in fishes is switched on or off by environmental factors. Not surprisingly, environmental sex differentiation is ubiquitous among fishes. Capture and aquaculture are important food-production sectors. Overexploitation has progressively decreased reproductive output through reductions in fecundity and/or egg size by advancing age/size at sexual maturity. However, it does not disrupt sex differentiation process. Unequal access to food rather than crowding in aqua-farms disrupts differentiation and skews sex ratio in favor of males.

Of over 60% of the 100,000 man-made chemicals that are in routine use worldwide, some mimic endogenous hormones. For example, the widely used endosulfan in developing countries and pills in developed countries disrupt endogenous hormones and feminizes the fishes. For the first time, this book views endocrine disruption from the point of labile period covering egg, embryo, alevin, larval and juvenile stages and non-labile differentiated adult stage. It shows that sex can be reversed during labile period by endocrine disrupting chemicals (EDCs) but not in adults. However, the EDCs induce reproductive impairment in adults.

With the consumption of enormous quantities of fossil fuels by industries and transportation systems, carbon dioxide level has risen from 250 ppm in 1750 to 380 ppm in 2010 and is projected to increase to 560 ppm by 2050. As a result, the global climate is gradually warming aquatic habitats, which serve as a sink for CO_2; consequent to acidification, the pH of oceans is slowly but progressively decreasing. Briefly, the climate change has elevated temperature, reduced pH and depleted oxygen levels in our waters. Describing the occurrence and distribution of hypoxic dead zones, the book shows that synthetic sites of serotonin and aromatase in the brain-pituitary-gonadal axis are disrupted by hypoxia and thereby masculinize the genetic females. The major pathway through which high temperature disrupts sex differentiation is by the development of abnormal germ cells → reduction in germ cells → repression of aromatase expression → elimination of oocytes by apoptosis. The overriding paternal genetic factors on the effects of elevated temperatures and the presence of temperature-insensitive individuals, families, strains and species of fishes are brought to light. These negative effects emphasize a need for conservation of strains and species of fishes. The book summarizes and evaluates the available *ex-situ* techniques for conservation of fish genome.

The book is a comprehensive synthesis of over 1,024 relevant publications collected from widely scattered 268 journals and other 84 literature sources. Fishes have retained sexual plasticity as long as possible to ensure wider genetic diversity and greater reproductive success. The possible existence of temporally separated two labile periods in many primary gonochorics is indicated for the first time. The analysis of the cited publications in this book has discovered new from old, suggests directions for future research and has led to propose new conclusions.

Wanting to accomplish this treatise on Sexuality, Sex Determination, Genetic Sex Differentiation, Endocrine Sex Differentiation and Environmental Sex Differentiation in Fish was aching me for the last five years. With this volume, they are delivered to the users, thanks to Science Publishers and CRC Press

April, 2014 T.J. Pandian
Madurai-625014

Acknowledgements

It is with great pleasure that I wish to thank Drs. C.A. Strussmann and E. Vivekanandan, who have critically read the manuscript of this book and offered many useful suggestions. I gratefully appreciate Dr. R. Jayabaskaran and R. Koteeswaran, who have very patiently searched relevant publications and brought them to me. I wish to thank Drs. A. Gopalakrishnan, W.S. Lakra and A.G. Ponniah, Directors of Fisheries Institutions of the Indian Council of Agricultural Research for providing the best library for fisheries science and excellent service to visitors. I gratefully appreciate my students, Drs. A. Balasubramani, R. Bhaskaran, C.J. David, T. George (S. Africa), M.A. Haniffa, S. Kavumpuruth, S. Kirankumar, R. Koteeswaran, L.A. Marian (Australia), P. Murugavel, S. Palanichamy, S.G. Sheela (USA), S. Varadaraj (USA) and E. Vivekanandan, whose publications from my laboratory have indeed helped me to launch this book. I extend sincere thanks to Prof. T. Balasubramanian for encouragement. I wish to record my sincere thanks to the Department of Science and Technology, New Delhi, and Annamalai University for support. The manuscript of this book was prepared by Ms. C.K. Chitra Prabha, M. Sc., M. Phil. and I wish to thank her for patience, cooperation and excellent work.

To reproduce figures and tables from published domain, I need to thank many. Firstly, I wish to thank my guru Prof. O. Kinne for providing me copies of his journals. I gratefully appreciate the open door policy and record my sincere thanks to the American Society of Ichthyologists and Herpetologists (Copeia), Current Science Association (Current Science), The Fishery Society of British Isle/John Wiley (Journal of Fish Biology), Food and Agricultural Organization, Rome, and The Ichthyological Society of Fisheries Science (Ichthyological Research). I am equally thankful to the editors/publishers: CRC Press, Fisheries Bulletin, Oxford Journals (ICES Journal of Marine Science) and Science Publishers. I welcome and gratefully appreciate the open door policy of Central Marine Research Institute, PLoS ONE and Fish and Fisheries, and wish to record my grateful appreciation to Drs. A. Gopalakrishanan, P.L. Munday, F. Piferrer and C.A. Strussmann. I also record my sincere thanks to Dr. E. Vivekanandan and R. Koteeswaran for consenting to include their unpublished data. For advancing our knowledge in this area by their rich contributions, I thank all my fellow scientists, whose publications are cited in this book.

T.J. Pandian

Contents

1

Introduction

The objective theme of this book is to comprehensively elucidate the negative effects of anthropogenic activity on skewing of sex ratio in fishes by disruption of sex differentiation process and impaired reproductive output through precocious sexual maturity and consequent decline in recruitment. More than 98% fishes are gonochores (Pandian, 2011, p. 18), and are divided into (i) primary and (ii) secondary gonochores. The remaining 2% fishes are hermaphrodites, and in them, sexuality ranges from simultaneous to sequential and serial hermaphroditism (see later). Sex is a luxury and a system, where two individuals are required for reproduction, frequently involving sexually selected traits and complicated courtship behavior leading to mating (Schultheis et al., 2009). Hence it costs time and energy but ensures recombination to generate genetic diversity (see Pandian, 2012, p. 17). The reproductive strategy of fishes seems to sustain and increase reproductive success and/or genetic diversity rather than to maintain the Mendelian sex ratio of one female to one male.

1.1 Genetic basis

In mammals sex is decisively determined by a single *Sry* gene. Among vertebrates, the only other decisive sex determining single gene is *Dmy/ Dmrt1bY* present in *Oryzias latipes* (Matsuda et al., 2002; Nanda et al., 2002) and *O. curvinotus* (Matsuda et al., 2003). A report on sex determining gene located on W chromosome is awaited. For gene(s) harbored in W chromosome overrides that of Y chromosome in *Xiphophorus maculatus* (see Table 1.6, see also Pandian, 2011, p. 35). However, the *Dmy* is not the sex determining gene in other fishes including those belonging to genus

Oryzias (Kondo et al., 2003). In other than these two medaka species, sex determination remains indecisive, and a host of sex differentiation genes follows the event of sex determination. Interestingly, many genes at the top of the genetic cascade are not conserved, while the downstream genes are homologous in a wide spectrum of fishes (Herpin and Schartl, 2011).

Sex differentiation genes: In the platyfish *Xiphophorus maculatus,* Zhang et al. (2011) have recognized 2,250 male-enriched and 2,304 female-enriched contigs with various degrees of difference. Among these contigs, 22 and 23 are female and male predominant transcripts, respectively. With these contigs, eight female predominant and nine male predominant contigs are recognized as differentially expressing sex genes. Santos et al. (2007) have found that the gonadal transcriptomes differ substantially between sexes in zebrafish. Microarray analysis of individuals has revealed the consistent expression of about 8,769 genes in the gonad, of which 7,976 and 7,060 are consistently expressed in ovaries and testes, respectively. Of them, statistical analyses have identified 2,940 and 1,370 genes that are differently expressed in ovaries and testes, respectively. Among the genes differentially expressed between ovaries and testes in the array dataset, 53 genes have been further identified to show a 10-fold difference in expression between sexes. A magnitude of 3,000-fold difference in expression of some genes between ovary and testis is likely to play a crucial role in the context of sex specific gonadal function. Gene ontological analysis has shown that *sox1lb, sox2la,* and *sox3l* are overexpressed in ovaries; in contrast the absence of elevated expression of *sox9a* and the relative higher expression of *amh* and its receptor induce development of testis in the secondary gonochoric zebrafish. Hence it is the co-action of *sox9a* and *amh* that masculinizes 50% of the presumptive males that have earlier developed an ovary.

Not only in the primary sex differentiation process, but also in the artificially induced sex reversal process, many genes are involved. Characterizing 2,474 genes that are either up-regulated or down-regulated in the Androgen-Induced Female (AIF) *Oncorhynchus mykiss*, Baron et al. (2007) have selected 1,133 unique annotated genes that are associated with at least one gene ontology, and grouped these genes into 13 k-means clusters of temporally correlated expression profiles. Among them, cluster 6 contains genes with increasing expression profiles and commences from day 16 in AIF but from day 27 in normal female and thereby accounts for the observed precocious testicular development in AIF. Clusters 7 and 8 are of particular interest, as they are up-regulated only in AIF. Cluster 11 contains genes that are down-regulated in AIF from day 27 but in females from day 12. The genes in cluster 13 are all down-regulated in AIF from day 27.

Sex chromosomes: In fishes, the cytological sex determining mechanisms of hereditary transmission by chromosomes as carriers of genes have received much attention. Yet, of 30,000 and odd fish species, cytogenetic and/or genetic analysis made thus far are limited to about 1,700 species. Classical karyotypic analysis did not show any dimorphic difference between male and female (e.g., *Oreochromis niloticus*, Majumdar and McAndrew, 1986). Attempts were made to identify the X and Y chromosomes by synaptonemal complex analysis of meiotic chromosomes. Carrasco et al. (1999) reported the absence of pairing in the terminal part of the largest chromosome pair in 25% of the pachytene preparations obtained from XY males, while the normal pair was found in both XX and YY genotypes of homogametic individuals. The inhibition of pairing of these large chromosomes was considered to indicate the accumulation heterochromatin namely, the sex determining region (Griffin et al., 2002). However, interspecific hybridization (Pandian, 2011, Chapter 3), gynogenesis, androgenesis (Pandian, 2011, Chapter 4, 5) and/or progeny testing following sex reversal by hormone treatment (Pandian, 2013, Chapter 7) have demonstrated the existence of male or female heterogametism in many more fishes.

From genotyping, genome-wide linkage analysis and recombination estimates of an accurately developed SNP genetic map of zebrafish, Bradley et al. (2011) have shown that sex determination is indeed a complex process; sex determination in zebrafish, a secondary gonochore, does not use sex chromosomes in the conventional homogametic/heterogametic mode. Further analyses have led Bradley et al. to suggest that the *Dmrt1* and *cyp2la2* are the candidate genes of sex determination and are harbored on chromosomes 5 and 16, respectively. The female associated allele 5 'CUGCUACAGAU-3' would yield to lower *Dmrt1* expression relative to allele 5'CUGCUGCAGAU-3' in the developing gonadal primordium. *cyp2la2* encodes 21-hydroxylase biosynthesizing corticosteroids. Reduced 21-hydroxlase activity results in shunt of precursors toward steroid synthesis. Corticosteroid can induce female to male sex change (see also, p. 200, 201), as in zebrafish and protogynic hermaphrodites.

1.2 Genes meet environment

In mammals, sex is decisively determined at fertilization by a single *Sry* gene harbored on a morphologically distinguishable chromosome. The expression of a cascade of sex differentiation genes is not amenable to changes in environmental factor(s). Hence, sex differentiation proceeds according to genetic determination. In majority of fishes (> 89%, see Pandian, 2011, p. 23), sex is indecisively differentiated by a host of genes harbored on many chromosomes (Fig. 1.1) during a protracted labile period (see Pandian, 2013, Section 7.3) after hatching, and the period may last for a short (e.g., *Betta splendens*) or longer (e.g., *Anguilla anguilla*) duration. In fact, the secondary gonochores commence with a differentiated or non-

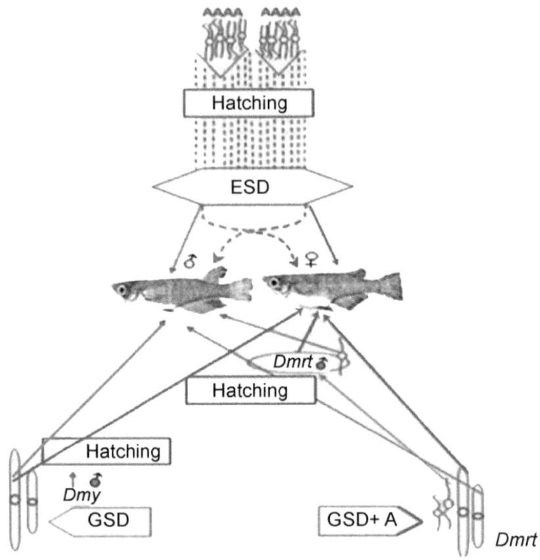

Fig. 1.1. Ontogenetic pathways of sex differentiation in fishes. Note the dimorphic sex differentiation is commenced before hatching in GSD but after hatching in GSD+A pathway. In GSD+A pathway, morphologically distinguishable sex chromosomes are identifiable and sex differentiation is altered by minor autosomal genes in some individuals alone. In the ESD pathway, one or other combinations of genes harbored on autosomes determine sex in all individuals but the genetic cascade of sex differentiation is mostly under the control of one or other environmental factor.

Color image of this figure appears in the color plate section at the end of the book.

functional intersex; their genotypic and phenotypic sexes differentiate in parallel lines for a short (e.g., *Danio rerio*) or longer (e.g., *Pterogymnus lanario*) duration but only one sex, as decided by genes and/or environment, becomes functional. Unlike in mammals, the expression of the cascade of sex differentiation genes is readily amenable to changes in environmental factor(s) and differentiation may result in phenotypic sex as opposite of genotypic sex. These events namely sex determination and differentiation are clearly and temporally separated in mammals but not in fishes. This may be a reason why many fishery scientists consider these two events together and some of them indiscriminately use one for the other. This book, however, recognizes these two events (see p. 263) and considers the latter under: (1) Genetic Sex Differentiation (GSD), (2) GSD + overridden by Autosomal genes (GSD + A) and (3). Environmental Sex Differentiation (ESD) (Fig. 1.1).

(1) In Genetic Sex Differentiation (GSD) pathway, sex differentiation in demersal fishes like *Hippoglossus hippoglossus* my solely depend on genes that are not subjected to wide changes in environmental factors (Hughes

et al., 2008). (2) In the second Genetic Sex Differentiation plus overriding autosomal (GSD + A) pathway, to which *Oreochromis niloticus* and many others belong, sex is determined after hatching by *Dmrt* harbored on morphologically distinguishable chromosome Y (Figs. 1.1, 1.2). However, during the ontogenetic pathway of sex differentiation, one or more minor but overriding autosomal gene(s) may derail the differentiation process in some individuals resulting in phenotypes of the opposite sex, while many individuals may realize the genetic sex (see p. 20). The overriding minor autosomal gene(s) may express inherently, as in *Puntius tetrazona* (Kirankumar and Pandian, 2003) or may be induced to express by one or other environmental factors like temperature (Baroiller et al., 1995). (3) In the third pathway (Figs. 1.1, 1.2), to which majority (89%, Pandian, 2011, p. 23) of fishes belong, sex is determined by a combination of many genes (see Conover and Heins, 1987), each with small additive effects (Kosswig, 1964), randomly distributed in some or all morphologically indistinguishable chromosomes but after hatching. While passing through an extended ontogenetic pathway, the genetic sex may (or may not) be changed to the opposite phenotypic sex by some autosomal gene(s), whose

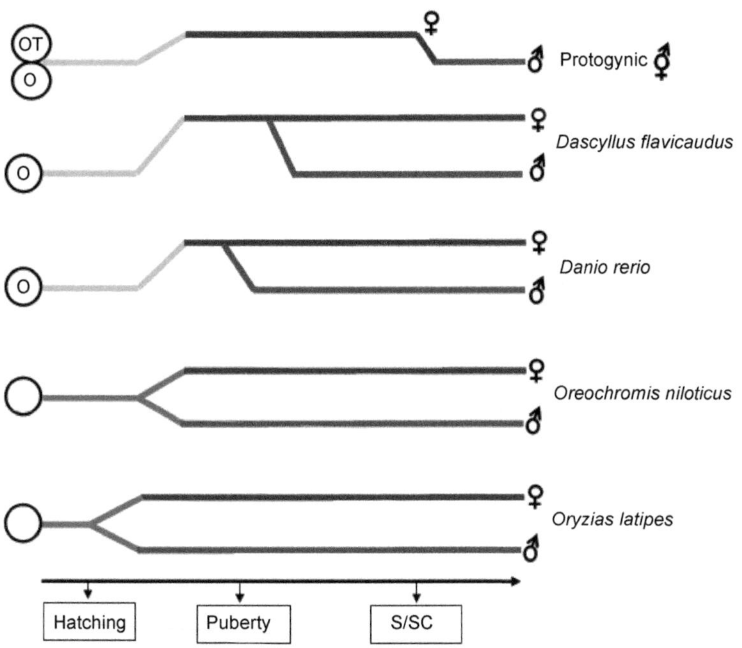

Fig. 1.2. Ontogenetic pathways of sex differentiation in selected fishes. O = Ovary, OT = Ovotestis, S/SC = Senesence/Sex change, Red = Female, Blue = Male, Green = Undifferentiated phase, Yellow = Differentiated but non-functional phase.

Color image of this figure appears in the color plate section at the end of the book.

expression is under the control of environmental factor(s). Hence, the pathway may be called Environmental Sex Differentiation (ESD) and is much more widespread in fishes than presently recognized.

Of more than 1,700 fish species that have been cytogenetically characterized, 188 species belonging to 86 families, i.e., 11% fish species are reported to have cytologically and/or cytogenetically distinguishable heteromorphic sex chromosomes (see Pandian, 2011, p. 23). Of them, 67 and 33% are male (XX/XY) and female (ZZ/ZW) heterogametics, respectively. Incidentally, morphologically distinguishable multiple sex chromosomes system has been identified in 9% fishes, in which karyotypes have been described (Pandian, 2011, p. 27). At present, it is not known how they operate. The presence of male and female heterogametic sex determination systems as well as simple and multiple sex chromosomes clearly indicates the absence of a common, universal sex determining chromosome system in fishes.

Baroiller et al. (2009) identified some of the following environmental factors that control the expression of sex differentiation genes harbored on autosomes. Of them, density (i.e., crowding), pH, temperature and social status are important. In primary gonochoric adult fishes, social induction alters the status but not sex (Pandian, 2013, Chapter 5). However, it induces protogynic sex change in adult secondary gonochores and both protogynic and protandric sex change in hermaphrodites (Pandian, 2013, Chapter 6). Briefly, experimental removal of the territorial male or harem master/mistress initiates immediate and dramatic sex change. But isolated housing of one or more, equal or unequal sized individuals of the same sex elicits benign, slower sex change. More importantly, man-made chemicals are known to disrupt endocrine sex differentiation process. Off from these, sex ratio of fishes is skewed in favor of one or the other sex due to inherent genetic factors to ensure reproductive success and/or genetic diversity, as in (i) social structure, (ii) mating system, (iii) parental care and (iv) hermaphroditism. Some reproductive strategies of fishes, in which the balanced Mendelian sex ratio is not maintained at unity are described later.

1.3 Social structure

Among pelagic spawners, group formation is a common phenomenon that considerably alters sex ratio. For example, sex ratio of surgeonfish *Acanthurus leucosternon* is 0.64♀ : 0.36♂; the surgeonfish population consists of 'solos', 'pairs' and 'trios', i.e., 1♂ + 2♀ (Robertson et al., 1979). In *Neocirrhites typus*, the population consists of 45% 'pairs', 11% 'trios' and 0.7% 'tetras', i.e., 1♂ + 3♀ (Donaldron, 1989). Moyer (1990) lists social organization consisting of trios in nine pomacanthid species: *Centropyge bicolor*, *C. bispinosus*, *C. ferrugatus*, *C. interruptus*, *C. multispinis*, *C. shepardi*,

C. tibicen, C. viroliki and *Holocanthus tricolor.* Hawkfishes live in a kind of loose social organization. In the hawkfish *Cirrhitichthys falco*, groups are described with (i) one male plus 5–7 females and (ii) one male plus one sneaker male, two juveniles and seven females. In another hawkfish *C. aprinus*, a male dwells with two females in Nyake-jima, Japan, four in Papua New Guinea and seven at Great Barrier Reef (Donaldson, 1987). More field-oriented investigations are required on social organizations of fishes to know how the organization alters sex ratio.

Harem: Another social structure that is responsible for female-biased sex ratio is the presence of a stabilized harem. Among the social structures of fishes, harems are important and are prevalent; for example, 20 cichlids (18%) out of 122 species inhabiting the Lake Tanganyika are haremic (Kuwamura, 1997). This may also be true of protogynics; more than 25 species (≈13%) (see Table 26, Pandian, 2010, p. 48) out of about 200 protogynics are haremic. In harems of gonochoric and protogynic, a dominant territorial male holds two to nine females and one to seven females/hermaphrodites, respectively. Consequently, the mean male ratios are just 0.24 and 0.22 in a gonochoric and protogynic harem, respectively (Table 1.1). Briefly, a territorial male holds 3.2–3.6 females in his harem. Interestingly, Spence and Smith (2005) offered many females to a male zebrafish (*Danio rerio*) and found that a male can effectively 'handle' 2–3 females at a time and display the best reproductive performance. In his harem, a territorial male *Lamprologus ocellatus*, in which sex differentiation of helpers is inhibited by aggression, holds seven brooding females and nine helpers. Of 20 haremic cichlids, five of them hold one sub-adult male as helper; surprisingly, in one of them the helper belongs to another cichlid species (Kuwamura, 1997). In *Julidochromis marlieri*, a polyandrous harem like system is reported; of eight females observed, two of them hold two males each with a helper (Yamagishi and Kohda, 1996). Likewise, the harem mistress of *Amphiprion alkallopsis* holds one male and many sub-adults and juveniles, whose sex differentiation is inhibited. Hence the sex ratio in these harems can not be easily estimated.

Sakai (1997) has described alternate strategies of harem formation in a model fish *C. ferrugatus*. However, not much is known about the early stages of harem formation. That the male ratio is around 0.22–0.24 in both hermaphroditic and gonochoric haremics indicates the existence of a genetic mechanism to reduce the male ratio. Further, on loss of the harem master or mistress from a sequential hermaphroditic harem either naturally or experimentally, the α-female or male changes sex and holds the harem. In a gonochoric harem, rather a social structure, the α-female/sub-adult male of secondary gonochorics like *Crenicara punctulata* and *Matriaclima* cf. *livingstoni* changes sex/social status (see Pandian, 2013, p. 138) to hold the social structure (Table 1.2). The freshwater haremic cichlids may prove

Table 1.1. Male ratio in haremic gonochoric and protogynic fishes (© 2010, Pandian, T.J., Sexuality in Fishes. Science Publishers/CRC Press, reproduced by Permission of Taylor and Francis Group, LLC, a division of Informa plc).

Family/Species	Male sex ratio	Females (no/harem)
	Gonochoric harem	
Acanthuridae		
Acanthurus gahhm	0.33	2.0
Ctenochaetus striatus	0.25	3.0
C. strigosus	0.20	4.0
Pomacanthidae		
Holacanthus clariorensis	1+	
H. passer	0.11	7–9
Ctenopyge argi		
8 m depth	0.13	7.0
12 m depth	0.29	2–3
14 m depth	0.33	2.0
Ostraciidae		
Lactoria fornasimi	0.20	4.0 + helper
L. diaphanous	0.33	2–4
Ostracion cubicus	0.33	2–4
Labridae		
Cymolutes torquatus	0.33	2-4
Cichlidae		
Pelvicachromis pulcher	0.33	3.0
Mean	≈0.24	≈3.2
	Protogynic harem	
Labridae		
Labroides dimidiatus	0.14–0.26	3–6
Xyrichtys martinensis	0.22	3–6
X. pentadactylus	0.20	1–7
Scaridae		
Sparisoma cretense	0.25	3.0
Serranidae		
Cephalopholis roseus	0.27	1–6
*Serranus baldwini**	0.20	1–7
*S. fasciatus**	0.18	1–8
Pomacanthidae		
Chaetodonotplus mesoleucus	0.40	1–2
Centropyge ferrugatus	0.25	3.0
Pomacanthus imperator	0.22	2–5
Chaetodontidae		
Chaetodon fromblii	0.29	4.0
C. trifasciatus	0.20	2–3
Pyglopites diacanthus	0.17	5.0
Parapersidae		
Parapersis snyderi	0.13	3–10
Mean	≈0.22**	≈3.55**

*holds hermaphrodites in his harem, **values taken from Pandian (2010).

an excellent model to study the genetic and/or environmental basis for reduction in male ratio.

Interestingly, the mating system within a species may change from place to place; *Halichores maculipinna* is haremic in Florida but not in Panama; on the other hand, *H. garnoti* is haremic in Panama but not in

Table 1.2. Social mediated sex change/reversal in adult gonochoric fishes. J = Juvenile. ♀-sm = sub-male possessing male's body color and aggressive behavior (© 2013 from Pandian, T.J. Endocrine Sex Differentiation in Fish, CRC Press, Boca Raton, reproduced by permission of Taylor and Francis Group, LLC, a division of Informa plc).

Social components	**Reported observations**
Crenicara puntulata (Carruth, 2000)	
(i) In natural territory ♂ + ♀♀♀♀	On removal of territorial male, the dominant α female changes sex to male
(ii) Group rearing ♂ + ♀sm♀sm♀♀♀	2-years old females differentiate into sub-males and display male's body color and aggressive behavior. On removal of male, one of the sub-males changes to male
(iii) J reared in isolation	Passes through female phase to become an adult male at the age of 7–10 months
Matriaclima cf. *livingstoni* (Stauffer and Ruffing, 2008)	
(i) Six tanks each housing ♀♀♀	One female from a tank changes to male and fertilized eggs of one female, "resulting in three large broods"
(ii) Two tanks each housing ♀♀♀♀♀♀	One female from a tank changes to male, as "indicated by a female holding eggs"
Macropodus opercularis (R. Koteeswaran pers com)	
(i) J reared alone	Matures into male
(ii) ♂♂ reared individually	Changes to female
(iii) Reared in group of 10 females or more than 19 females	2–3 females change to males

Florida (Robertson, 1981). Harems are limited to a section of population in *Sparisoma radians* and *Cryptotomus roseus* (Robertson and Warner, 1978). The mating system of the coral-dwelling damselfish *Dascyllus marginalis* may range from a solitary male to a pair, harem and promiscuous group depending on the coral colony size (Fricke, 1980). In the Nigerian cichlid *Pelvicachromis pulcher*, the red morphs are haremic but the yellow ones are not. Its sex ratio differs from 0.67♀ : 0.33♂ in the red morphs at the upper wet region of the Sombreiro River to 0.5♀ : 0.5♂ in the yellow morphs at the lower dry region (Martin and Taborsky, 1997). Indeed genes meet the environment in deciding sex ratio in these mating systems.

1.4 Mating system

Fishes display all the four recognized mating systems (Table 1.3). Among monogamous fishes, sex ratio may not be altered. The exclusive monogamy is characterized by site fidelity and low mobility as well as no parental care of eggs and/or fry; it is mostly limited to some coral reef fishes (Barlow, 1984). Sex ratio may deviate from unity in biparental monogamy involving parental care, as this monogamy does not last longer than one reproductive cycle in many paternal egg-guarding gobiids (e.g., *Valenciennea longipinnis*,

Table 1.3. Mating systems leading to intense selection (source: Searcy and Yasukawa, 1995, modified).

Mating system	No. of successful mates/ mating ♂	No. of successful mates/ mating ♀	Gender subjected to intense selection	Example
Monogamy	One	One	Neither	*Typhlogobius californiensis*
Polygamy				
a) Polygyny	Many	One	Male	*Lamprologus ocellatus*
b) Polyandry	One	Many	Female	*Gambusia holbrooki*
c) Polygyandry	Many	Many	Either or neither	*Gasterosteus aculeatus*

Takegaki and Nakazona, 1999) and maternal mouth brooding cichlids (e.g., *Sarotherodon galilaeus*, Fishelson and Hilzerman, 2002). Polygamic fishes may either be polygynic or polyandric; in the most common polygynic system, male-male competition occurs and males are subjected to intense selection by female. The reverse is true in the less common polyandric system. However, polygamy is much more common among fishes. Consequently, multiple paternity or maternity, which may alter sex ratio, is not uncommon. For example, eggs spawned by a single female *Gadus morhua* are fertilized by at least three attendant males (Hutchings et al., 1999). The nest-holders are indeed polygyandrics, as 5–15% of the tended eggs are not sired by the nest holder and the eggs tended may be from 3–6 mothers (see Pandian, 2011, p. 14). Pouch brooders like *Syngnathus floridae* and *S. typhle* receive embryos from two and three females, respectively. As their sperm can be stored for 2–4 months duration (Love et al., 2002), the viviparous females selectively fertilize their eggs from stored sperm of more than one male. The rockfish females of *Ditrema temmincki* and *Neoditrema ransonneti* gestate embryos sired by two males (Takegaki et al., 2008). Not surprisingly, 71–100% females in populations of marine viviparous fishes gestate embryos sired by more than one male. Multiple paternity is also not uncommon among freshwater poeciliids: 52% females of *Poecilia latipinna* (Travis et al., 1990), 56% females of *Gambusia affinis* (Chesser et al., 1991) and 100% females of *G. holbrooki* (Zane et al. 1999) gestate embryos sired by more than one male. Barring exclusive monogamy, the mating systems in fishes seem to ensure genetic diversity through multiple paternity/maternity.

Alternate mating strategy (AMS): Energy required for the normal development and functioning of ovary is six to 18 times more than that required for the development of testis (Diana, 1983). This wide difference in reproductive investment between sexes has led to intense male-male competition, although this conventional male role is dynamic and may reverse with unusual female-biased sex ratio (e.g., Berglund et al., 1986; Forsgren et al., 2004). With intense competition, males of lower competitive ability, adopt

AMS to make use of a bad situation (Taborsky, 2001) and produce as many offspring as possible. The share of fertilization success by parasitic male ranges from 3.5% in *Xiphophorus nigrensis* to 36% in *Symphodus tinca* (see Taborsky, 1994). Incidentally, males of some AMS-displaying fishes like the Mediterranean labrid *Symphodus* spp. and Mexican blennoids *Auxiclinus* spp. pass through sneaker → satellite → territorial phases. Oliveira (2006) classified fishes displaying AMS into (i) plastic alternative phenotypes and (ii) fixed alternative genotypes. The plastic transformants can change from sneaker to satellite and from satellite to territorial, but the territorial can not revert to satellite or satellite to sneaker. Despite passing through two or three parasitic male phases, the male ratio is around 50% in four species belonging to *Symphodus* (Warner and Lejeune, 1985). Contrastingly, the Mexican blennoids *Auxiclinus nigricaudus* and *A. carminalis* (Tripterigiidae), in which males also pass through sneaker to satellite and to territorial phase, the male ratio is 66% in *A. nigricaudus* and 80% in *A. carminalis* (Neat, 2001). In other AMS-displaying fishes, territorial and parasitic male characteristics are genetically fixed, hatched as parasitics or territorials and are classified as the fixed sex-linked genotypes (see Pandian, 2013, Section 5.4). Irrespective of the presence of different genotypes like jacks and hooknose among salmonids, cuckolders and parental among centarchids, the sex ratio remains almost equal. With haremic bourgeois and parasitic dwarf morphotypes in *Lamprologus callipterus* (Taborsky, 2001) as well as haremic red and non-haremic yellow morphs of the Nigerian cichlid *Pelvicachromis pulcher*, sex ratio is considerably altered. For example, the male ratio is 0.33♂ and 0.50♀ in the red and yellow morphs (Martin and Taborsky, 1997). Hence Oliveria's classification may not include changes in male ratio as a criterion.

Operational Sex Ratio (OSR) is the ratio of sexually active males to females ready to mate in a given space and time (Emlen and Oring, 1977). Some examples are given to show how time, space and sexually active males and females are important in the context of OSR. Anabantid males build bubble nests to accommodate and tend the sired embryos. In *Betta splendens*, a bubble nest is built by an adult male once every 4–5 days, indicating that the duration of breeding cycle of a male is 4–5 days; however, that of a female is 14–15 days (Kirankumar and Pandian, 2002). Hence a male fighting fish may theoretically be able to attend on three females. From the spatial point of view, sex ratio of the rose bitterling *Rhodeus ocellatus* is 1.0♀ : 1.2♂. The female bitterling oviposits through inhalant siphon into the interlamellar space of the gills of its symbiotic partner the mussel *Anadonta woodiana*. The mussel irrigates the attached (to the lamella) developing bitterling embryos under a much secured habitat. On arrival of a female bitterling at the mating site namely the mussel, many 'parasitic' males may join the territorial male, resulting in an OSR of 1.0♀ : 3.3♂ (Kanoh, 2000).

In northern anchovy *Engraulis mordax*, the sex ratio in catches ranged from 0.50♀ : 0.50♂ to 0.11♀ : 0.89♂ but it was 0.28♀ : 0.72♂ at the active spawning sites of Oregon and Washington during July (Laroche and Richardson, 1980). Hamilton et al. (2008) recorded spatial variations of sex ratio of the greenhumped parrotfish *Bolbometopon muricatum*; the ratios were 0.53♀ : 0.43♂, 0.51♀ : 0.49♂ and 0.22♀ : 0.78♂ for the populations of the Solomon Islands at Munda, Nasabunga and Tetepara, respectively. Seghers (1973) reported that the adult ratio of the guppy *Poecilia reticulata* from the Paria River in the West Indies varies spatially from 0.65♀ : 0.35♂ to 0.60♀ : 0.40♂ and temporally from 0.60♀ : 0.40♂ to 0.47♀ : 0.53♂. These examples show that the spatial and temporal variations in sex ratio are not uncommon among marine and freshwater fishes. Rather interesting sex ratios were reported for the Finnish *Perca fluviatilis* by Olin et al. (2012). By number, the ratio was 0.41♀ : 0.59♂/ha but by weight (kg/ha) it was 0.51♀ : 0.49♂. Consideration of the weight based OSR may challenge the OSR theory, as fecundity increases with body size in many fishes (see Fig. 2.5).

The viviparous poeciliids like *P. sphenops* and *P. velifera* realize the maximum reproductive potential, when they receive the first impregnation as virgin at the age of about 7 month and the second at the age of 12 months. These two impregnations are adequate for their reproductive life span of 20 and 25 months, respectively (George and Pandian, 1997). In *P. reticulata* also, Jirotkul (1999) found that the females are ready to mate as virgin and then once or twice after giving birth. The viviparous fishes can store sperm for periods of 2 to 4 months and selectively fertilize their eggs using the stored sperm from more than one male. Hence the number of receptive viviparous females at a given time is very low, resulting in the very strongly male-biased OSR. However, the sex ratio of many viviparous fishes is female-biased, providing from 2–3 females to every male (e.g., 2.02 females for every male *Jenynsia multidentata*, Mai et al., 2007) and as many as 11 females to every male (e.g., *Poeciliopsis baenschii*, Zuniga-Vega et al., 2012). For this reason, Jirotkul (1999) has re-defined the OSR stating that OSR is the number of sexually active males divided by the total number of sexually active adults of both sexes. He has also assessed the reproductive success by the number of copulations rather than eggs or progenies.

"Operational sex ratio (OSR) is determined by three factors: (i) the adult sex ratio in the population, (ii) the spatial and temporal distribution of the two sexes and (iii) their potential reproductive rates. The rate at which a sex can potentially reproduce depends on its investment in gamete replenishment and parental care. Even in species with no parental care, the OSR is usually skewed because females are limited by the number of eggs. Thus, males typically compete among each other for mating, whereas the females are selective of mating partners. Competition among males for mating takes various forms including (i) contest, (ii) alternative mating tactics, (iii) mate guarding and (iv) sperm competition (Spence and Smith, 2005)" (Pandian, 2010, p. 51).

1.5 Parental care

Some experimental observations made from a simultaneous offer of more females or males to a single opposite sex partner within a limited space seem not quite in agreement with the OSR concept but rather indicate the importance of parental care. Reproductive performance of fishes with no parental care or with parental care has been assessed in a few fishes by manipulating OSR with simultaneous offer one or more females to a male and one or more males to a female. Interesting observations are reported for (i) oviparous fishes with no parental care (e.g., *Danio rerio*, Delaney et al., 2002; Spence and Smith, 2005), oviparous females with (ii) paternal parental care (e.g., *Pseudorasbora pava*, Katano and Maekawa, 1997), (iii) paternal gestation (e.g., *Syngnathus typhle*, Berglund, 1993) and (iv) viviparous fishes (e.g., *Poecilia reticulata*, Jirotkul, 1999). The number of fertilized eggs or progenies arising from an experimental mating has been taken as a criterion to assess the reproductive performance, although the number of copulations per male alone has been considered to assess the reproductive success in viviparous fishes. The reported observations clearly show that (1) A male can effectively 'handle' only two or at the most three females to produce maximum number of fertilized eggs in an oviparous fish with no parental care (Table 1.4). (2) Conversely, three males are required to produce maximum progenies from eggs spawned by a single female in fish with paternal parental care. (3) In oviparous fish with paternal gestation, two males are required to produce maximum number of progenies from eggs transferred by a female. 4. The reproductive performance of viviparous fish is limited by the availability of receptive females (for more details see Pandian, 2010, pp. 51–54).

Extending his investigations, Jirotkul (2002) provided simultaneous opportunity to *P. reticulata* to mate and copulate with OSR of 2♀ : 4♂, 3♀ : 3♂, 4♀ : 2♂ and 5♀ : 1♂, i.e., the female OSR increased from 0.33 to 0.50, 0.67 and to 0.83. He estimated the opportunity for sexual selection, being the variance in the total number of copulation events per male divided by the square of the mean copulation events. The opportunity for sexual selection was positively correlated with increased female OSR, i.e., the opportunity for copulation was more than four times greater, when the female OSR was increased from 0.33 to 0.83.

"In the pipefish *S. typhle*, known for paternal gestation of embryos, Berglund (1993) manipulated sex ratio with 4♂ (3 small plus one large): 2♀ (one small plus one large) in the first male-biased OSR series and in the second female-biased OSR series, the combination was 2♂ (one

Table 1.4. Effect of altered OSR and mode of reproduction on egg production in selected teleostean fishes (© 2010, Pandian, T.J., Sexuality in Fishes. Science Publishers/CRC Press, reproduced by Permission of Taylor and Francis Group, LLC, a division of Informa plc).

Mode of reproduction/Species	Experimental design	Eggs/ copulations (no)	Male reproductive success
	Oviparous		
Danio rerio	i) 2 ♂ + 1 ♀	52	high upto 1♂ + 2 ♀♀
	ii) 1 ♂ + 2 ♀	70	
	iii) 10 ♂ + 5 ♀	42	
	iv) 5 ♂ + 10 ♀	36	
	Oviparous: ♂ egg guarding		
Pseudorasbora parva	i) 1 ♂ + 1 ♀	1996	high with 3 ♂♂ + 1♀
	ii) 3 ♂ + 1 ♀	2053	
	iii) 10 ♂ + 1 ♀	1717	
	iv) 1 ♂ + 3 ♀	1170	
	v) 1 ♂ + 10 ♀	632	
	Oviparous with ♂ gestation		
Syngnathus typhle	i) 4 ♂ + 2 ♀	85	high with 2 ♂♂ + 1♀
	ii) 1 ♂ + 4 ♀*	58	
	Viviparous		
Poecilia reticulata	♂ biased		low with ♂ bias
	♀ biased		high with ♀bias

*male took 702 min in courting against 121 min in the male bias series.

small and one large): 4♀ (3 small plus one large) to study the effect of size and OSR. In the male-biased OSR series, the males did not select the large or small females, took shorter durations to court (121 min) and mated for longer (440 min) duration than in the female-biased series. Consequently, each male in the first series received more (85) eggs and got his pouch filled to 93% capacity, whereas in the second series, each female produced 58 eggs only and filled the male's pouch only to 63% of its capacity. With increased availability of females, the male spent longer duration of 702 min to select and court rather than to spermiate and acquire eggs. Thus in the oviparous fish with male gestation, sex ratio of 2♂ : 1♀ increases the progeny output" (Pandian, 2010, p. 54).

"Table 1.4 represents a comprehensive summary to show the effect of experimentally manipulated OSR and mode of reproduction on the reproductive success of fishes, for which relevant information is available. In oviparous species with no parental care, which constitute about 64%

of teleosts, the reproductive output of a male increases with increasing availability of females, but only upto a maximum of two-three females, beyond which the increased availability of females itself has a negative effect on the male reproductive success. Incidentally, the frequency of courting attempts terminating with successful spawning in *Cirrhitichthys falco*, a haremic protogynous hermaphroditic fish was 76%, when a male courted 3 females against 64%, when it courted 5 females (Donaldson, 1987)" (Pandian, 2010, p. 55). Viviparous fish constitute 19% of teleosts; for them the availability of upto five females increases the copulation frequency and hence may increase reproductive output.

However, in oviparous fish with male guarding or gestating (as in syngnathids) the eggs, the presence of two–three males for every female results in (i) provision of a wider choice to the females, (ii) courting males to quickly select competing females and, (iii) higher reproductive output. Incidentally, male guarding/gestating species constitute 11% of teleosts; in these fishes, it is the availability of males that limits the reproductive output. But in oviparous species with no parental care and in viviparous species (64 + 19 = 83% teleosts), it is the availability of females per male that limits reproductive output. Exceptionally, the Mediterranean wrasse *Xyrichthys novacula* male holding 12 females in his harem fertilized 56,777 eggs/d during the spawning season, in comparison to 24,482 and 36,864 eggs fertilized daily by males holding four and six females, respectively (Marconato et al., 1995, see also Fig. 2.16). No information is yet available with regard to 4% teleosts, in which both male and female afford protection to the eggs. Further research work is required, however, to confirm this generalization. Briefly, 83% fishes are polygynic and 11% polyandric. Hence, the scope for genetic diversity in fishes is wider.

1.6 Hermaphroditism

Among vertebrate taxa, sex change in fishes is a unique, fascinating but an intriguing phenomenon. Instantaneously, this phenomenon poses a series of challenging questions regarding the presence of sex chromosome systems, sex determining genes and consequently sex ratio. A few secondary gonochores and almost all the hermaphrodites undergo natural sex change or artificially induced sex reversal (see Pandian, 2013, p. 138). Among hermaphroditic fishes, the sequentials are divided into (i) female to male sex changing protogynics and (ii) male to female sex changing protandrics (see Fig. 1.3). In the monandric protogynics, secondary males arise indirectly from sex changing females but in the diandrics, primary males also arise directly, in addition to the secondary males arising from sex changing females. A similar dichotomy has been recognized in the protandrics and the corresponding terms are monogyny and digyny. The

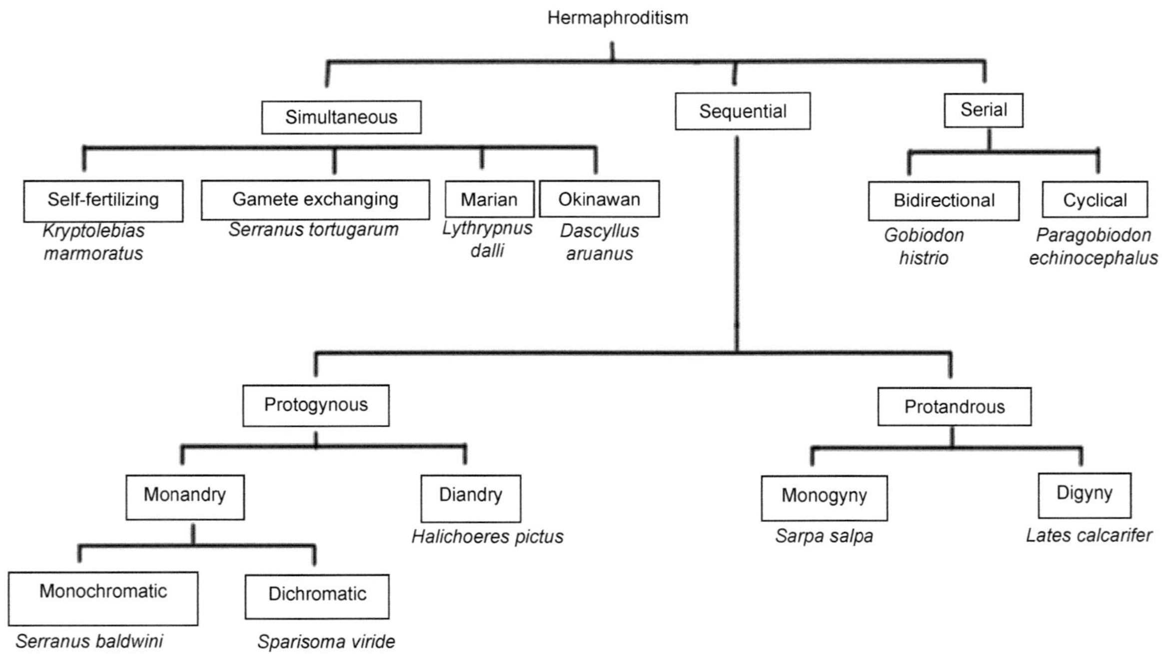

Fig. 1.3. Patterns of functional hermaphroditism in fishes (from Pandian, Sex Determination in Fish. Science Publishers/CRC Press, Boca Raton, reproduced by permission of Taylor and Francis Group, LLC, a division of Informa plc).

monondrics are further divided into monochromatics with all mature (initial phase, IP) males displaying the same dull body color as the female and dichromatics, in which (terminal phase, TP) males change their body color to bright terminal ones (see Pandian, 2013, p. 142). Pandian (2013, p. 146) has described many routes through which male and female phenotypes change sex naturally or reverse artificially (Fig. 1.4). Thus the changes in morphotypes are intrasexual among primary gonochores (see p. 4–5) but are intersexual among hermaphrodites. Hence the sex determination and differentiation processes are several times more flexible in hermaphrodites than in primary gonochores.

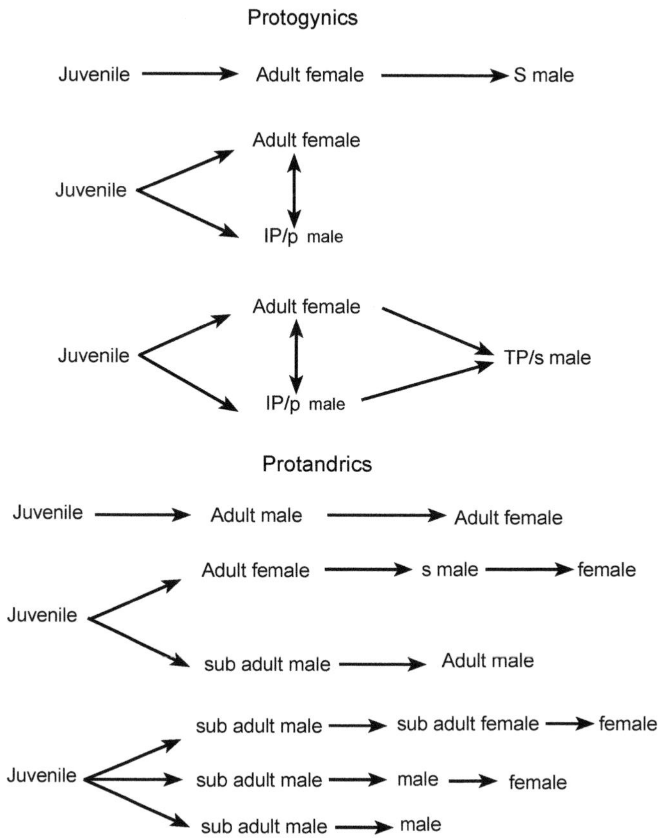

Fig. 1.4. Pathways of ontogenetic sex differentiation and sex change in protogynic and protandric sequential hermaphrodites to show the plasticity in sex differentiation . IP = Initial Phase, TP = Terminal Phase, p = Primary, s = Secondary (© 2013 from Pandian, T.J. Endocrine Sex Differentiation in Fish, CRC Press, Boca Raton, reproduced by permission of Taylor and Francis Group, LLC, a division of Informa plc).

In the protogynics, female-biased sex ratio is to be expected. Protogynic females undergo natural sex change, when the fish attains a body weight of 80% of their maximum body size, corresponding to 80–85% of their life span (see Pandian, 2012, p. 113). Consequently, with increasing number of sex changing females, male ratio of monochromatic monandric protogynic fishes increases from 0.09 to 0.31 (see also Table 2.11). Interestingly, the female ratio for three protandrous hermaphrodites is also 0.31 (see Pandian, 2010, p. 112). Among dichromatic monandrics, the male ratio is 0.29, which includes IP and TP male ratios of 0.23 and 0.06 (see Table 2.12). With 4 times larger in number and 2.5 times higher gonado somatic index (GSI) than their respective TP male, the IP male partially compensates the reduced male ratio. In diandrics, the primary (PM) and secondary (SM) male ratios are 0.10 and 0.16 during the final stage, respectively (Table 2.12). Hence the cumulative male ratio of diandrics is 0.26 only (see Table 1.5). The primary (PM) male shows nearly four times higher GSI than the secondary (SM) male. Pandian (2010, p. 109) recognized four types of protogynics like monochromatism, dichromatism, diandry and harem to understand the effect of mating systems on male ratio. From a comprehensive analysis of available information, he indicated that male ratio decreases from 0.31 in monochromatic monandrics to 0.22 in haremics (Table 1.5). Investigations on the mechanisms of sex determination and differentiation may prove rewarding.

The increased male ratio in the presence of parasitizing primary males increases genetic diversity at the cost of sperm economy. In the diandric *Coris julis*, 80% of paired matings occur between a female and secondary male leaving 20% matings being parasitized by primary males (see Taborsky, 1994). In the diandric bluehead wrasse *Thalassoma bifaciatum* mating is followed by pair spawning between a female and a secondary (S) male or group spawning between a female and a secondary male along with as many as 20 primary males. In both cases, 97–98% fertilization success is achieved, though at the cost of 50 times more sperm by group spawning males (Shapiro et al., 1994; Marconato et al., 1997). In pair spawning, sperm

Table 1.5. Male ratio and number of females per male in protogynous hermaphroditic fishes (from Pandian, 2010, updated).

Protogynic hermaphrodites	Male ratio	Females/male
Monandric monochromatics	0.31	2.23
Monandric dichromatics	0.29	2.45
Diandric dichromatics	0.26	2.85
Haremic protogynics	0.22	3.55
Haremic gonochorics	0.24	3.16
Monogynic protandrics	0.31*	

*female ratio.

economy is achieved but group spawning increases genetic diversity among the progenies.

1.7 Model fishes

Here it is elucidated how (i) changes in constellation of chromosomes alters male to female heterogamety in poeciliids, (ii) autosomal genes override sex differentiation in cichilids and, (iii) environmental factors like temperature alters phenotypic sex ratio in the European sea bass.

Poecilidae, a family of Cyprinodontiformes, is comprised of about 200 species belonging to 22–29 genera. With an exception of *Tomeurus gracilis* (Hrbek et al., 2007), all poeciliids are viviparous. By virtue of small size and short generation time, many species belonging to the genera *Poecilia* and *Xiphophorus* have been the favorites for experimental studies (Schultheis et al., 2009). Table 1.6 summarizes the available information on the genetic mechanism of sex differentiation in poeciliids, in which deviations in sex ratio are also reported to be genetically governed. For example, a deviation reported for *X. helleri* is 0.26♀ : 0.74♂ (Kallman, 1984). The species sequence listed in Table 1.6 broadly indicates the possible course of origin and evolution of sex differentiation mechanisms in poeciliids but it does not claim who has originated from whom. To achieve genetic diversity by skewing sex ratio in favor of either male or female through the described genetic mechanisms (GSD + A, Fig. 1.1), all these poeciliids have involved

Table 1.6. Genetic sex differentiation mechanisms in selected species belonging to *Poecilia* and *Xiphophorus* (source: Kallman, 1984; Kavumpurath and Pandian, 1992, 1993; George and Pandian, 1995; Schultheis et al., 2009; Pandian, 2011).

Species	Sex chromosomes	Sex determination and differentiation
P. reticulata	X, Y	♂ heterogamety confirmed by FISH. Autosomal factors override to generate XX ♂ and XY ♀; YY ♂ and YY ♀ are viable
P. cortezi	X, Y	XX ♂ exists due to autosomal modifiers
P. velifera	X, Y	XX ♂ exists due to autosomal modifiers
X. xiphidium	X, Y	Autosomal modifiers present
X. montzumae	X, Y	Occasional XY ♀, XX ♂, YY ♂ viable
X. millerii	X,Y,Y^1	2 Y chromosomes. XY^1 ♀, XY ♂; YY^1♂ viable
X. nigrens	X, A, Y	XY ♂ but XX with 2 autosomal allelies A and a. XXAA ♀, XXaa ♂. Among XXAa, 95% ♀ and 5% ♂
X. maculatus	X, Y, W	XX, XW and YW ♀, XY and YY ♂ Autosomal modifiers present
X. helleri	Y, W	YW ♀, YY ♂
P. sphenops	W, Z	♀ heterogamety shown by FISH; ZZ ♀ viable

a role by one or more autosome(s). However, two different trends are distinctly notable: (1) With the exception of *P. sphenops*, genetic mechanism for deviation of sex ratio in *Poecilia* is limited to autosome(s) alone. (2) Contrastingly, some *Xiphophorus* species have progressively reduced the importance of X chromosome and its replacement by W chromosome. 2a. Thus, the Y chromosome is duplicated in *X. millerii*, 2b. To weaken X chromosome, AA/Aa/aa autosomes are added in *X. nigrens*, (2c). The more powerful W chromosome is included, while still retaining X in *X. maculatus* and (2d) Complete replacement of X chromosome with W in *X. helleri*. On the other hand, *P. sphenops* seems to have at one stroke replaced the XX/XY male heterogamety by ZW/ZZ female heterogamety.

From his long term experiments, Kallman (1984) has elaborately described a typical sex determination resulting in unexpected deviations in sex ratios of six species belonging to *Xiphophorus*. Of them, *X. maculatus* possesses both male determining Y chromosome and female determining W chromosome. With regard to the mechanisms, through which the autosomal factors bring about the deviations in favor of male ratio in *X. maculatus*, the following are summarized. The summary may also provide an idea about the complicated role played by autosomal factors in other fishes. 1. To bring about a skewed sex ratio in favor of the male, the autosomal factors interact specifically and independently with the W chromosome, irrespective whether the second chromosome is an X or a Y. They do not interact with the X chromosome to induce WX individual to develop into a male. 2a. the autosomal regulatory gene interacts with the W chromosome to cause the transcription of male-determining gene in 86% of all WX individuals but it does it differently with the X chromosome to activate male determining gene to rarely produce 6% males. 2b. these interactions suggest a similarity of the sites on both W and X chromosomes. But the differences between these chromosomes are the presence of (i) a suppressor gene in W for the Y-linked male determining gene and (ii) differently controlling elements for the male-determining gene on the chromosomes. 2c. WW zygote may develop into male only, when the suppressor genes in both WW chromosomes are blocked. It can happen, only when all the three genes are closely linked to one another and form a functional unit.

Cichlidae: In aquaculture, the short generation time and frequent reproduction causes stunted growth in tilapias. Undesired reproduction in mixed sex population causes up to 70% of total harvest being small fish of no commercial value (FAO, 2010). To produce all-male tilapia, several protocols are available and they may be grouped under (i) inter-specific hybridization (see Table 3.2), (ii) endocrine sex reversal and (iii) androgenesis. Using endocrine sex reversal technique, YY supermales have been generated in *Oreochromis mossambicus* (Varadaraj and Pandian, 1989) and *O. niloticus*

(Scott et al., 1989). However, crosses between the YY supermale and XX female tilapia usually produced unexpected female progenies, though at low frequencies (e.g., 3–12% in *O. mossambicus*, Pandian, 1993). In *O. niloticus*, the reported frequencies range from 0 to 7%, when YY male of the Egyptian strain was crossed with six different dames (Mair et al., 1991) and from 0 to 20%, when YY male of the Thai strain was crossed with 12 different dames (Tuan et al., 1999). The LG1 marker developed by Lee et al. (2003) correctly predicted the sex of 95% of individuals in two of the three families of *O. niloticus*. However, there was no association between LG1 and sex in the third family. Eshel et al. (2010) found that two linkage groups namely LG1 and LG23 are associated with sex differentiation in *O. niloticus* and suggested that LG23 harbored male-associated alleles. Mair et al. (1997) also attributed the occurrence of female progenies of YY males to several autosomal genes. Using sex-specific molecular markers, these unexpected female progenies have been shown as genotypic males (e.g., Kirankumar and Pandian, 2004; Kirankumar et al., 2003; David and Pandian, 2006; however, see also Williamson et al., 2008). Hence the unexpected sex ratio of tilapia has a genetic basis. Interestingly, a cross between YY supermale and YY sex reversed Δ neosuperfemale *O. niloticus* eliminated phenotypic female progenies (Muller Belecke and Horstgen-Schwark, 2007), clearly indicating that two YY chromosomes are required to silence the expression of modifying autosomal genes.

From single pair matings, strong evidence for strain-dependent genetic variation in male ratios has been reported for the Auburn strain (71 families, male ratio ranging from 0.31 to 0.77, Shelton et al., 1983), Egypt Swansa strain (57 families, male ratio ranging from 0.35 to 0.75, Mair et al., 1991) and Thai Chitralada strain (95 families, male ratio ranging from 0.15 to 1.00, Tuan et al., 1999). Figure 1.5 shows a much wider range of male ratio for the Thai strain than the Egyptian strain and peaks in the distribution is indicative of at least one autosomal recessive gene influencing sexual phenotype (Beardmore et al., 2001). Hence all these strain-dependent variations in male ratio are genetic-based (see also Chapter 7).

There are publications to show that these genetic variations in sex ratios in *Oreochromis* spp. are inherited. In *O. niloticus* single pair mating between 'selected males' generated predominantly male broods (85–100%), irrespective of the female parent was a selected or control one. Control males generated broods with normal sex ratios (47–57%), again irrespective of type female parent used (Wohlfarth and Wedekind, 1991). From a study involving 1077 full-sib families produced over six generations, Lozano et al. (2013) reported that across all generations, there was a low but significant 'additive' genetic component for male ratio with inheritability estimates of 0.12. Selection for increased male ratio based on family means may lead to an increased frequency of XX males. If a normal (XX x XY) mating produces

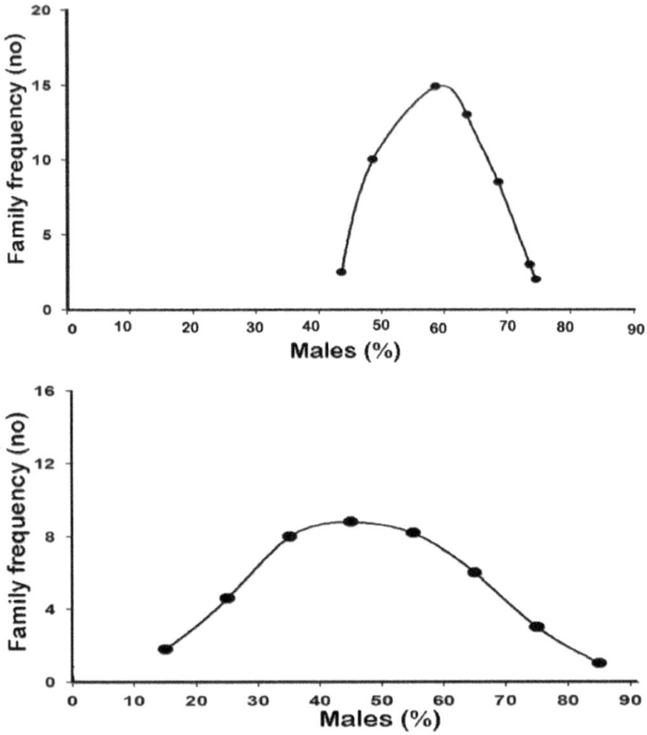

Fig. 1.5. Free hand drawing to show the number of families displaying deviated male ratio in Egypt Swansa (upper panel) and Chitralada strain (lower panel) of *Oreochromis niloticus* (source: Mair et al., 1991; Tuan et al., 1999).

a significantly skewed sex ratio, it can be also due to difference in ability of 'male' or 'female' sperm.

Social mediated sex differentiation and change occur in cichlids like *Crenicara puntulata* and *Matriaclima* cf. *livingstoni* (see Pandian, 2013, p. 138). There is a claim that the fast growing juvenile dominance determines the differentiation of male sex in the Midas cichlid *Amphilophus citrinellus* (Francis and Barlow, 1993). However, Oldfield (2007) has experimentally shown that the larger body size typically observed in adult male Midas cichlid is due to post-maturational growth rather than that of juvenile differentiating into male. That seems to be true for sea bass but with the female growing faster.

Sea bass: The European *Dicentrarchus labrax* is one of the most important marine aquacultured species in the Mediterranean region. Thanks to large scale investments and accumulation of information on reproduction, nutrition and growth of the sea bass, its production has enormously increased since 1980s (120,000 t in 2010). The sea bass males outnumber the females. The male sex ratio ranges from 0.67 to 0.80 (Piferrer et al., 2005). However, the females are larger than males and attain sexual maturity at the age of 2–3 years, long after attaining marketable size. Associated with retarded growth, males attain sexual maturity on or before attaining the age of 2 years, coinciding with the marketable size. To study the causes for the deviation in sex ratio from unity and sex dependent differences in age and marketable size, a large volume of information has been published on genetic, endocrine and environmental role on sex differentiation process of the sea bass (e.g., Zanuy et al., 2001).

Vanderpuatte et al. (2012) analyzed the sex ratios in wild sea bass from 13 populations consisting of 4,889 individuals collected from major parts of the distribution range of the bass. They reported that the sex ratio in wild populations was biased toward female (0.594). In fact, the female ratio increased from 0.52 in smaller fish (< 30 cm total length TD) to 0.70 in larger bass (>40 cm TL). From an elaborate experimental study, Saillant et al. (2003) showed that density of the bass had no effect on sex ratio. This finding suggests that the high density culture usually practiced in aquaculture operations is not responsible for the observed male biased sex ratio. Repeated size grading from 84 to 199 day post-fertilization (dpf), the duration representing the labile period, during which sex is differentiated in the bass, had also no effect on sex ratio. Clearly, neither the density nor the size difference alters sex ratio of the bass in aquaculture systems. In a study, Gorshkov et al. (2003) estimated the parental effects from diallel crossings at 24°C. Within each full-sib families, the female ratio averaged to 50% but significantly varied from 20 to 68% among families. Natural spawning in wild populations occurs during Febraury (Saillant et al., 2003), when prevailing temperature is >17°C. However, the male-biased sex ratio in the bass is reported mostly from the laboratory (e.g., Saillant et al., 2003) and aquaculture systems, where the prevailing temperature exceeds 17°C. Hence it is likely that the high temperature induces male-biased sex ratio through parental effects (see Chapter 7, 6).

References

Barlow, G.W. 1984. Patterns of monogamy among teleost fishes. Arch Fischwiss, 35: 75–123.

Baroiller, J.F., D. Chourrout, A. Fostier and B. Jalabert. 1995. Temperature and sex chromosomes govern sex ratios of the mouthbrooding cichlid fish *Oreochromis niloticus*. J Exp Zool, 273: 216–223.

Baroiller, J.F., H. D'Cotta and E. Saillant. 2009. Environmental effects on fish sex determination and differentiation. Sex Dev, 3: 118–135.

Baron, D., J. Montfort, R. Housgatte et al. 2007. Androgen-induced masculinization in rainbow trout results in a marked dysregulation of early gonadal gene expression profiles. BMC Genomics, 8: 357. doi: 10.1186/1471-2164-8-357.

Beardmore, J.A., G.C. Mair and R.I. Lewis. 2001. Monosex male population in finfish as exemplified by tilapia: applications, problems and prospects. Aquaculture, 197: 283–301.
Berglund, A. 1993. The operational sex ratio influences choosiness in a pipefish. Behav Ecol, 5: 254–258.
Berglund, A., G. Rosenqvist and I. Svensson. 1986. Reversed sex roles and parental environment in zygotes of two pipefish (Syngnathidae) species. Mar Ecol Prog Ser, 29: 209–215.
Bradley, K.M., J.P. Breyer, D.B. Melville et al. 2011. An SNP based linkage map of zebrafish reveals sex determination loci. Genes/Genomes/Genetics, 1. doi: 10.11534/93.111 000190.
Carrasco, L.A.P., D.J. Penman and N.R. Bromage. 1999. Evidence for the presence of sex chromosome in the Nile tilapia (*Oreochromis niloticus*) from synaptonemal complex analysis of XX, XY and YY genotypes. Aquaculture, 173: 207–218.
Carruth, L.L. 2000. Freshwater cichlid *Crenicara punctulata* is a protogynous sequential hermaphrodite. Copeia, 2000: 71–82.
Chesser, R.K., M.W. Smith and M.H. Smith. 1991. Biochemical genetics of mosquitofish. 3. Incidence and significance of multiple paternity. Genetics, 74: 77–81.
Conover, D.O. and S.W. Heins. 1987. The environmental and genetic components of sex ratio in *Menidia menidia* (Pisces: Atherinidae). Copeia, 1987: 732–743.
David, C.J. and T.J. Pandian. 2006. Cadaveric sperm induces intergeneic androgenesis in the fish *Hemigrammus caudovittatus*. Theriogenology, 65: 1048–1070.
Delany, M., C. Follet, N. Ryan et al. 2002. Social interaction and distribution of female zebrafish (*Danio rerio*) in a large aquarium. Biol Bull, 203: 240–241.
Diana, J.S. 1983. Growth, maturation and production of northern pike in three Michigan Lakes. J Fish Res Bd Can, 36: 481–487.
Donaldson, T.J. 1987. Social organization and reproduction behavior of the hawkfish *Cirrhitichthys falco* (Cirrhitidae). Bull Mar Sci, 41: 531–540.
Donaldson, T.J. 1989. Facultative monogamy in obligate coral-dwelling hawkfishes (Cirrhitidae). Env Biol Fish, 26: 395–302.
Eshel, O., A. Shirak, J.I. Weller et al. 2010. Fine mapping of a locus on linkage group 23 for sex determination in Nile tilapia (*Oreochromis niloticus*). Anim Genet, 42: 222–224.
Emlen, S.T. and L.W. Oring. 1977. Ecology, sexual selection and the evolution of mating systems. Science, 197: 215–223.
FAO. 2010. Cutured Aquatic Species Information Programme. J.E. Rakocy (ed.). Fisheries and Aquaculture Department, FAO, Rome.
Fishelson, L. and F. Hilzerman. 2002. Flexibility in reproduction styles of male St. Peter's tilapia *Sarotherodon galilaeus* (Cichlidae). Env Biol Fish, 63: 173–182.
Forsgren, E., T. Amundsen, A.A. Borg and J. Bjelvenmark. 2004. Unusually dynamic sex roles in a fish. Nature, 429: 551–554.
Francis, R.C. and G.W. Barlow. 1993. Social control of primary sex differentiation in the Midas cichlid. Proc Natl Acad Sci USA, 90: 10673–10675
Fricke, H.W. 1980. Control of different mating systems in a coral reef fish by one environmental factor. Nature, 266: 830–832.
George, T. and T.J. Pandian. 1995. Production of ZZ females in the female-heterogametic black molly *Poecilia sphenops* by endocrine sex reversal and progeny testing. Aquaculture, 136: 81–90.
George, T. and T.J. Pandian. 1997. Interspecific hybridization in poeciliids. Indian J Exp Biol, 35: 628–637.
Gorshkov, S., H. Meiri, H. Rosenfeld et al. 2003. Parental effects on sex ratio in progeny of the European sea bass (*Dicentrarchus labrax*). Isr J Aquacult, Bamidgeh, 55: 265–273.
Griffin, D.K., S.C. Harvey, R. Campos-Ramos et al. 2002. Early orgins of the X and Y chromosomes: lessons from tilapia. Cytogenet Genome Res, 99: 157–163.
Hamilton, R.J., S. Adams and J.H. Choat. 2008. Sexual development and reproductive demography of the green humped parrotfish (*Bolbometopon nuricatum*) in the Solomon Islands. Coral Reefs, 27: 153–163.

Herpin, A. and M. Schartl. 2011. *Dmrtl* gene at the cross roads: a widespread and central class of sexual development factors in fish. FEBS J, 278: 1010–1119.
Hrbek, T., J. Seckinger and A. Meyer. 2007. A phylogenetic and biogeographic perspective on the evolution of poeciliid fishes. Mol Phylogenet Evol, 43: 986–998.
Hughes, V., T.J. Benfey and D.J. Martin-Robichaud. 2008. Effects of rearing temperature on sex ratio in juvenile Atlantic halibut *Hippoglossus hippoglossus*. Env Biol Fish, 81: 415–419.
Hutchings, J.A., T.D. Bishop and C.R. McGregor-Shaw. 1999. Spawning behavior of Atlantic cod *Gadus morhua*: evidence of mate competition and mate choice in a broadcast spawner. Can J Fish Aquat Sci, 56: 97–104.
Jirotkul, M. 1999. Operational sex ratio influences female preference and male-male competition in guppies. Anim Behav, 58: 287–294.
Jirotkul, M. 2002. Operational sex ratio influences the opportunity for sexual slection in guppies. J Fish Biol, 56: 739–741.
Kallman, K.D. 1984. A new look at sex determination poeciliid fishes. pp. 95–171. *In*: B. Turner (ed.). *Evolutionary Genetics of Fishes*. Plenum, New York.
Kanoh, Y. 2000. Reproductive success associated with territoriality, sneaking and grouping in male rose bitterling *Rhodeus ocellatus* (Pisces: Cyprinidae). Env Biol Fish, 57: 143–154.
Katano, O. and K. Maekawa. 1997. Reproductive regulation in the female Japanese minnow *Pseudorasbora pavo* (Cyprinidae). Env Biol Fish, 49: 197–205.
Kavumpaurath, S. and T.J. Pandian. 1992. Production of YY male guppy (*Poecilia reticulata*) by endocrome sex reversal and selective breeding. Asian Fish Soc, 57: 265–276.
Kavumpaurath, S. and T.J. Pandian. 1993. Production of YY female guppy *Poecilia reticulata* by endocrome sex reversal and progeny testing. Aquaculture, 116: 183–189.
Kirankumar, S. and T.J. Pandian. 2002. Effect on growth and reproduction of hormone immersed and masculinized fighting fish *Betta splendens*. J Exp Zool, 293: 606–616.
Kirankumar, S. and T.J. Pandian. 2003. Production of androgenetic tiger barb *Puntius tetrazona*. Aquaculture, 228: 37–51.
Kirankumar, S. and T.J. Pandian. 2004. Interspecific androgenetic restoration of rosy barb using cadaveric sperm. Genome, 47: 66–73.
Kirankumar, A., V. Anathy and T.J. Pandian. 2003. Hormonal induction of supermale golden rosy barb and isolation of Y chromosome specific markers. Gen Comp Endocrinol, 134: 62–71.
Kondo, M., I. Nanda, U. Hornung et al. 2003. Absence of the candidate male sex determining gene *dmrt1/Y* of medaka from other fish species. Curr Biol, 14: 1664–1669.
Kosswig, C. 1964. Polygenic sex determination. Experientia, 20: 190–194.
Kuwamura, T. 1997. The evolution of parental care and male mating systems among Tanganyikan cichlids. pp. 59–86. *In*: H. Kawanabe, M. Hori and M. Nagoshi (eds.). *Parental Care and Mating Systems*. Chukyo University Press, Yagota, Nagoya.
Laroche, J.K. and S.L. Richardson. 1980. Reproduction of northern anchovy *Engraulis mordax* off Oregon and Washington. Fish Bull, 78: 603–618.
Lee, B.Y., D.J. Penman and T.D. Kocher. 2003. Identification of a sex-determining region in Nile tilapia (*Oreochromis niloticus*) using bulked segregant analysis. Anim Genet, 34: 379–383.
Love, M.S., M. Yoklarich and L. Thornstein. 2002. *The Rockfishes of the North east Pacific*. University of California Press, Berkeley, CA, USA.
Lozano, C.A., B. Gjerde, J. Odegard and H.B. Bensten. 2013. Heritability estimates for male proportion in the GIFT Nile tilapia (*Oreochromis niloticus* L.) Aquaculture, 372-375: 137–148.
Mai, A.C.G., A.M. Garcia, J.P. Vieira and M.G. Mai. 2007. Reproduction aspects of the one-sided live-bearer *Jenynsia multidentata* (Jenyns, 1842) Cyprinidontiformes in the Patos Lagoon estuary, Brazil. Pan-Am J Aquat Sci, 2: 40–46.
Mair, G.C., A.G. Scott, D.J. Penman et al. 1991. Sex determination in the genus *Oreochromis* 1. Sex reversal, gynogenesis and triploidy in *O. niloticus* (L.). Theor Appl Genet, 82: 144–152.

Mair, G.C., J.S. Abucay, D.O.F. Skinbinski et al. 1997. Genetic manipulation of sex ratio for the large scale production of all-male tilapia *Oreochromis niloticus* L. Can J Fish Aquat Sci, 54: 396–404.
Majumdar, K.C. and B.J. McAndrew. 1986. Relative DNA content of somatic nuclei and chromosomal studies in three genera: *Tilapia, Sarotherodon* and *Oreochromis* of the Tilapiini. Genetics, 63: 165–168.
Marconato, A., V. Tessari and G. Martin. 1995. The mating system of *Xyrichthys novacula*: sperm economy and fertilization success. J Fish Biol, 47: 292–301.
Marconato, A., D.Y. Shapiro, C.W. Petersen et al. 1997. Methodological analysis of fertilization rate in the bluehead wrasse *Thalassoma bifasciatum*: pair versus group spawns. Mar Ecol Prog Ser, 161: 61–70.
Martin, E. and M. Taborsky. 1997. Alternative male mating tactics in a cichlid *Pelvicachromus pulcher*; a comparison of reproductive effort and success. Bahav Ecol Sociobiol, 41: 311–319.
Matsuda, M., T. Sato, Y. Toyazaki et al. 2002. *Dmy* is a Y-specific DM-domain gene required for male development in the medaka fish. Nature, 417: 555–563.
Matsuda, M., T. Sato, Y. Toyazaki et al. 2003. *Oryzias curvinotus* has *DMY*, a gene that is required for male development in the medaka *O. latipes*. Zool Sci, 20: 159–161.
Moyer, J.T. 1990. Social and reproductive behaviour of *Chaetodontoplus mosoleucus* (Pomacanthidae) at Bantayan Island Phillipines with notes on pomacanthid relationships. Jap J Ichthyol, 36: 459–462.
Muller-Belecke, A. and G. Horstgen-Schwark. 2007. A YY male *Oreochromis niloticus* strain developed from an exceptional mitotic gynogenetic male and growth performance of genetically all-male progenies. Aquacult Res, 38: 773–775.
Nanda, I., M. Kondo, U. Kornung et al. 2002. A duplicated copy of *Dmrt1* in the sex determination region of the Y chromosome of the medaka *Oryzias latipes*. Proc Natl Acad Sci USA, 99: 11778–11783.
Neat, F.C. 2001. Male parasitic spawning in two species of triple fin blenny (Tripterigiidae) contrasts in demography, behavior and gonadal characteristics. Env Biol Fish, 61: 57–64.
Oldfield, R.G. 2007. Behavioral interaction, body size and sex determination in the Midas cichlid *Amphilophus citrinellus*. J Fisher Internatl, 2: 242–249.
Olin, M., J. Jutila, H. Lehtonen et al. 2012. Importance of maternal size on the reproductive success of perch *Perca fluviatilis* in small forest lakes: Implications for fisheries management. Fisher Mgmt Ecol, 19: 363–374.
Oliveira, R.F. 2006. Neuroendocrine mechanism of alternative reproductive tactics in fish. *In*: *Fish Physiology: Behaviour and Physiology of Fish*, Elsevier, Amtserdam, The Netherlands, 24: 297–357.
Pandian, T.J. 1993. Endocrine and chromosome manipulation techniques for the production of all male and all female population in food and ornamental fishes. Proc Ind Natl Acal Sci, 59B: 549–566.
Pandian, T.J. 2010. *Sexuality in Fishes*, Science Publishers, Enfield/CRC Press, Boca Raton, USA, p. 208.
Pandian, T.J. 2011. *Sex Determination in Fish*. Science Publishers, Enfield/CRC Press, Boca Raton, USA, p. 282.
Pandian, T.J. 2012. *Genetic Sex Differentiation in Fish*. CRC Press, Boca Raton, USA, p. 213.
Pandian, T.J. 2013. *Endocrine Sex Differentiation in Fish*. CRC Press, Boca Raton, USA, p. 303.
Piferrer, F., M. Blazquez, L. Navarro and A. Gonzalez. 2005. Genetic, endocrine and environmental components of sex determination and differentiation in the European sea bass (*Dicentrarchus labrax* L.). Gen Comp Endocrinol, 142: 102–110.
Robertson, D.R. 1981. The social and mating systems of two labroid fishes *Halichoeres maculipinna* and *H. garnoti*. Mar Biol, 64: 327–340.
Robertson, D.R. and R.R. Warner. 1978. Sexual patterns in the labroid fishes of the western Carribean. 2. The parrotfishes (Scaridae). Smithsonion Contributions to Zoology, No. 255.

Robertson, D.R., N.V.C. Polunim and K. Leighton. 1979. The behavioral ecology of three Indian Ocean surgeonfishes (*Acanthurus lineatus*, *A. leucosternon* and *Zebrasoma scopas*): their feeding strategies and social and mating systems. Env Biol Fish, 4: 125–170.

Saillant, E., A. Fostier, P. Haffray et al. 2003. Effect of rearing density, size grading and parental factors on sex ratios of the sea bass (*Dicentrarchus labrax* L.) in intensive aquaculture. Aquaculture, 221: 183–206.

Sakai, Y. 1997. Alternative spawning tactics of female angelfish according to two different cantexts of sex change. Behav Ecol, 8: 372–377.

Santos, E.M., V.L. Wormann, G.C. Paull et al. 2007. Molecular basis of sex and reproductive status in breeding zebrafish. Physiol Genomics, 30: 111–122.

Scott, A.G., D.J. Penman, J.A. Beardmore and D.O.F. Skibinski. 1989. The YY supermale in (*Oreochromis niloticus*) and its potential in aquaculture. Aquaculture, 78: 237–251.

Schultheis, C., A. Bohne, J.N. Volff and D. Galiana-Arnoux. 2009. Sex determination diversity and sex chromosome evolution in poeciliid fish. Sex Dev, 3: 67–77.

Searcy, W.A. and K. Yasukawa. 1995. *Polygyny and Sexual Selection in Red winged Blackbirds*. Princeton University Press, Princeton, NJ.

Seghers, B.H. 1973. Analysis of geographic variation in the anti-predator adaptation of the guppy *Poecilia reticulata*. Ph.D. Thesis, University of British Columbia, Vancouver.

Selim, K.M., A. Shinomiya, H. Otake et al. 2009. Effects of higher temperature on sex differentiation and germ cell population in medaka *Oryzias latipes*. Aquaculture, 289: 340–349.

Shapiro, D.Y., A. Marconato and T. Yoshikawa. 1994. Sperm economy of a coral reef fish. Ecology, 75: 1334–1344.

Shelton, W.L., F.H. Meriwether, K.J. Semmens and W.E. Calhoun. 1983. Progeny sex ratios from intraspecipic pair spawning of *Tilapia aurea* and *T. nilotica*. pp. 270–280. *In*: L. Fishelson and Z. Yaron (eds.). Proc Internatl Symp Tilapia Aquacult. Nazareth, Isreal.

Spence, R. and C. Smith. 2005. Male territoriality mediates density and sex ratio effects on oviposition in the zebrafish *Danio rerio*. Anim Behav, 69: 1317–1323.

Stauffer, R.J., Jr. and R.A. Ruffing. 2008. Behaviorally induced sex reversal of *Matriaclima* cf. *livingstoni*. Copeia, 2008: 618–620.

Taborsky, M. 1994. Sneakers, satellites and helpers: Parasitic and co-operative behavior in fish reproduction. Adv Stud Behav, 23: 1–100.

Taborsky, M. 2001. The evolution of bourgeois, parasitic and cooperative behavior in fish reproduction. J Hered, 92: 100–110.

Takegaki, T. and A. Nakazona. 1999. Reproductive behavior and male fidelity in the monogamous goby *Valenciennea longipinnis*. Ichthyol Res, 46: 115–123.

Takegaki, M., K. Sakai and N. Taniguchi. 2008. Direct evidence of multiple paternities in natural population of viviparous Japanese surfperch by allelic markers of microsatellite DNA loci. Fish Sci, 74: 976–782.

Travis, J., J.C. Trexler and M. Mulvery. 1990. Multiple paternity and its correlates in female *Poecilia latipinna*. Copeia, 1990: 722–724.

Tuan, P.A., G.C. Mair, D.C. Little and J.A. Beardmore. 1999. Sex determination and the feasibility of genetically male tilapia production in the Thai-Chitralada strain of *Oreochromis niloticus* (L.). Aquaculture, 173: 257–269.

Vanderputte, M., E. Quillet and B. Chatain. 2012. Are sex ratio in wild European sea bass (*Dicentrarcus labrax*) populations biased? Aquat Living Resour, 25: 77–81.

Varadaraj, K. and T.J. Pandian. 1989. First report on production of supermale tilapia by integrating endocrine sex reversal with gynogenetic technique. Curr Sci, 58: 434–441.

Warner, R.R. and P. Lejeune. 1985. Sex change limited by parental care: a test using four Mediterranean labrid fishes genus *Symphodus*. Mar Biol, 87: 89–99.

Williamson, K.S., R. Phillips and B. May. 2008. Characterization of a chromosomal rearrangement responsible for producing apparent XY-female fall-run chinook salmon in California. J Hered, 99: 483–490.

Wohlfarth, G. and H. Wedekind. 1991. The heredity of sex determination in tilapias. Aquaculture, 92: 143–156.
Yamagishi, S. and M. Kohda. 1996. Is the cichlid fish *Julidochromis marlieri* polyandrous? Ichthyol Res, 43: 469–471.
Zane, L., W.S. Nelson, A.G. Jones and J.C. Avise. 1999. Microsatellite assessment of multiple paternity in natural population of a live-bearing fish *Gambusia holbrooki*. J Evol Biol, 12: 61–69.
Zanuy, S., M. Carillo, A. Felip et al. 2001. Genetic, hormonal and environmental approaches for the control of reproduction in the European sea bass (*Dicentrarchus labrax* L.). Aquaculture, 202: 187–203.
Zhang, Z., Y. Wang, S. Wang et al. 2011. Transcriptome analysis of female and male *Xiphophorus maculatus* JP 163A. PLoS ONE, 6: e18379. doi: 10.1371/journal pone.0018379.
Zuniga-Vega, J.J., A.L. Hermandez-Roses, A. Molina-Moctezuma et al. 2012. Population abundance and sex ratio of the viviparous fish *Poeciliopsis baenschii* (Poeciliidae) throughout its range in Western Mexico. West North Am Naturlist, 72: 357–368.

2

Overexploitation by Fisheries

Introduction

This chapter describes the negative effects of overexploitation by capture and recreational fishing on decreased Reproductive Life Span (RLS) and reduced reproductive output. Irrespective of overexploitation or under-exploitation, fishing is not known to disrupt sex differentiation but it regulates age/size at which sexual maturity is attained. Broadly, the chapter has two major sections: (i) overexploitation-induces precocious sexual maturity resulting in reduced recruitment and collapse of stocks of a fish or even a fish species of gonochores and (ii) male and size-specific capture fishing induces precocious maturity and sex change in females, and sperm economization in males of protogynic fishes. Due to overexploitation, the Pacific mackerel disappeared in 1933 and Californian sardine in 1949, and the Norwegian and Icelandic herring waned during 1950s followed by South African pilchard in 1960, Peruvian anchovies in 1962 and George Bank's herring in 1967 (Beverton, 2000). Nevertheless, the depletion of a stock or species may be caused by overexploitation (e.g., *Gadus morhua*, Hutchings and Reynolds, 2004; *Alosa alosa* in the Rhine River, Rougier et al., 2012) or aquatic pollution (e.g., *A. alosa* in Morroco, Rougier et al., 2012), which shall be described in another chapter, or loss of spawning and nursery habitats (e.g., *A. sapidissima* in the Atlantic coast of USA and Canada, Rougier et al., 2012). Hence depletion of a species can be traced to overexploitation in an aquatic system but to pollution in another. Evidently, overexploitation by fisheries alone can not be held responsible for the depletion and/or collapse of a stock or species fish. Rather overexploitation, pollution and habitat degradation may be individually and/or jointly responsible for collapse of several stocks. Incidentally, there are ongoing debates, whether a stock or species by capture fisheries is really depleted or not. For example, questioning the gear and habitat selected by Myers and Worm (2003) to report the decline of tuna populations, Hampton et al. (2005) showed that biomass decline and fishing impacts are much less severe. Interestingly, the annual landings of the pelagic planktivore *Sardinella longiceps* and benthic carnivores *Upeneus supravittatus, Trichiurus lepturus* and *Saurida undoquamis* is 2,402 tonne at Chennai (Southeast coast of India). From a study on trophodynamics, Gomathy (2013) assessed the approximate food consumption of these fishes as 59,673 t, of which the amount of commercially important fishes and crustaceans consumed by the three benthic carnivores was 23,815 t. Remarkably, predatory fishes seem to consume nearly 10 times more commercially important fishes than they are being captured. Hence, the prey-predator relationship in an ecosystem may also determine the impact of overexploitation of fish stocks.

2.1 Fisheries scenario

Historically, fishes constitute an important source of human food and nutrition. An attempt is made to provide a comprehensive picture of the levels of fisheries and related activities up to 2010 (Table 2.1). Of 90 million tonnes (mt) captured world over, just 11.2 mt arise from freshwater. Contributing 87% of the capture, marine fisheries constitute a major resource. Fisheries provide not only food but also gainful employment. Following the collapse of cod fishery, for example, the number of sailors and fishing

Table 2.1. Fishing scenario of the world and India (source: Mostly from FAO, 2012, others are from Vivekanandan et al., 2013a,b, personal communication A. Gopalakrishnan, Kochi, Funge-Smith and Kennely, 2013).

Particulars	World	India
Total catch (mt)	90	3.8
Species captured (no)	800	296 + 68*
Species collapsed (no)	9	-
Stock collapsed (no)	102	-
Fishing personnel (no × 6^{10})	3.8	1.0
Craft		
Non-mechanized (no)	-	50,618
Mechanized (no)	83,000†[1]	143,872
Capture area (m km^2)	?	2.02
Mean capture depth (m)	200	100
Capture effort (t/fisher/year)	7.0–8.0†[2]	3.2
Fuel used for fishing (l/t fish)	680	393
CO_2 emission (t CO_2/t fish)	1.7	1.0

*Marine ornamental fish,†[1] trawlers from southeast Asia alone, †[2] European fisher.

vessels have been halved during the past 20–25 years in European countries, especially France (Villisante, 2010). Annual fishing capacity ranges from 2–3 t for a fisher from developing countries to 7–8 t for a fisher of Europe and North America; for example, fishing capacity of an Indian fisher is just 3.2 t. Understandably, overexploitation of marine fishery resources especially by developed countries has led to almost total collapse (99.9%) of the Atlantic cod (Hutchings and Reynolds, 2004). Besides the family of Antarctic rock cods, FAO (2010) lists nine depleted fish species that are beyond recovery (Table 2.2). Developed countries are more responsible for depletion/collapse of these world fishery resources. For example, the catches of *Notothenia rossil* alone exceeded 100,000 t per season from 1960s to 1970s around South Georgia and the Kerguelen Islands (Fallon and Stratford, 2003). The number of European sturgeon *Acipenser sturio* has been depleted from 10,000 in 1990 to about 200 in 2010 (Gessner et al., 2011). Both condition factor and CPUE (catch per unit effort) of the Antarctic toothfish *Dissostichus mansoni* dramatically decreased in 1992 following heavy fishing since 1972 (Ainley et al., 2013). Hutchings and Reynolds (2004) have listed > 80% depletion of 102 stocks out of 193, for which reliable data are available for periods longer than 25–73 years.

With progressively increasing use of more and more efficient crafts and gears, as well as fossil fuel for their operation especially during the last 60 years, the global capture of fishes has impressively increased and stabilized at 90 mt (Fig. 2.1). However, a developing country like India, with 3.8 mt from its seas in 2012, has fixed a target of 8.0 mt by 2025. With increasing area and depth up to 200 m of fishing, a huge volume of seawater is filtered annually to capture about 90 mt of marine fishes (Table 2.1).

The number of mechanized boats including trawlers increased, for example from 6,708 in 1961 to 143,872 in 2010 in India (Table 2.1). In a seminal publication, Vivekanandan et al. (2013) have determined the fuel used for

Table 2.2. Depleted species by developed countries (source: Food and Agricultural Organization of the United Nations, 2010. General Situation of World Fish Stock, Reproduced with permission).

Species	Main fishing countries
Gadus morhua	Canada, USA, Greenland
Melanogrammus aeglefinus	Canada, USA
Petrus rupestris	South Africa
Atractoscion aequidens	South Africa
Clupeonella cultriventris	Russia, Ukraine
Alosa pontica	Bulgaria
Thunnus thunnus	France, Italy, Turkey
T. maccoyii	Japan, Taiwan, South Africa
Chaenocephalus gunnari	United Kingdom, S. Korea
Nototheniidae	United Kingdom

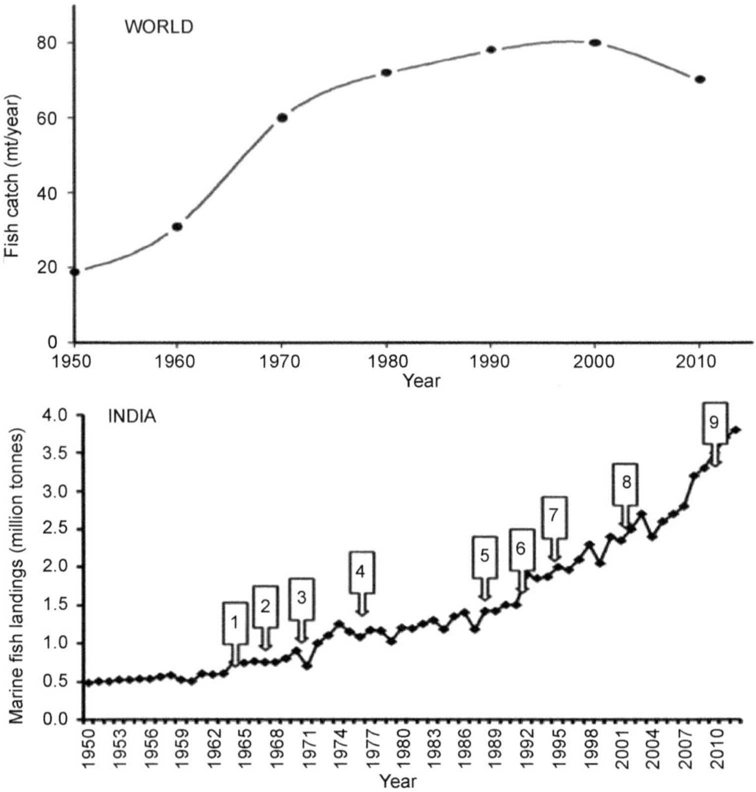

Fig. 2.1. Upper panel: Global trend in capture fisheries as function of year (source: Food and Agricultural Organization of the United Nations, 2012. The State of World Fisheries and Aquaculture, Reproduced with permission). Lower panel: Milestones of increases in Indian capture fisheries (Vivekanandan per com). 1. Introduction of mechanization, 2. Introduction of synthetic gear, 3. Construction of fishing harbors, 4. Trawlers and shrimp fishery become popular, 5. Motorization of artisanal craft, 6. Introduction of ring-seine, 7. Expansion of fishing to distant grounds, 8. Expansion of motorized craft and ring-seine and 9. Introduction of fishery for tunas.

fishing as 680 l/t fish and 393 l/t fish for the world and India, respectively. The carbon dioxide output from the use of fossil fuel for marine fishing is in the range of 1.7 t CO_2/t fish captured for the world and 1.0 t CO_2/t fish for India. Considering 42.4 mt fossil fuel (≈2.4% of the global oil consumption) used by fishing vessels and the consequent CO_2 emission (134 mt CO_2 for the world and 3.6 mt CO_2 for India) (Table 2.1), 'green fishing' is being emphasized in recent years.

Recreational fishing, essentially a quiet, benign and often solitary activity, involves selectively high-value carnivores, which are already depleting. The catch by more than 65 million anglers from Europe (22 countries), USA, Canada and Australia is estimated as 2 mt (Cooke and Cowx, 2006). Introduction of exotic fish for sport is exemplified by rainbow trout now in 82 countries and still spreading along with the associated expensive angling gear and related infrastructure (Cambray, 2003). Around Trinidad in the Caribbean, recreational fishing accounts for 11% of the commercial catch (Mike and Cowx, 1996). The Mediterranean Island Majorca captures 31% of fish production at the highest trophic level (Morales-Nin et al., 2005). During 1990s, the Atlantic States Marine Fisheries Council has reported that recreational capture of *Morone saxatalis*, *Pomatomus saltarix*, *Coryphaena hippurus*, *Sebastes marinus* and *Tautoge onitis* has exceeded those of commercial fisheries (Beal et al., 1998). Post et al. (2002) documented instances of decline in fish stocks of *Salvelinus namayacush*, *Sander vitreus*, *Esox lucius* and *Oncorhynchus mykiss*. In Spain, angling has decreased the breeding stock and egg production relative to unexploited rivers (Almodovar and Nicola, 2004). However, there are no reports on depletion of either fish or fish stock or reproductive impairment due to recreational fishing.

2.2. Sexual maturity

Depletion of a commercial fish stock or species is recognized by fisheries managers from the (i) drop in commercial landings, (ii) mean size landed and (iii) catch effort estimated in number of fish (iiia) trapped per trap/hour or (iiib) captured per unit effort (see Table 2.14). Reproductive characteristics such as size/age at sexual maturity and fecundity are considered by scientists (e.g., Kjesbu et al., 2010) to trace the reproductive step(s), through which a stock/species is depleted. The most important reproduction steps that are affected by overexploitation in capture fisheries are: (i) age and/or size at sexual maturity and (ii) fecundity. In gonochoric fishes, survival, growth, age or size at sexual maturity, fecundity, RLS, mating patterns and life history strategies are important factors that ultimately determine recruitment. Sexual maturity or puberty is the developmental stage, at which an individual fish becomes capable of sexually reproducing for the first time. It commences sometime after sexual differentiation (termination of labile period in primary gonochores, see Pandian, 2013, p. 221) and culminates at the first spermination in male and ovulation leading to spawning of the non-hydrated demersal eggs or hydrated pelagic eggs by female (Taranger et al., 2010).

Many commercially important food-fishes are large, live long and are characterized by high juvenile mortality. Inhabiting 260,000 km^2 area of the Pacific Ocean, the breeding population of the Japanese sardine *Sardinops melanogaster* spawn 5,130 trillion eggs annually; of these, less than 0.1% juveniles alone are recruited (Watanabe et al., 1996). In many pelagic fishes, both male and female attain sexual maturity at the same size, for example 46–48 cm Fork Length (FL) in *Coryphaena hippurus* (Schwenke and Buckel, 2008) and 10–11 cm Standard Length (SL) in *Engraulis mordax* (Laroche and Richardson, 1980). However, in many others, females require understandably a longer duration to attain sexual maturity, as adequate nutrients have to be stored to meet the vitellogenic requirement. For example, the mullet (*Mugil cephalus*) females begin to mature at the age of 2 years at 29.1 cm Total Length (TL) and 100% of females in a population attain sexual maturity at the age of 4 years and size 40 cm TL, against males beginning to attain maturity at the age of the one year and size of 24.8 cm TL; 100% of males mature at the age of 2 years and size of 30 cm TL (McDonough et al., 2005). In the viviparous rockfish *Sebastes brevispinis*, 60 and 40% of males and females mature at the age of 13 and 17 years, respectively (Stanley and Kronlund, 2005). Among freshwater fishes too, the age at which the sexual maturity is attained, for example, in the perch (*Perca fluviatilis*) is delayed in females by one to two years (Olin et al., 2012). In this perch, both age and size at sexual maturity increase with increasing latitude with corresponding decrease in temperature. However, Gonado Somatic Index (GSI) decreased linearly from about 29% at 35 °N (≈21°C) to about 18% at 70 °N (≈ 3°C) (Fig. 2.2). Whereas the GSI may compensate predation or fishing pressure in tropical and temperate waters, it may not be in Arctic and Antarctic waters. Hence fishing pressure at the Arctic and Antarctic waters may inflict depletion of a stock or species much more rapidly than in tropical waters (see Table 2.10). Notably, female red brand

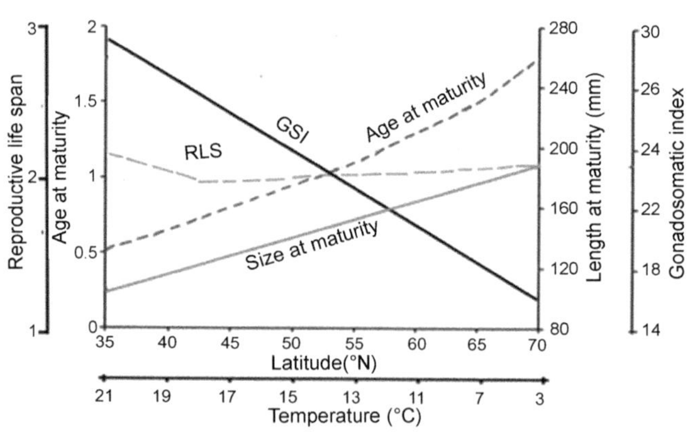

Fig. 2.2. Decreasing trends in body length and age at sexual maturity estimated for 75 populations of Eurasian perch *Perca fluviatilis* as function of latitude and corresponding temperature. Figure also shows the trends of GSI and RLS as functions of latitude and temperature (source: Heibo et al., 2005).

Color image of this figure appears in the color plate section at the end of the book.

fish *Cepola macropthalma* attains maturity at the age of 5 years and size of 42.0 cm TL in the highly predated southern Euripos strait of the Aegean Sea, against 8 years and 58.7 cm TL in the northern strait (Stergiou, 1999). The age at maturity of cod *Gadus morhua* ranges from the age 2 years in the warmer Irish Sea to 3–4 years in relatively cooler North Sea and ≈8 years in the cold North east Arctic Sea (Nash et al., 2010). Similarly, the size at 50% maturity for the Greenland halibut *Rheinhardtius hippoglossoides* varies from 48 cm at Gulf of St. Lawrence during 1978–1981 to 78.5 cm at eastern New Foundland (Cooper et al., 2007). Obviously, apart from fishing, gender, temperature and predation may also alter the age and size of sexual maturity.

According to the theory of life history, the populations that encounter high or variable adult mortality invest in reproduction at an early age (i.e., short generation time), while those that experience high or variable juvenile mortality, reproduction is delayed, i.e., their generation time is relatively longer (Wooton, 1992). For a description on morphological, ecological and behavioral adaptation to escape from predation, Wooton (1998) may be consulted. However, quantitative studies on mortality caused by predation in fishes characterized by shorter generation time are limited. Reznick et al. (2006) evaluated the reproductive performance of the laboratory-born guppy *Poecilia reticulata* from the Oropuche and Yarra drainages. Exposing the guppy to four different combinations of predation and feeding levels, they estimated generation time, RLS, inter-brood interval and progeny production. From Table 2.3, the following may be summarized: 1. Experiencing high predatory mortality of adults, the guppy matured at an early age 83 days against 103 days, i.e., shorter generation time and produced more number of progenies (25–28 against 16–21) by (a) shortening the inter-brood interval (26–28 days against 30–31 days) and (b) extending the RLS from 462–627 days in low-predation series to

Table 2.3. Effect of combinations of predation and feeding on reproductive performance of *Poecilia reticulata* drawn from Oropuche and Yarra populations (compiled from Reznick et al., 2006).

Predation level	Feeding level	Generation time (day)	Reproductive life span (day)	Inter-brood interval (day)	Progeny (no)
			Oropuche		
Low	Low	103	627	31	21
Low	High	91	524	30	17
High	Low	98	817	29	28
High	High	83	800	28	28
			Yarra		
Low	Low	96	572	31	19
Low	High	91	462	30	16
High	Low	87	641	27	25
High	High	79	646	26	26

641–817 days in the high predation series. 2. High feeding at low predation level not only shortened the RLS but also reduced progeny production; at low predation level, high feeding reduced the progeny production. Hence predation causing high adult mortality plays a more crucial role than food availability in enhancing reproductive performance of the fish. Another factor that alters mortality and RLS is the latitude and corresponding temperature. Investigating 75 Eurasian populations of perch *P. fluviatilis*, Heibo et al. (2005) reported that RLS tends to decrease with increasing latitude; consequently, the high mortalities may not be compensated (Fig. 2.2). Hence, juvenile and/or adult mortality due to increased predation or fishing may not be compensated by the delayed sexual maturity and the decreased RLS beyond 45° N corresponding to ≈17°C.

2.3 Fecundity

The total number of oocytes contributing to fecundity is assured by waves of oogonial proliferation and subsequent oocyte recruitment (see Pandian, 2013, p. 24). Fecundity is decisively an important factor in recruitment. In this treatise series, many aspects of fecundity have been described (Pandian, 2010, Section 1.3). However, a re-visit to fecundity in the context of recruitment has become necessary. For familiarization, a few terms related to fecundity are explained below: 1. Batch Fecundity (BF) or a clutch is the number of eggs produced per spawning. It is a function of body size (L) (Fig. 2.5). Because BF is related to the volume of space available in the body cavity to accommodate the ripe ovaries ($F = aL^b$), geometry suggests that length exponent 'b' would be 3.0. Sampling 238 species, the exponent for the relationship between BF and L is found to be 3.1 (Wooton, 1998), although the exponent ranges widely, for instance, from 2.57 to 7.12 in *Lethrinus nebulosus* (Ebisawa, 1990). Figures 2.3a and 2.3b provide representative examples to show the levels, to which spatial and temporal changes alter the exponent in *Rheinhardtius hippoglossoides* (Cooper et al., 2007) and *Nemipterus japonicus*. In fact, *N. japonicus* is subjected to intensive fishing pressure in the southwest Bay of Bengal; while smaller and younger (< 23 cm TL, 147 g, 21 month) females respond by increasing fecundity (Fig. 2.3b) the larger and older females (26 cm TL, 217 g, 28 month) are unable to increase fecundity to the expected level (Vivekanandan, 2013a). 2. Season fecundity is the batch fecundity multiplied by the number of spawning during a breeding season. 3. Life time fecundity is the season fecundity multiplied by the number of years during RLS. 4. Relative Fecundity (RF) is the number of eggs spawned per unit body weight (e.g., *Perca fluviatilis*, Heibo et al., 2005) or ovary-free body weight (e.g., *Engraulis mordax*, Laroche and Richardson, 1980) of fish. In *E. mordax*, for example, RF is estimated as 720 eggs per g body weight and 829 eggs per g ovary-free body weight. Irrespective of this

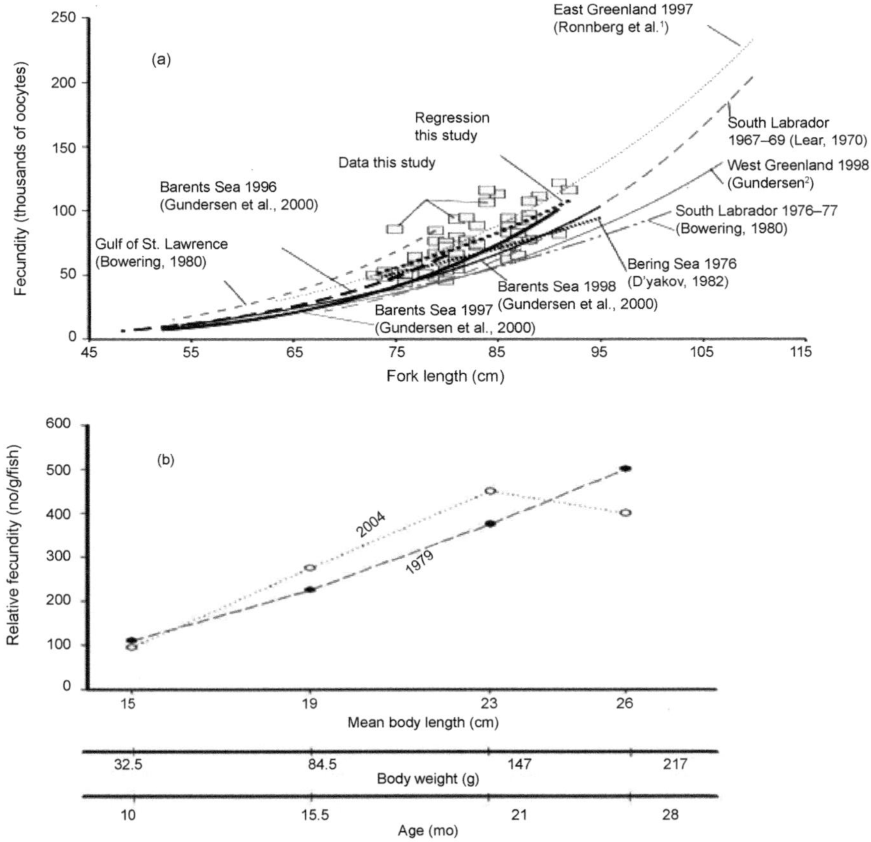

Fig. 2.3. (a) Fecundity as a function of body length of Greenland halibut *Rheinhardtius hippoglossioides,* as reported by different authors during different years (from Cooper et al., 2007). For references cited in Fig. 2.3a, Cooper et al. may be referred. (b) Fecundity as functions of body length, weight and age of threadfin bream *Nemipterus japonicus* landed from southwest Bay of Bengal (compiled from Vivekanandan, 2013a).

difference, it provides a scope for comparative analysis of the reproductive performance of stocks from different geographic areas and during different years, as well as between species. For example, a 3-kg weighing carpenter sea bream *Argyrozona argyrozona* produces five-fold more eggs per kilogram body weight than a fish of 1-kg (Brouwer and Griffiths, 2005). A large (60 cm FL) female red snapper *Lutjanus compechanus* spawn as many eggs as 212 small (42 cm FL) females (Sadovy, 1996). Besides, larger females produce larger eggs in *Salmo trutta* (Ojanguren and Brana, 1996) and larger larvae in *Gadus morhua* (Pepin et al., 1997). 5. Realized fecundity is the number of eggs actually spawned and is usually less than that of potential fecundity (oocyte number) due to different quanta of follicular atresia. For example, potential and realized fecundities of *Pleurogrammus monoptergius* are 41,994 and 30,644 eggs, respectively, indicating that a substantial fraction (27%) of oocytes undergo atresia (McDermott et al., 2007). 6. Reproductive potential is a measure of the capacity of a population to produce a viable eggs and larvae that may be recruited to the fishery (Trippel, 1999).

Even at interspecies level, the minimum batch fecundity in many tunas increases linearly with increasing age at sexual maturity (Fig. 2.4). Fecundity is correlated exponentially with body length (Fig. 2.5a) and is linearly correlated with body weight (Fig. 2.5b). In long living large fishes, a significant relation is also found between fecundity and age, but beyond a specific age, fecundity begins to decrease due to senescence (see Pandian,

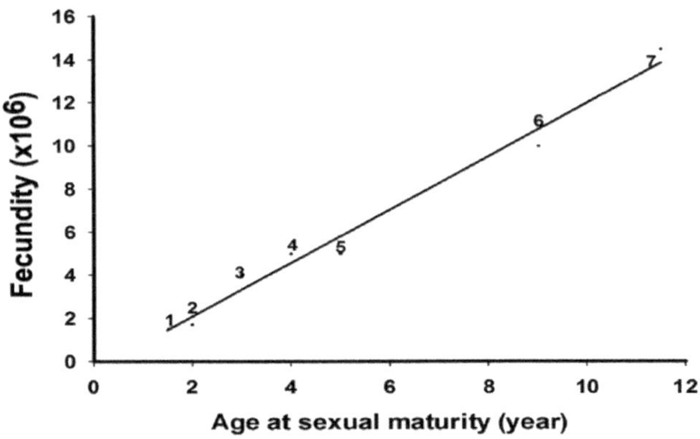

Fig. 2.4. Fecundity as a function of age at sexual maturity in selected tunas. 1. *Auxis thazard*, 2. *Euthunnus affinis*, 3. *Thunnus albacares*, 4. *T. thunnus*, 5. *T. obesus*, 6. *T. orientalis* and 7. *T. maccoyii* (source: Kumar, 2012).

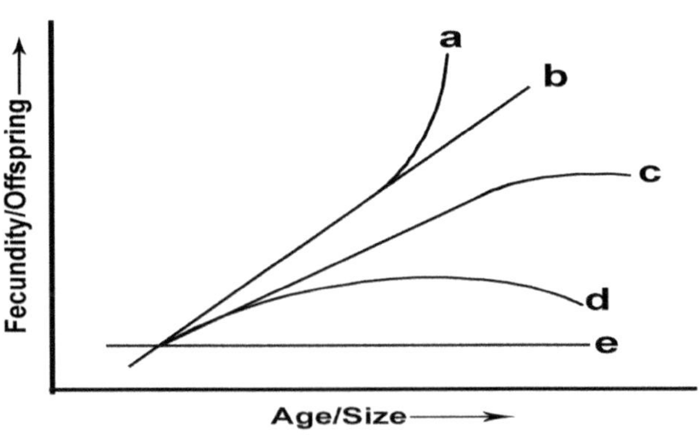

Fig. 2.5. Fecundity as functions of (a) body length, (b) weight, (c) age (e.g., *Hoplostethus atlanticus)* and (d) age (e.g., *Sebastes alutus*) and (e) body length (e.g., oviparous fish *Cyprinodon radiosus*) and matrotrophic viviparous fish *Heterandria formosa* (from Pandian, 2010, modified and added).

2012, p. 114) (Fig. 2.5c). For example, the net annual fecundity, after initially increasing with advancing age, progressively decreases from 120,000 eggs in the age class of 20 years to 3,000–4,000 in the 70–77 year age classes in *Sebastes alutus* (Fig. 2.5d). *Coregonus pidschian* also shows a positive relationship with age up to 15–20 years and decreasing thereafter. However, fecundity remains independent of females body size in many fishes (e.g., oviparous fish: *Cyprinodon radiosus*, Mire and Millet, 1994; viviparous fish: *Heterandria formosa*, see Pandian, 2010, p. 11, Fig. 2.5e). It may also decrease with size and age, as in *E. australis* and *Clupea bentincki* (Fitzhugh et al., 2012, see also Fig. 2.5c). In many fishes, however, both BF and RF decrease with ageing due to senescence (see Pandian, 2011, pp. 82–83).

With reference to the spawning strategy, Hunter et al. (1992) have recognized determinate and indeterminate spawners. Elaborating it, Witthames et al. (2009) have grouped them into: 1. Synchronous determinate total spawner (e.g., *Clupea harengus, Sebastes mentella*), 2. Synchronous determinate *batch* spawners (e.g., *Gadus morhua, Pleuronectes platessa*) and 3. Asynchronous indeterminate spawners (e.g., *Merluccius merluccius, Scomber scombrus*). Some life history traits of these three groups are summarized in Table 2.4. In the first two types, the follicles are in the process of synchronous development but in the first the demersal eggs are spawned in a single batch and in the second the pelagic eggs are spawned batch by batch, as eggs in successive batches are hydrated at different intervals. In asynchronous indeterminate spawners, the follicular size distribution is continuous. Whereas determinate fecundity is readily estimated using pre-spawning ovaries, that of indeterminate species can be estimated from the daily egg production method (DEPM, see Kjesbu et al., 2010), which

Table 2.4. Life history traits of marine fishes and their relevance to fecundity and spawning (compiled from Kjesbu et al., 2010).

Traits	Type 1	Type 2	Type 3
Breeding type**	High capital breeder	Low capital breeder	Income breeder
Breeder	Synchronous	Synchronous	Asynchronous
Fecundity	Determinate	Determinate	Indeterminate
Relative fecundity*	175	430	-
Food	Planktivores	Carnivores	Piscivores
Reserves	Muscles, Viscera	Liver	Liver
Spawning area	Demersal	Pelagic	Pelagic
Spawning period	Days	Weeks	Months
Behavior	Shoaling	Solitary	Solitary
Example	Herring *Clupea harengus*	Cod *Gadus morhua*	Hake *Merluccius merluccius*

*(egg/g), **For explanation, Pandian (2013, p. 58) may be consulted.

includes batch fecundity and percentage female spawning per day, as well as duration of spawning season. Spawning season lasts for a few days to weeks in species inhabiting higher altitudes but is prolonged for months in temperate and tropical fishes (see Fitzhugh et al., 2012). Understandably, sampling procedures for the quantification of fecundity differ for species with determinate and indeterminate fecundity (see Yoda and Yoneda, 2009). Hence quantification of fecundity ranges from the 'simple' herring to the 'semi-intricate' cod and to the 'highly intricate' hake. Sub-samples from the ovaries are subjected to histological (Emerson et al., 1990) and auto-diametric methods (Thorsen and Kjesbu, 2001) to estimate realizable fecundity. The reliability of these methods has been evaluated (Witthames et al., 2009). Carter et al. (2009) and Yoda and Yoneda (2009) have suggested methods to increase accuracy of quantification of fecundity; for example, ovarian samples of the coral grouper *Plectropomus leopardus* have to be collected by spear filling during the new moon in austral spring alone (Carter et al. 2009). Considering these complications for quantification of fecundity, especially in the context of a on board situation, Daily Egg Production (DEP, e.g., Kjesbu et al., 2010, see also Tables 4.22, 4.23) and spawning stock biomass (SSB, e.g., Hussy, 2011) are taken as measures to assess the reproductive potential of a fish or its stock.

Interestingly, potential fecundity of low capital breeders like plaice and cod is significantly correlated with hepato-somatic index, implying that determinate fecundity depends on stored energy reserves from the liver. In threadfin bream *N. japonicus*, Vivekanandan (2013a) found a parabolic relationship between body size and condition factor of the bream landed in 2004, implying that the larger females are unable to store adequate nutrients and energy; hence its fecundity is less than that expected (see also Fig. 2.3b). On the other hand, the indeterminate fecundity of income breeder like *Scomber japonicus* is significantly correlated with body condition but not with body size (Yamada et al., 1998), as it draws the required nutrients for vitellogenesis from their liver. In an extreme case of income breeder like the anchovies of the Solomon Islands, the indeterminate fecundity is positively correlated with zooplankton density of the waters (Milton et al., 1995). In fact Mangel and Smith (1990) suggested that plankton may provide a useful means of estimating the size of spawning biomass of some pelagic fishes (see also Gaughan et al., 2004). In indeterminate spawners, fecundity seems to be more dependent on nutrient sources rather than size or age of fish. Hence fecundity may vary with age, size and body source, from where the nutrients are drawn for vitellogenesis.

2.4 Spawning stress

Examining the FAO capture fisheries production database for 1959–1994 in terms of trophic levels of the catch, Pauley et al. (1998) found a clear shift in global fisheries from large piscivorous fishes to small planktivorous fishes, a process now called "fishing down the marine food web" (cf. Vivekanandan et al. 2005). The reduced or non-availability of mid-trophic level feed organisms appears to be critically more important for larval survival and recruitment rather than fecundity. A fact that has not so far been recognized is the spawning stress-induced mortality, especially fishes that produce hydrated oocytes. Spawning stress-induced mortality of adult females may significantly reduce Spawning Stock Biomass (SSB) in the ensuing years. Realizing its importance, an account is provided for the first time.

Natural mortality of fishes comprises of the loss of individuals in a population due to predation and parasites, and others like senescence leading to death. Sex specific natural mortality may also inherently alter sex ratio. However, reports on direct estimates of natural mortality, especially sex specific natural mortality of fishes are rare. The natural mortality is estimated often as the difference between total mortality (e.g., estimated by the catch-curve of the age frequency) and fishing mortality (e.g., estimated by mark-recapture). For the unexploited population, the mortality can be estimated by catch-curve analysis of the population age frequency. Estimating natural and fishing mortality of the walleye *Sander vitreus* in Escanabe Lake, Wisconsin during 1956–2009, Hansen et al. (2011) have confirmed the earlier reports, i.e., the relative vulnerability of the fish to natural mortality progressively decreases with increasing size/age, say, from unit 1.0 in 0+ age class to 0.2 in 16+ age class.

The dolphinfish *Coryphaena hippurus*, an oceanic epipelagic fish found in tropical and subtropical waters, is a fast growing fish attaining body weight of 30 kg (2 m FL) in three years. The dolphinfish is an income breeder and characterized by an indeterminate fecundity pattern. It spawns twice a year, the first one from May to July and the second from November to January; each spawning coincides with gonado somatic index (GSI) of 4.4. The hydrated eggs are spawned at an interval of 1.2–3.3 days, 27–74 times during each of the 90 day-long spawning season. From Aleja-Plata et al. (2011), it is noted that 1. The batch fecundity increases from 45,022 hydrated oocytes (1.3 mm diameter) in a mature small (49 cm FL) female to 1,930, 245 hydrated oocytes in a large (129 cm FL) female and 2. But the female ratio is decreased progressively from 0.76 in small size class (30 cm FL) to 0.13 in the largest (140 cm FL) size class (Fig. 2.6) indicating that the larger females suffered heavy mortality due to spawning stress. Unfortunately, Aleja-Plata et al. have not provided data on mortality of the dolphinfish. Interestingly, Schwenke and Buckel (2008) have estimated the age and growth relationship

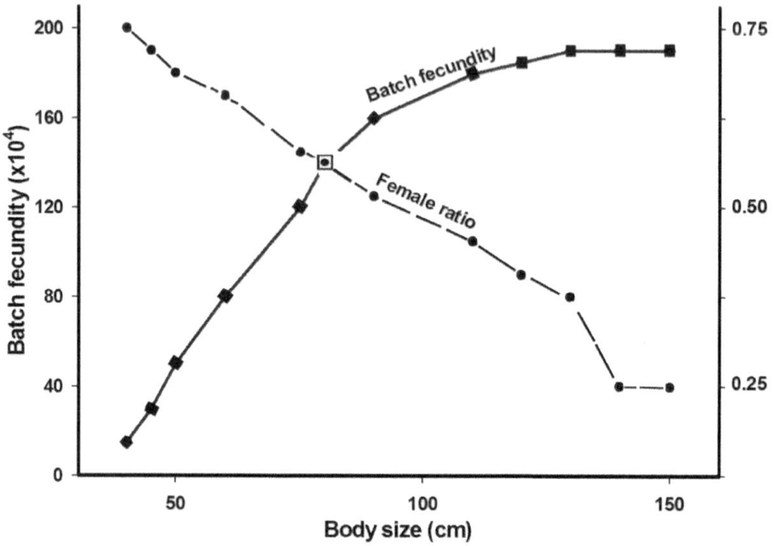

Fig. 2.6. Relationships of body size with batch fecundity and female ratio in dolphinfish *Coryphaena hippurus* (compiled from Aleja-Plata et al., 2011).

of 194 male and 217 female *C. hippurus* collected between May 2002 and May 2004 off North Carolina. Fifty percent females attain sexual maturity at the size of 46 cm FL and age of 6 months, and may undertake about three spawnings during the first and second year of their life. Figure 2b of Schwenke and Buckel (2008) shows the total absence of females (Fig. 2.7), indicating the spawning stress-induced mortality of females. In bigeye tuna *Thunnus obesus* too, Sun et al. (2013) found a progressive decrease in female ratio from 140 cm to 160 cm body size. In threadfin bream *N. japonicus* too, a demersal spawner characterized by indeterminate (?) fecundity, the female ratio decreases from 0.5 in 9.5 cm total length (TL) size class to 0.0 in 28.0 cm TL size class, perhaps due to spawning induced stress; however, males of 30 cm TL size class continued to be captured from southwest Bay of Bengal (Vivekanandan, 2013b). In *Labeo calbasu*, that spawn planktonic eggs in the Indian rivers, Dwivedi et al. (2011) have also reported a decrease in female ratio from 0.92 at age 2+ to 0.5 at age 5+. More publications on the subject, especially for viviparous fishes are needed. However, an indirect evidence for the parturition induced mortality in the quillback rockfish *Sebastes maliger* is reported. Rodgveller et al. (2012) reported the capture of gravid female quillback rockfish at 30–75 m depth from three different locations during 2006–2008. According to their Fig. 4, 89 gravid females have been captured. Of them 68 (76.4%) females were aged between 10 and 30 years; another 16 (18.8%) belonged to the age group of 35–45 years classes and the remaining 5% alone belonged to 50+ age classes (Fig. 2.8). Clearly, the

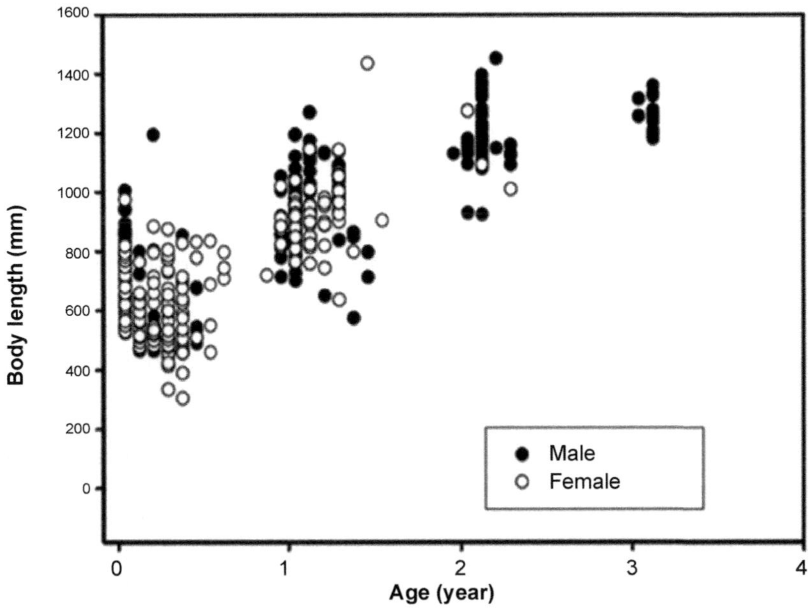

Fig. 2.7. Incidence of female (open circle) and male (closed circle) dolphinfish *Coryphaena hippurus* as a function of age. The fishes were collected off the coast of North Carolina during 2002–2004. Note the decreasing number of females from 2 year class and their total absence in 3 year class (from: Schwenke and Buckel, 2008).

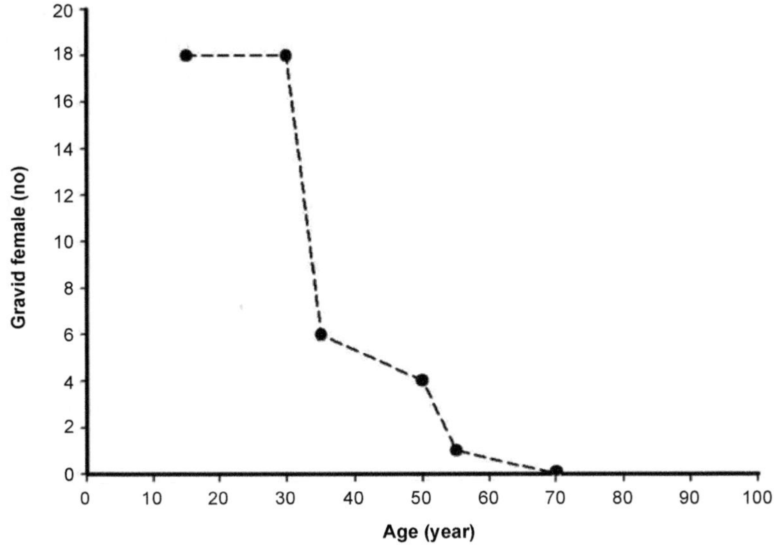

Fig. 2.8. Incidence of quillback rockfish *Sebastes maliger* as a function of age. Note the drastic decrease in female frequency from 30 year class onwards (source: Rodgveller et al., 2012).

older females, especially > 30 year classes suffered progressive increased mortality perhaps due to parturition stress. Hence spawning stress may be an important cause for the decrease in female ratio.

The Atlantic cod *G. morhua,* a low income breeder (Pandian, 2013, p. 48), is characterized by determinate fecundity pattern, and draws nutrients from the liver for vitellogenesis. It spawns for weeks during the spawning season, which lasts from May–July every year. Feeding activity of the cod is suppressed for 2–3 months before spawning and is completely ceased a month before spawning (Skjaeraasen et al., 2004). Morgan et al. (1999) have described the reproductive stress in the cod. GSI of the cod increases from 12.9 to 22.3 and to 25.9 in females of the 2nd, 3rd and 4th year classes, respectively (Table 2.5). More than 25% of the body weight of the 4+ year-old female consists of hydrated oocytes alone, which are spawned within weeks during the spawning season. An ovulated oocyte of the cod is 8.2 times larger in volume and 7.2 times heavier than a pre-hydrated oocyte (Finn et al., 2000). Hence retention of the ovulated, heavily hydrated oocyte batch within ovarian lumen leads to substantial increase in GSI and expansion of the belly. Examination of many post-spawning females has shown bruises and wounds particularly around the genital pore. Thanks to Arnason and Bjornsson (2012), experimental data on sex-specific mortality of the cod are available. Rearing cod of two year class for two years, they estimated separately the cumulative and sex specific natural mortality. The average annual spawning stress-induced mortality was 21 and 36% of the reproductive total mortality for males and females, respectively. Of this respective mortality, mortality during spawning season alone increased from 7 to 19 and to 34% in the females belonging to the 2nd, 3rd and 4th year classes, respectively (Table 2.5). The spawning-related stress in fishes that ovulate heavily hydrated oocytes may cause high mortality of females and consequent change in sex ratio. The described spawning stress induced female specific mortality, one for indeterminate (dolphinfish) and the other for determinate (cod) fecundity patterns, are the only available examples. For some reason, many authors, who have considered many aspects of reproductive performance of depleted fishes, for example cod (Kjesbu et al., 2010), have not taken note of the spawning induced stress leading to

Table 2.5. Spawning mortality and gonado somatic index of *Gadus morhua* reared from November 2002 to November 2004 (compiled from Arnason and Bjornsson, 2012).

Age (year)	Spawning mortality as % of total annual mortality		Gonado somatic index	
	Male	**Female**	**Female**	**Male**
2	-	7	12.9	5.8
3	5	19	22.3	8.6
4	7	34	25.9	11.1

sex specific mortality. Another fact that has also not so far been considered is the severe inbreeding within declining/depleted breeding population in formerly abundant marine fishes (O'Leary et al., 2013).

2.5 Collapse of fishery

Published fecundity estimates for overexploited fishes are rare and often based on limited samples and restricted size and age groups. Decline in catch of a fish species or its stock is primarily caused by depletion in SSB and the consequent decrease in Daily Egg Production (DEP) (Corten, 2013). Confirming the Allee effect, rate of population growth declines beyond recovery, when SSB falls below a threshold level of abundance. As much is known about the collapse of the Atlantic cod, *Gadus morhua* is chosen to explain how overexploitation of fishery affects reproductive performance by advancing age/size at sexual maturity and fecundity. With the history of sustained fishing for 500 years (Hutchings and Reynolds, 2004), the estimated annual harvest of cod was 0.2 mt (Xu et al., 2012) and the harvest during 1950–2000 was more than 41.3 mt from the Baltic Sea alone (Zeller et al., 2011). As indicated, the cod is a semi-intricate species for quantification of fecundity. Added to it, its distribution extends from the saline waters of Canadian Atlantic to low saline waters of eastern Baltic (Hussy, 2011). The minimum salinity required to ensure fertilization is 15–16‰ and 11–12‰ for the high saline western and low saline eastern Baltic cod stocks, respectively (Nissling and Westin, 1997). Due to lower buoyancy at lower salinities, eggs sink and fail to develop to hatching (von Westernhagen, 1970). The stocks of the Irish and North Seas distinctly differ from those of Barrents and Islandic Seas. The Irish stock matures at the age of 1.6 years and size of 36 cm (Table 2.6); both age and size at maturity follow a North-South gradient with the first maturity at an earlier age and much smaller length in the south in comparison to the north. This North-South gradient

Table 2.6. Selected reproductive traits of European cod *Gadus morhua* (compiled from Thorsen et al., 2010).

Particulars	Irish Sea	North Sea	Barents Sea	Islandic Sea
Temperature (°C) at 100 m depth	10	-	4	-
Mean size (cm)	55	69	82	94
Age at 80 cm (y)	4	5	8	8
Age at maturity (y)	1.6	3.8	7.2	6.6
Size at maturity (cm)	36	62	78	76
Condition factor	1.27	-	0.89	-
Oocyte diameter (µm)	570	537	644	672
Egg size (µm)	558*	540	646–603**	618–534***

*Value for 2003, **from 1986 to 2006, ***from 1995 to 2000.

also holds true for the condition factor. Correspondingly, the matured oocyte size ranges from 537–570 μm in the warmer southern waters to 644–672 μm in the colder northern waters; i.e., the oocyte size of northern cod stocks is 16% larger than that of the southern cod stocks (Thorsen et al., 2010).

Nash et al. (2010) quantified selected reproductive traits of *G. morhua*, whose stocks in the North Sea, Barents Sea and Northeast Arctic Sea are overexploited at different levels. 1. Fishing mortality (F) in these stocks increased from 0.2 in 1945 to 1.6 in 2005. 2. Correspondingly, total stock biomass also decreased from about 4.5 X 10^6 t in 1945 to 1.5 X 10^6 t in 2005 and 3. During this period, age diversity decreased from 0.75 H to 0.4–0.6 H. Hence overexploitation of larger sized fish resulted in (i) significant difference in age/size composition of the stock, (ii) SSB and (iii) reduced age/size at sexual maturity. For example, the size composition of the

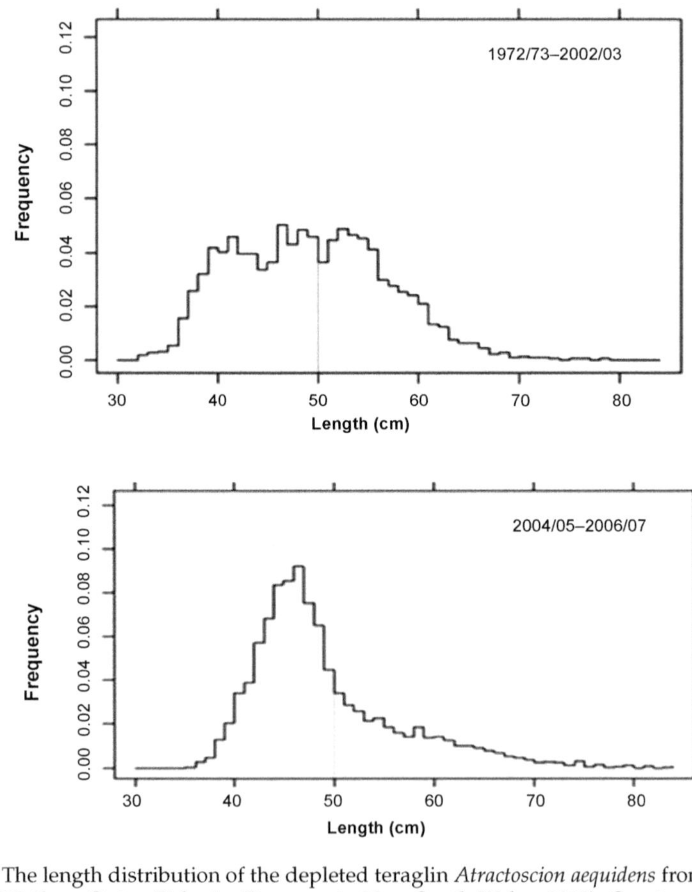

Fig. 2.9. The length distribution of the depleted teraglin *Atractoscion aequidens* from 1972/73 to 2006/07 (from Status Fisheries Resource in New South Wales, 2010). Courtesy: Status of Fisheries Resource in New South Wales: Industry and Investment of New South Wales, 2010.

depleted teraglin *Atractoscion aequidens* (see Table 2.2) landed since 1970s shows that the large fishes (>50 cm FL), which contributed more than 50% of the catch in 1972/1973, were almost no more available by 2006/2007 (Fig. 2.9). As a consequence, the age/size at sexual maturity is advanced, as older and larger individuals are being captured. In the depleted stock of *Chaenocephalus gunnari*, the age at maturity was advanced to 3 years from 5–6 years (Fallon and Stratford, 2003). With progressively increasing fishing pressure, the SSB (expressed in numbers) of the Baltic cod also decreased. For example, the time series of SSB and Recruitment (R), despite wide fluctuations, significantly decreased from 50,000 females (≈ 40,000 t) in 1970 to about 18,000 (≈20,000 t) in 2008; correspondingly, the number of cod recruited also declined from 200,000 in 1971 to about 75,000 in 2009 (Fig. 2.10). Similarly, the abundance of the spawners of the allis shad *Alosa alosa* in southwest France, whose fishery is also depleted, decreased from 285,918 females in 1994 to 6,429 in 2008. Correspondingly, the number of juveniles recruited also decreased from 34,120,025 in 1995 to 308,280 in 2009 (Rougier et al., 2012). The SSB of the European hake *Merluccius merluccius* of the Bay of Biscay decreased from 48.7 t in 1995 to 24.1 t in 2004. In fact, the number of females with hydrated ovaries encountered by Nash et al. (2010) also decreased from 22.7% in 1995 to 10.4% in 2001 (cf. Section 2.4). But the effective spawners of *A. alosa* remained 44% of the

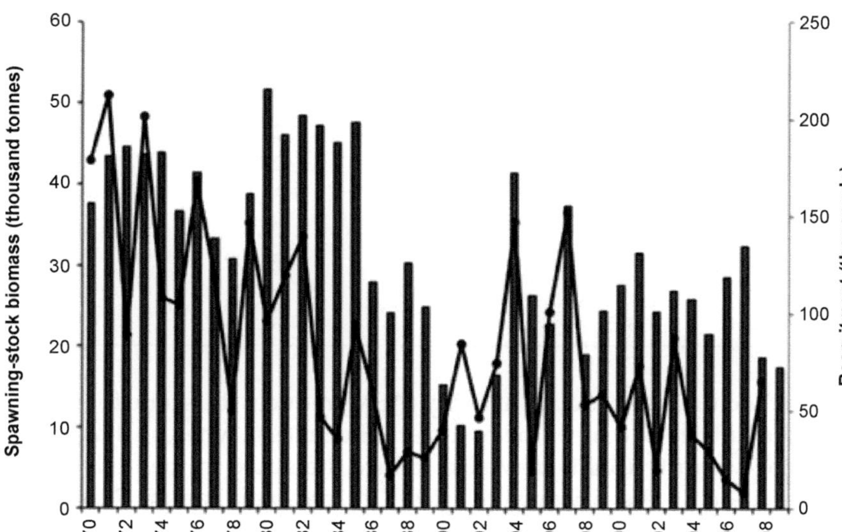

Fig. 2.10. Spawning Stock Biomass (SSB) of Baltic cod *Gadus morhua* from 1970 to 2007 and recruitment from 1971 to 2008. Bars represent SSB and line represents recruitment (from Hussy, 2011. Permission by International Council for Exploration of the Sea).

potential spawners during the entire period from 1991 to 2008 (Rougier et al., 2012). In the southwest Bay of Bengal, the number of female spawners of *Nemipterus japonicas* decreased from 4,917 during 1980s to 2,355 during 2001–2005 (Vivekanandan, 2013b).

Age at Sexual Maturity (ASM) of the Northeast Arctic Sea cod is advanced from 10 years in 1940 to 7 years in 2000 and size at sexual maturity (SSM) from 100 cm to 75 cm. Values reported by others increased from 2 years and 40 cm in the southern warmer parts (George Bank and the Baltic) to 7 years and 75 cm in the northeastern colder parts (see Nash et al., 2010). There are also other publications to confirm the reports on cod. For example, the Pacific stocks of the orange roughy *Hoplostethus atlanticus* are captured since 1979 but that of the Atlantic from 1991 only. Nevertheless, its stock in Northwest Ireland of the Atlantic underwent severe depletion and the landings fell from 3,502 t in 1991 to 138 t in 2008 (Minto and Nolan, 2006). Being subjected to intensive fishing since 1979, the SSM is advanced to 26–35 cm in different Pacific stocks, in comparison to 38 cm of the Atlantic stock (Table 2.7). Hence advancing ASM and/or SSM is a signal by a stock suffering intense exploitation (Trippel, 1995).

Table 2.7. Size range captured, size at maturity and egg weight of *Hoplostethus atlanticus* (compiled from Koslow et al., 1995; Minto and Nolan, 2006, modified).

Area	Size range (cm)	Size at maturity (cm)	Relative fecundity (no/kg)	Egg weight (µg)
		Australian Pacific		
Challenger Plateau	25–40	26	27,270	27
Cook Canyon	25–42	29	27,180	27
Richtie Banks	30–40	32	28,550	29
Chathan Rise	30–42	32	31,480	31
Eastern Tasmania	32–45	32	31,085	31
South Australia	32–45	35	33,339	-
		Atlantic		
Porcupine Bank	36–51	38	33.376	33

Advancing age at sexual maturity (ASM) reduces the scope for storing adequate nutrients to meet the cost of vitellogenesis, whereas that of Size at Sexual Maturity (SSM) may also reduce the space in the body cavity to accommodate the ripening ovaries. On overexploitation, RLS, ASM, SSM, age diversity (H) and SSB are all decreased (Tables 2.8, 2.9). With advancement of ASM and/or SSM, a fish may opt to (a) reduce both the number and size of eggs, (b) reduce the egg numbers alone or (c) reduce the egg size alone. Unfortunately, relevant information on the options of the depleted fishes listed in Table 2.2 is not available. Table 2.9 summarizes

Table 2.8. Effect of overexploitation on reproductive impairment in gonochoric and protogynic fishes.

Group characteristics	Reported/inferred changes
	Gonochoric fishes
Tropical, Small , Short life span (>2 years), Lecithotrophic, Non-superfetatious, e.g., *Poecilia reticulata* Reznick et al. (2006)	1. Advances age at sexual maturity 2. Increases RLS, 3. Increases progeny output
Medium sized, Short life span (14 months), Demersal eggs, e.g., *Nemipterus japonicus* Vivekanandan (2013b)	1. Advances SSM and ASM 2. Reduces egg size
Long life span (5 years), Determinate fecundity, e.g., *Coilia mystus,* He et al. (2011)	1. Reduces egg size but with increase in RF
Large, Relatively long life span Low capital breeder, Determinate fecundity, Hydrated pelagic eggs, e.g., *Gadus morhua* Nash et al. (2010)	1. Advances age at maturity by 2–3 years 2. Reduces age diversity 3. Decreases RLS due to spawning stress-induced mortality, 4. Reduces SSB (Fig. 2.10)
Large indeterminate fecundity, Income breeder, Hydrated pelagic eggs, e.g., *Merluccius merlucccius* Murua et al. (2010)	1. Advances size at maturity? 2. Reduces daily egg production
Large, Relatively short life span, Indeterminate fecundity, Hydrated pelagic eggs, e.g., *Coryphaena hippurus* Aleja-Plata et al. (2011)	1. Reduction in RLS due to spawning stress-induced mortality 2. Reduced fecundity
Large, Long living > 180 years Determinate fecundity, e.g., *Hoplostethus atlanticus* Minto and Nolan (2006)	1. Advances size at sexual maturity 2. Reduces RF (Table 2.7) 3. Reduces fecundity 4. Reduces egg size
	Protogynic fishes
Low income breeders, Hydrated pelagic eggs, e.g., *Pagrus pagrus,* (Harris and McGovern, 1997), *Epinephelus niveatus* (Wyanski et al. 2003)	1. Reduction in mean size of landed fish indicating of large females with high fecundity are lost, e.g., 1.1 kg/porgy in 1972 to 0.7 kg/porgy in 1978 in porgy; ≈ 72 cm in 1970 to 55 cm in 1990 in grouper 2. Advance age and/or size at maturity, e.g., by 0.5 year and 5.5 cm in grouper and 5 cm in porgy 3. Consequently, RLS of female is reduced by 50 % in grouper and 5 cm growth period in porgy. RLS of male is also reduced by 33% 4. Significant reduction in female stock biomass, e.g., reduction to 20% in porgy (Table 2.14) and corresponding reduction in fecundity 5. Significant reduction in male number

Table 2.9. Effects of overexploitation on egg size and/or fecundity of some fishes.

Species/Authors	Reported observations
Reduction in relative fecundity (RF) and egg size	
Hoplostethus atlanticus Cold temperate, Long life span of >180 years, Determinate fecundity, Hydrated pelagic eggs. Minto and Nolan et al. (2006)	Egg size of Atlantic orange roughly decreased from 33 μg at less overexploited Porcupine Bank stock to 27 μg at relatively more overexploited Cook Canyon stock. RF also decreased from 33 eggs/g to 27 eggs/g in these stocks (Table 2.7)
Reduction in RF only	
Merluccius merluccius Cold temperate, Large, Indeterminate income breeder, Hydrated pelagic eggs. Murua et al. (2010)	In the Bay of Biscay, RF of the European hake decreased from 187 eggs/g ovary free fish in 1995 to 133 eggs/g in 2004, in response to 12% decrease in size at sexual maturity
Reduction in egg size followed by increased fecundity	
Nemipterus japonicus Tropical, Short life span of 14 month Vivekanandan (2013b)	In the southwest Bay of Bengal, egg size of threadfin bream decreased from 0.12 mg in 1979 to 0.10 mg in 2004, however, with marginal increase in fecundity
Coilia mystus Warm temperate, 5 year life span, Determinate fecundity. He et al. (2011)	In the Yangtze Estuary, China, egg size of the anchovy decreased from 1.4 mg in 1979 to 1.1 mg in 2007. However, RF increased from 712 eggs/g to 871 eggs/g during this period
Gadus morhua Cold temperate, > 8 year life span, Determinate low capital breeder, Hydrated pelagic eggs. Thorsen et al. (2010)	In the Islantic Sea, egg size of the European cod decreased from 618 μm in diameter during 1995 to 534 μm in 2006. However, RF of Baltic cod increased from 640 eggs/g in 1987 to 850 eggs/g in 1996 (Kraus et al., 2000)

available information on the patterns of response by overexploited fishes to advancement of ASM and/or SSM. Pattern 1: Reduction in both Relative Fecundity (RF) and egg size (e.g., *H. atlanticus*), Pattern 2: Reduction in RF only (e.g., *M. merluccius*), Pattern 3: Reduction in egg size. Notably, reduction in egg size is followed by a marginal (e.g., *N. japonicus*) or significant (e.g., *C. mystus*, *G. morhua*, see Kraus et al., 2000) increase in RF. Vivekanandan (2013a,b) has provided adequate information to estimate the effect of overexploitation on recruitment of *N. japonicus*. Despite beginning to suffer overexploitation, as shown by (i) ASM and (ii) reduction in egg size, the threadfin bream compensates overexploitation by increased recruitment (Table 2.10). Clearly, tropical fishes like *Poecilia reticulata* and *N. japonicus*, characterized by a short life span of < 2 years are able to compensate predation/exploitation. However, the temperate fishes, characterized

Table 2.10. Effect of 25-years of fishing threadfin bream *Nemipterus japonicus* on fecundity, spawners and recruitment in southwest Bay of Bengal (compiled from Vivekanandan, 2013a,b).

Parameter	1980	2004
Relative fecundity (egg/g)	303	319
Absolute fecundity (no)	69,847	72,909
Estimated spawners (no)	4917	2355
Recruitment (10^3/spawner)	220	500
Total recruitment (10^6)	1.08	1.18

by a longer life span of more than two years, are not able to compensate overexploitation, as their GSI is likely to decrease with increasing latitude (see Fig. 2.2). The temperate fishes characterized by life span of 5–8 years (*C. mystus, G. morhua*) experience reduction in egg size but partly compensated by increased fecundity (Table 2.9). The other temperate fish *M. merluccius* with a life span longer than 8 years experiences reduction in RF alone. This is true of the Peruvian hake *M. gayi* (Ballon et al., 2008). With about 3 years of life span, the tropical dolphin fish, whose female ratio decreases with advancing age owing to spawning stress, experiences reduction in SSB. Oocyte hydration significantly reduces SSB of pelagic spawners. Incidentally, energy cost of oocyte hydration in warm tropical and cold temperate fishes may prove to be impressive. Interestingly, *H. atlanticus* with a life span of more than 180 years (Minto and Nolan, 2006) suffers from reduction in both RF and egg size (Fig. 2.11). Remarkably, overexploited fishes characterized by determinate fecundity (e.g., *G. morhua*) reduced egg

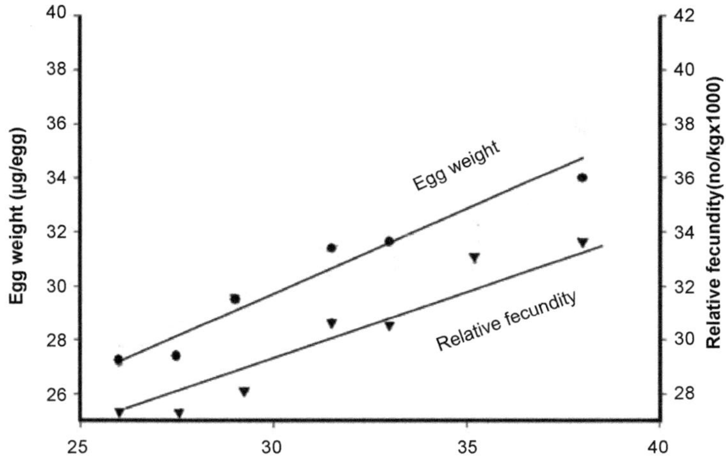

Fig. 2.11. Fecundity and egg size as function of size at sexual maturity in Atlantic and Pacific stocks of *Hoplostethus atlanticus* (drawn from data reported in Table 2.7).

size with slight increase in fecundity but that with indeterminate fecundity (e.g., *M. merluccius*) reduced egg size with no increase in fecundity. Briefly, overexploited fishes characterized by determinate fecundity (e.g., *G. morhua*) reduce egg size but that of indeterminate fecundity (e.g., *M. merluccius*) reduced fecundity. Briefly, overexploitation disrupts reproductive potential of a stock or species by (i) reductions in SSB and (ii) age diversity as well as (iii) advancing ASM and/or SSM and thereby (iv) fecundity and/or egg size (Tables 2.8, 2.9). Further, the ability to compensate overexploitation seems to depend on life span of fish and latitudes of its geographical distribution.

2.6 Sex change

In vertebrates, fishes are the only taxa that display sex change and provide an opportunity to understand the very plastic process of sex differentiation. Sex change occurs in many hermaphroditic fishes and also in a few secondary gonochoric fishes (Table 1.2). In hermaphroditism, both male and female reproductive function is expressed simultaneously or sequentially in a single individual. In fishes, three patterns of functional hermaphroditism have been recognized: simultaneous, sequential and serial (Fig. 1.3). Simultaneous hermaphrodites function as male and female at the same time (e.g., *Kryptolebias marmoratus*) or within a short span of time (e.g., *Serranus tortugarum*). They do not undergo natural sex change. But the sequential and serial hermaphrodites undergo natural sex change; the sequential change sex only once in a single direction during their life time but the serials do it more than once in either direction. Sex changing patterns of sequential have been described elsewhere (see p. 17). It is also shown that all hermaphroditic fishes are capable of undergoing natural sex change in either direction; only the number of times a hermaphrodite undergoes sex change is not frequent among simultaneous hermaphrodites, is minimum among sequential and is more frequent among serials (Pandian, 2013, p. 215). Figure 1.4 shows the directions and number of times natural sex change or artificially induced sex reversal occurs in protogynic and protandric fishes.

A fraction of protogynic *Pagrus pagrus* females does not change sex during its entire life span (Kokokris et al., 1999); the same is true of protandric *Diplodus vulgaris* (Pajuelo et al., 2006). On the other hand, administration of optimal MT doses to 2-years old juvenile *Epinephelus suillus*, 1+ year old juvenile *E. marginatus* and 5-months old juvenile *E. malabaricus* are reported to directly differentiate into male and produce sperm; hence they do not pass through the female phase (see Pandian, 2013, p. 159). Many protandric clownfishes, when reared in groups of two, pair formation takes place sooner or later, the large one directly differentiates into female and does not pass through the male phase (see Pandian, 2013, p. 160).

In hermaphroditic fishes, sex change has evolved independently in at least 23 teleost families and more than 350 species (Frisch, 2004). Nevertheless, it is a relatively rare phenomenon and occurs only in 4% teleost families, i.e., 23 out of 450 families and includes about 1% species alone. Among the sex changing hermaphrodites, 74% are protogynic hermaphrodites. Clearly, this pattern of hermaphroditism has been the choice of natural selection; the choice per se decreases in the order of protogyny (74%) < protandry (20%) < serial (6%) (Pandian, 2012, p. 105). Not surprisingly, more publications are available on sex change in protogynics. Hence protogynics are more focused in this book. Table 2.11 lists the male ratios reported for monochromatics. With lowest male ratios of 0.06, 0.11 and 0.15 in *Mycteroperca bonaci*, *Pagurus auriga* and *Choerodon schoenleinii* may be depleted for want of adequate males in their respective populations. The ratios for the dichromatic monoandrics and diandrics are listed in Table 2.12. Among dichromatic monandrics, Initial Phase (IP)

Table 2.11. Increase of male ratio in monochromatic monandric protogynous hermaphroditic fishes (© 2010, Pandian, T.J., Sexuality in Fishes. Science Publishers/CRC Press, reproduced by Permission of Taylor and Francis Group, LLC, a division of Informa plc).

Species	♂ ratio
Serranidae	
Mycteroperca bonaci	0.06
Serranus baldwini	0.30
Epinephelus akaara	0.33
M. interstitialis	0.37
E. flavolimbatus	0.37
Hypoplectrodes maccullochi	0.54
Labridae	
Choerodon schoenleinii	0.15
Clepticus parrae	0.30
Thalassoma cupido	0.38
Cirrhilabrus temmincki	0.39
Sparidae	
Pagrus auriga	0.11
Pagrus pagrus	0.25
Pagellus erythrinus	0.33
Chrysoblephus puniceus	0.36
Gobiidae	
Coryphopterus glaucofraenatum	0.27
C. personatus	0.39
Gobiosoma multifasciatum	0.47
Lethrinidae	
Lethrinus rubrioperculatus	0.31
Pomacentridae	
Dascyllus reticulatus	0.23
Mean	0.31

*hermaphrodite to male

Table 2.12. Male ratio in monandric monochromatic and diandric hermaphroditic fishes (© 2010, Pandian, T.J., Sexuality in Fishes. Science Publishers/CRC Press, reproduced by Permission of Taylor and Francis Group, LLC, a division of Informa plc).

Species	♂ ratio	I.P./P.M.	T.P./S.M.
	Monandric dichromatics	I.P	T.P
Labridae			
Pseudolabrus celidotus	0.3	0.25	0.05
Scaridae			
Sparisoma viride	0.25	0.20	0.05
S. chrysopternum	0.26	0.21	0.05
S.rubripinne	0.33	0.25	0.08
Mean	0.285	0.23	0.06
	Diandrics	P.M.	S.M.
Labridae			
Halichoeres bivitattus	0.17	0.05	0.12
Suezichthys gracilis	0.17	0.05	0.12
Thalassoma bifasciatum	0.18	0.06	0.12
H. poeyi	0.23	0.09	0.14
H. pictus	0.24	0.12	0.12
H. marginatus	0.30	0.03	0.27
H. maculipinna	0.33	0.11	0.21
Scaridae			
Calomotus spinidens	0.20	0.06	0.14
Scarus vetula	0.25	0.14	0.11
S. croicenosis	0.53	0.26	0.27
Mean	0.26	0.10	0.16

I.P. = Initial Phase; T.P. = Terminal Phase; P.M. = Primary Male and S.M. = Secondary Male.

male ratios are four times higher than that of Termial Phase (TP) males. Hence, the smaller and younger IP males may play a more important role than TP males in fertilizing the eggs of (71.5%) females in the respective populations. Contrastingly, the secondary male ratios among diandrics are two times higher than that of primary males. The large and old secondary males may play a more important role in fertilizing the eggs spawned by (74%) females in their respective populations. Barring *Halichoeres marginatus*, *H. maculipinna* and *Scarus croicenosis*, the secondary male ratios of all other diandrics are less than 0.20. Hence overexploitation of larger secondary males of these diandrics may rapidly deplete their population size. Pandian (2010, p. 109) has estimated the male ratios as 0.31 in monandric monochromatics and 0.22 in protogynic haremics (see Table 1.5). Quite independently, Molloy et al. (2007) have also estimated the male ratio of 0.29 for 14 protogynic species.

Attempts have been made to estimate the cost and size at which protogynics change sex. In most gonochorics, fecundity increases with body size but it does not during the terminal body size (weight) and age (Fig. 2.5). When fishes attain 70% of their respective maximum body size of the total life span, they are approaching a more or less reproductively inactive menopause stage (Pandian, 2012, p. 113). The empirical estimation predicts that sequential hermaphrodites change sex, when they reach 80% of their respective maximum body size (Allsop and West, 2003). Taken together, the protogynic hermaphrodites switch to males, when they reach 70% of their maximum body size corresponding to 80% of their maximum life span (see Pandian, 2012, p. 113). Molloy et al. (2007) have estimated that the gonadal investments by sex changed males is just 20% of that of gonochores, and by protogynic females is about 50% of that of gonochores. Hence the protogynic sex changers invest less on gonads, perhaps to neutralize the cost involved in sex change (cf. Kazancioglu and Alonzo, 2009).

In natural populations, sex changing transitional constitutes a load on actively reproducing fraction of population and may require a period of 4 months in *Thalassoma bifasciatum* to 36 months in *Halichoeres marginatus* to complete sex change in 100% females (see Pandian, 2011, p. 171). The mean transitional load for 16 hermaphroditic species belonging to seven families is estimated as 10.4% of the population and that for percentage of life span, during which transitional persists is 32.1% at the mean male ratio of 0.33 (Pandian, 2010, p. 92). Male ratio seems to regulate the size of transitional load and percentage of the life span, during which the population has to bear the transitional load.

Hermaphroditic fishes have either an ovotestis or a gonad. Sadovy and Shapiro (1987) classified them into three groups. Group 1: Delimited ovotestis consisting of distinct heterologous zones of testicular and ovarian lobes, which are separated by a connective wall (Fig. 2.12), is present among both protogynics (e.g., *Pagrus pagrus*) and protandrics (e.g., *Acanthopagrus schlegeli*) (see Pandian, 2013, p. 144). In these sparids, the ovary is located dorsally and testis latero-ventrally. Figure 2.12:4b shows a section across the ovotestis to suggest the surgical scope for castration or ovariectomy. In fact castration of one-year old porgy successfully induced precocious sex reversal, comparable to the natural sex change in 2+ and 3 years old porgy (Wu et al., 2008). This surgery-induced precocious sex reversal is shown to be possible by virtue of the fact that Primordial Germ Cells (PGCs) and/or Spermatogonial Stem Cells (SSCs) and/or Oogonial Stem Cells (OSCs) are present in the testicular and ovarian zones (see Pandian, 2013, p. 217).

The undelimited type of gonads are further divided into two groups (Sadovy and Shapiro, 1987). In the undelimited type 1, the ovarian and

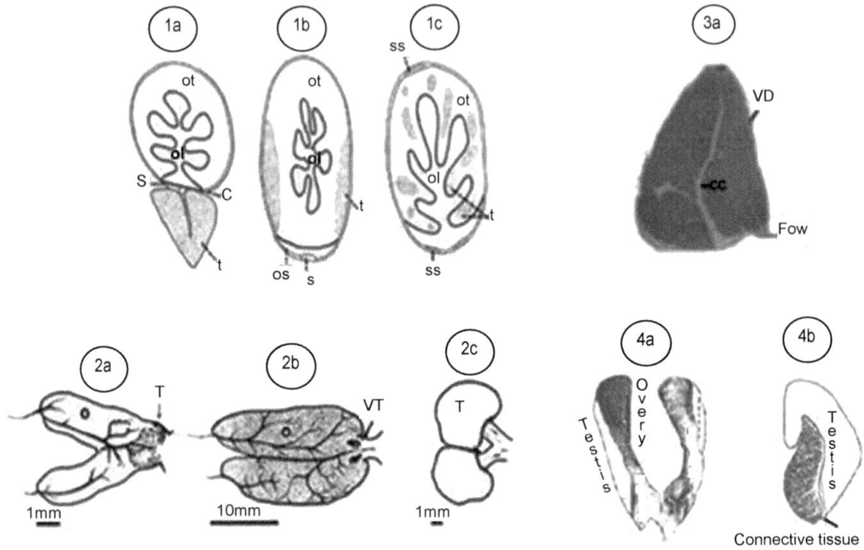

Fig. 2.12. Structural diversity in gonads of selected hermaphrodites. Changes in configuration of ovarian and testicular tissues of sequential hermaphrodites . 1a. Delimited type: e.g., *Pagrus pagrus*, b. Undelimited type 1: e.g., *Hypoplectrus puella*, c. Undelimited type 2: e.g., *Epinephelus guttatus*. c = connective tissue, ol = ovarian lumen, ot = ovarian tissue, t = testicular tissue, ss = sperm sinus, os = ovarian sinus, s = sperm duct (Sadovy and Shapriro, 1987). 2. Gonads of *Sufflamen chryspterus* a. gonads of juvenile with female coloration, b. a breeding female, c. a breeding male (from Takamoto et al., 2003), 3. Transverse section of testis a. *Lethrinus rubrioperculatus* (from Ebisawa, 1997). 4a. Schematic illustration to show the latero-ventral position of testes and dorsal position of ovaries in typical sparid. b. A cross section across the ovotesties to suggest the delicate surgery required for ovariectomy or castration (from Pandian, 2013, p. 144).

testicular tissues come in contact with each other but are never mixed (e.g., *Hypoplectrus puella*). In type 2, the ovarian and testicular tissues are intermingled during the course of sex change (e.g., *E. guttatus*). The undelimited gonads are found in almost all diandrics; in them, the gonad of the secondary sex is simply derived from the existing walls of the primary gonad. Presumably, the PGCs and/or SSCs/OSCs are attached to walls of the primary gonad; the entire basket of PGCs and/or SSCs/OSCs are lost, when these diandrics are gonadectomized. Hence *T. biafaciatum*, a diandric hermaphrodite is unable to regenerate its gonad following castration or ovariectomy (Godwin et al., 1996; Semsar and Godwin, 2003, 2004). Incidentally, the third type of gonad has been described in a blastid *Sufflamen chrysopterus* (Takamoto et al., 2003), in which natural sex change is accompanied by the diminishing ovarian component, and enlarging and maturation of dormant testis (Fig. 2.12:2a,b,c). The structure of this gonad provides excellent scope for ovariectomy and to test whether sex change from female to male can be induced.

2.7 Sperm economy

Due to fishing pressure on large sized males of protogynic fishes, there are many females for every male; for example, for every male black grouper (*Mycteroperca bonaci*), there are 15.4 females in Flordia, 33.3 in Cuba and 77.6 in Mexico (see Pandian, 2010, p. 90). Large males of *Thalassoma bifasciatum* spermiate 1,000 times more sperm than the number of eggs spawned (Warner et al., 1995, cf. with egg to sperm [1:1,000] ratio, Rurangwa et al., 1998). When a ratio of sperm to eggs ranges from 10^4 to 10^6, a single male can fertilize all the eggs in a population. However, when the ratio is decreased to 10^2, sperm limitation occurs, even in the absence of fishing or predation (Alonzo and Mangel, 2004). In protogynic stocks, overfishing of primarily large male members may reduce sperm availability even in a resilient species with high sperm production, and Fertilization Success (FS) and thereby reproductive output. Hence size- or age-selective fishing can inflict severe depletion in protogynic population through decreasing spermiating stock biomass. Hence Spawning-Per-Recruitment (SPR) and Spawning Stock Biomass-per-Recruitment (SSBR), commonly used to track per-recruitment production of larvae or eggs in gonochores (see Fig. 2.10), may not serve as good measures to protogynics (cf., e.g., Vaughan and Pragar, 2002). However, there is hardly any publication on the spermiating stock biomass of protogynics, comparable to spawning stock biomass of cod (Fig. 2.10) or allis shad (Rougier et al., 2012). Therefore, dependence on theoretical models addressing potential sperm limitation in protogynics has become the obvious choice.

For example, Alonzo and Mangel (2004) developed a model to show the effects of size/male selective fishing on the stock dynamics and sperm limitation in sex changing protogynics with the assumptions that: 1. The size of sex change remains fixed and is not socially mediated (see later) and 2. Considering 100% fertilization success occurs at the sperm to egg ratio of 1:1000. The model predicts that: (i) with the availability of 5 billion sperm, 95–98% of eggs spawned by 60 females may be fertilized but only about 85 and 75% of eggs of 300 and 700 females, respectively (Fig. 2.13). (iia) With increase in fishing mortality (F) from 0.5 to 3.0, the number of eggs produced remains equal (see next part); however, the FS per recruit is reduced by 20% (Fig. 2.14a). (iib) As fishing mortality increases, spermiating stock biomass-per-recruit decreases dramatically, while changes in spawning stock biomass is almost not detectable, i.e., 90% drop in male SSBR, compared with a 3% drop for female SSBR (Fig. 2.14b) (but see later). (iic) Mean sperm required per-recruit also decreases (Fig. 2.14c).

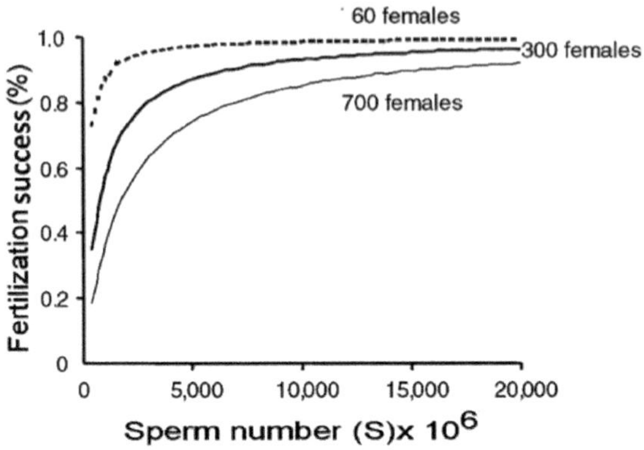

Fig. 2.13. Fertilization success (FS) as a function sperm number (s) 1 X 10^6 in the presence of 60, 300 or 700 protogynic females (from Alonzo and Mangel, 2004).

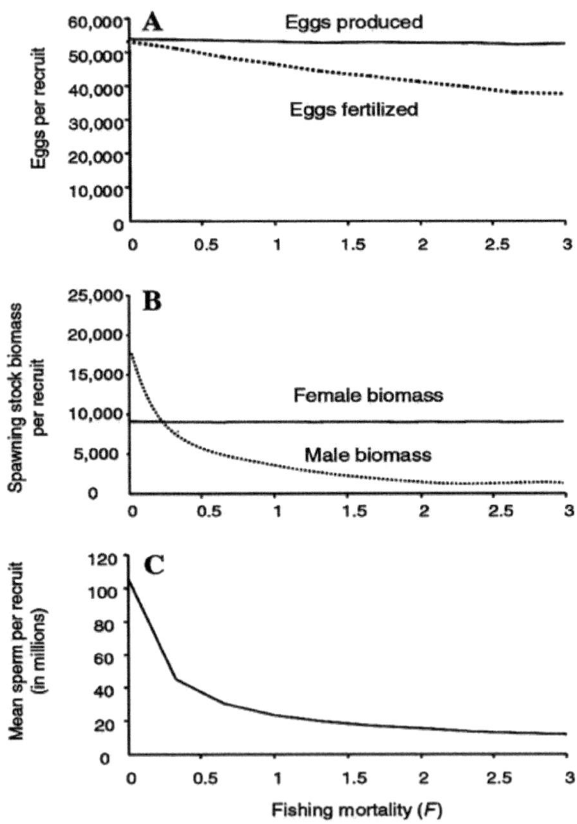

Fig. 2.14. Spawning-per-recruit measures as a function of fishing mortality (F) in the protogynic fishes (from Alonzo and Mangel, 2004).

Figure 2.15 represents a comparison between the number of eggs produced and fertilized in gonochoric and sex changing protogynic fishes as functions of increasing population size from left to right but increasing F from right to left. For an easier understanding, the following may be noted: 1. When F is negligible, 100% FS occurs in all the eggs produced by 825 females in sex changing protogynics but with 900 female (9% more than that of protogynics) gonochorics. 2. When F is 2.5, about 70% FS occurs in 675 protogynic females but FS is reduced to 45% in equal number of gonochoric females. Briefly, for every unit (1.0) of increase in F, FS is decreased by 8.5% in gonochores but by only 7.6% in sex changing protogynics. Beginning with an objective to show the sperm limitation, Alonzo and Mangel (2004) have but rightly shown that the sex changing protogynics have economized their sperm requirement, as they ensure, 65% FS for the egg produced, even at a higher F of 3.0 and at marginal reduction in female number than that for the gonochores. There is also evidence in support of this suggestion.

To demonstrate the sperm release and FS in the protogynic Mediterranean haremic wrasse *Xyrichthys novacula*, Marconato et al. (1995) collected all sperm and eggs released in individual spawn from territories holding four, six or 12 females. The haremic male undertook 'spawning rush' every day separately with each female from his territory during the

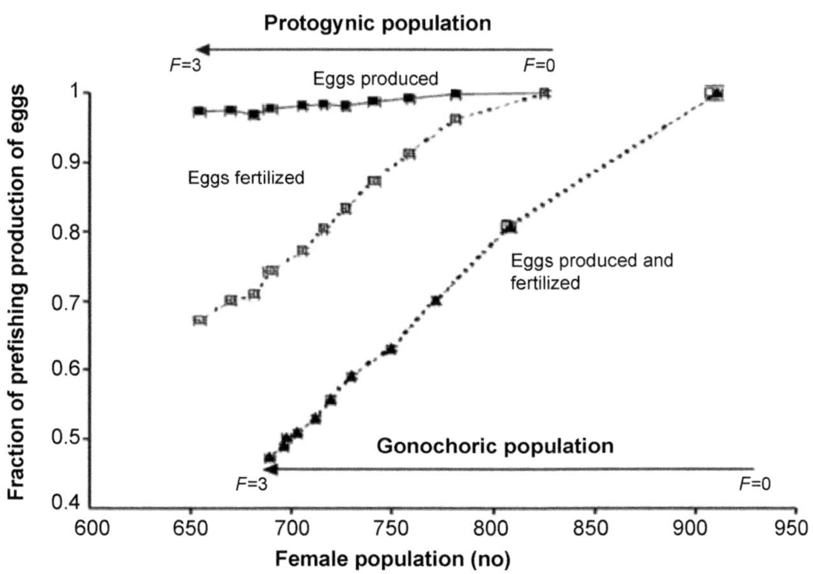

Fig. 2.15. Spawning-per-recruit (SPR) as a function of population size of protogynic (squares) and gonochoric (triangles) fishes. Mean egg production (filled) and fertilized (open) eggs per-recruit shown for a randomly mating population with one large mating group (from Alonzo and Mangel, 2004, modified).

breeding season from May to October. With increasing cumulative egg output by a female, the number of sperm released by the male increased from 5 X 10^6 sperm for 5 X 10^4 eggs to 15 X 10^6 sperm for 16 X 10^4 eggs, i.e., the sperm to egg ratio decreased from 1,000 sperm/egg in the former to 937 sperm/egg in the latter, but ensured 98% FS (Table 2.13). Remarkably, the number of eggs decreased dramatically from about 6,150 eggs spawned/♀ and 6,125 eggs fertilized/♀ in harems each with four or six females to 4,731 eggs spawned/♀ and 4,101 eggs fertilized/♀ in a harem with 12 females. Hence, the females in the harem with 12 inmates spawned significantly less number of eggs, which the respective territorial males were able to recognize and release sperm accordingly. This observation indicates that the male does indeed gauge his sperm release according to the number of eggs spawned by a female (Fig. 2.16, see also Pandian, 2010, p. 143). Despite decrease in

Table 2.13. Reported (bold numbers) and calculated (light numbers) estimates of egg and sperm released by haremic protogynic *Xyrichthys novacula* (compiled from Marconato et al., 1995).

Eggs and sperm	Total eggs spawned (x 10^4)	Eggs spawned (x 10^4/♀)	Sperm release (x 10^6)	Sperm /egg	Fertilization success (%)
Sperm released	**5**	-	**5**	1000	**98**
Sperm released	**16**	-	**15**	937	**98**
Eggs spawned by 4♀	**24.4**	6.1	24.5	1000	**98**
Eggs spawned by 6♀	**36.9**	6.1	-	-	95?
Eggs spawned by 12♀	**56.8**	4.7	49.4	863	**87**

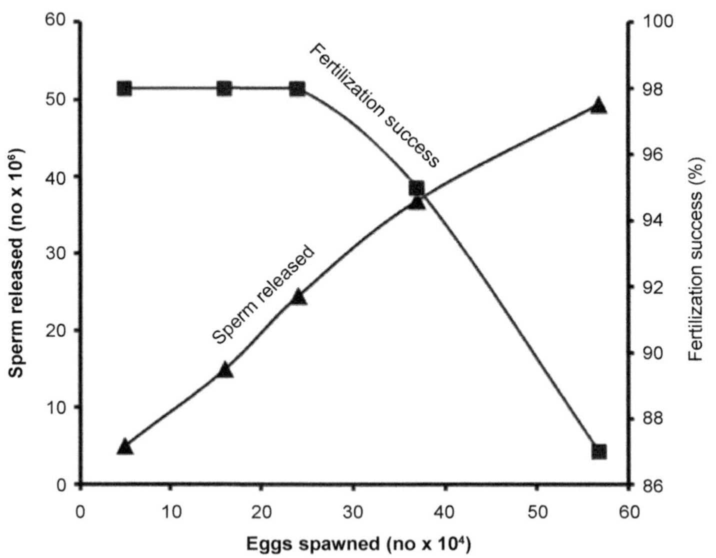

Fig. 2.16. With increasing number of eggs spawned by all females in a harem, the number of sperm released is also increased but fertilization success begins to decrease in *Xyrichthys novacula* (drawn from data reported in Table 2.13).

FS from 98 to 87%, the male with the largest harem consisting of 12 females fertilized 56,777 eggs every day, in comparison to 24,482 and 36,864 eggs fertilized by the males holding four and six females, respectively. Clearly, the protogynic males have the capacity to gauge the number of eggs awaiting fertilization and economize the quantum of sperm release accordingly. Secondly, aggregation behavior, which was not considered by Alonzo and Mangel (2004), may also result in sperm economization and ensure a higher FS. Breeding aggregations bring proximity between vents of mating partners to ensure higher FS and sperm economy, for example temporally and spatially predictable spawning aggregations in scamp *Epinephelus guttatus* and gag *E. microlepis* (McGovern et al., 1998) and *E. niveatus* (Wyanski et al., 2000) at the high risk of large males being captured. The longer residency at the aggregation site increases the vulnerability of reproductively active males to fishing (Rhodes and Trupper, 2008). Shapiro et al. (1993) suggested that another function of the annual spawning aggregation may enable a potential female to determine whether to continue as female or change to male. However, the aggregation groups in 10s and 100s are reduced from 17% in 1970s to 1% in 1990s in gag and from 38% in 1970s to 18% in 1990s in scamp (Coleman et al., 1996).

From their innovative investigation, Marconato et al. (1997) showed the comparative distance between the vents among mating partners during the event of pair-spawning and group spawning in *T. bifasciatum*. In both cases, 95% FS is achieved though at the cost of up to 50% more sperm by group spawning males. In pair spawning, sperm economy is achieved but in group spawning genetic diversity is enhanced among the progenies. Two points are to be noted: 1. The decreasing male number may impose intense inbreeding (cf. O'Laury et al., 2013) leading to reduced genetic diversity, an aspect which is yet to receive attention. Thirdly, Alonzo and Mangel (2004) assumed that time point of sex change remains fixed in exploited protogynic populations. However, available data described later seem to go against this assumption.

In contrast to sperm limitation in protogynics, reproductive performance of protandrics may encounter abundance of sperm produced by too many males, as their male ratio is ≈0.67 (Pandian, 2010, p. 112). In *Danio rerio*, the offer of more males to a female reduced the egg output, as competing males became increasingly aggressive against each other (Spence and Smith, 2005). This may be a reason why protandrics are not as abundant as progynics. However, the protandrics seem to have evolved the following strategies to avoid simultaneous presence of more than one male in mating system: (i) option to monogamic mating system, as in many clownfishes and (ii) holding only one male and retaining other members as undifferentiated juveniles in her harem, as in *Amphiprion alkallopsis* (Pandian, 2010, p. 114).

2.8 Social induction

To investigate the process of socially mediated sex change in sequential, two different experimental approaches have been made. 1. Removal of the haremic/dominant (e.g., *Labroides dimidiatus*, Sakai et al., 2001) or female from a harem or social unit (e.g., *Amphiprion alkallopsis*, Fricke and Fricke, 1977) or a monogamous pair (e.g., clownfishes) and 2. Isolated housing of one or more, or equal or unequal sized individuals of the same sex. The former induces immediate dramatic changes in behavior and in some morphological traits (e.g., body color) modification in haremic (e.g., *L. dimidiatus*, Nagashima et al., 2000) and non-haremic (e.g., *Lythrypnus dalli*, Black et al., 2011) in α-female which ultimately changes its gonadal sex to male. For the first time, Pandian (2013, pp. 166–167) summarized widely scattered observations on socially mediated sex change in sequential hermaphrodites, when one or more individuals of the same sex are kept together in isolation. Amazingly, the protogynics, which are considered to undergo female to male sex change only once in their life time, are shown to undergo social mediated sex change more than once in either direction.

To know the exact factor that induces sex reversal, Lorenzi et al. (2006) exposed goby (*L. dalli*) females to (i) physical absence of male, (ii) presence of a male behind a barrier that allowed visual and olfactory interactions but prevented physical interaction (iii) presence of a male behind a porous screen that allowed olfactory interaction but prevented physical and visual interaction. The α-female changed sex only in the physical absence of male. It is likely that the encountering frequency (Lutnesky, 1994) of a large male (Bullough, 1947) inhibits/delays sex change by a potential protogynic female. As a consequence of capture and removal of more and more larger males from natural population, smaller and younger females change sex to male, as they 'miss' the physical presence and encounters of larger males. An account on social induction of younger and smaller females to undertake natural sex change and its implication to reproductive performance has already been provided (Pandian, 2010, Section 10). Briefly, overexploitation of larger protogynics (i) reduces egg production by advancing age/size at sexual maturity and reducing the reproductive life span in females and (ii) economizes and/or limits sperm availability through reduction in number and biomass of males.

2.9 Reproductive life span

In protogynics like the red porgy *Pagrus pagrus*, increased fishing mortality (F) from 0.25 in 1972 to 0.80 in 1998, inflicted a series of negative effects and led to their depletion by (i) decline in commercial landings, as reported for *P. pagrus* and *Epinephelus niveatus* (Table 2.14), (ii) decreases in (a) mean

Table 2.14. Changes in reproductive features of commercial landings in overexploited protogynics during the period from 1972 to 2002 in the southeastern United States.

Features	Reported changes
Pagrus pagrus (Harris and McGovern, 1997; Vaughan and Prager, 2002)	
Fishing mortality (F)	Increased from 0.25 in 1972 to 0.8 in 1998
Commercial landings	Fell from 1.5 mt in 1971 to 0.2 mt in 1995; number decreased from 200,000 during 1970s to 100,000 after 1988
Mean size landed (kg/fish)	Decreased from 1.1 in 1972 to 0.7 in 1998
Largest size captured (cm)	57.5 in 1980s but 47.5 after 1988
Catch effort	
(i) (no/trap/hr)	Decreased from 10 in 1980 to 1.5 in 1995
(ii) (no/unit effort)	Decreased from 0.8 in 1992 to 0.1 in 1998
SSB (t)	Depleted from 2500 in 1972 to 500 in 1998 SSB ratio fell to 18%
Epinephelus niveathus (Wyanski et al., 2000)	
Landing in South Carolina (t)	Fell from 130 in 1982 to 40 in 1996
Mean size landed (cm)	65–80 during 1970 but 50–60 in the mid 1990s
Captured by snaper reeds	
(i) largest size landed (cm)	105 in 1979–1985 but 98 in 1993–1994
(ii) oldest landed (age)	22 years in 1979–1985 but 18 years in 1993–1994

size (b) largest and (c) oldest fishes; the mean size of captured red porgy decreased from1.1 kg/porgy in 1972 to 0.7 kg/porgy in 1998, which may have many implications to reduction in RLS of female and fecundity (cf. p. 36), (iii) reduction in catch per unit effort and (iv) decrease in spawning stock biomass (SSB). Table 2.14 also summarizes the reported changes snowy grouper (Epinephelidae) captured in southeastern states of USA. In *P. pagrus*, the SSB ratio of 18% is far below the 30% level used by the South Atlantic Fishery Management Council to define, when a species is overfished.

Reported values for the male ratio of many protogynics are summarized in Table 2.11 and 2.12. Only a few authors have reported decreases in male ratios of protogynics resulting from overexploitation. For example, the male ratio of the gag *Mycteroperca microlepis* is reported to have decreased from 0.17 during 1977–1980 to 0.02 in 1991–1993 in the Gulf of Mexico and from 0.20 during 1976–1982 to 0.06 in 1995 in the catches from U.S. Atlantic waters (McGovern et al., 1998; DeVries, 2007). In *P. pagrus*, the reported decrease in male ratio is from 0.40 in 1980s to 0.33 in 1990s (Harris and McGovern, 1997).

Remarkably, overexploitation is known to advance the age and/or size of not only sexual maturity but also sex changing time schedule in female protogynics. Consequently, these precocious events effectively reduce (i) the scope for storage of adequate nutrients to meet the demand by vitellogenesis; not surprisingly, many protogynics are income breeders (cf. Table 2.4) and (ii) the duration of RLS of a female. Though many publications hint at the reduced RLS in females of overexploited protogynics, relevant data on this aspect are still awaited. Nevertheless, an attempt was made to have an estimate of it. Wyanski et al. (2000) estimated size of L_{50} (length at 50% maturity) of *E. niveatus* collected during 1980s and 1990s as 54.1 cm and 48.6 cm, respectively (see also Fig. 1 of McGovern et al., 1998); the A_{50} (age at 50% maturity) was 4.92 years for females collected during 1980s. From their data on percentage of females attaining sexual maturity as function of age, the A_{50} for the female of 1990s was calculated as 4.43 years, i.e., overexploitation advanced maturity of *E. niveatus* female by 0.5 year in age and 5.5 cm in size. In Fig. 2.17 the reported values (Table 9 of Wyanski

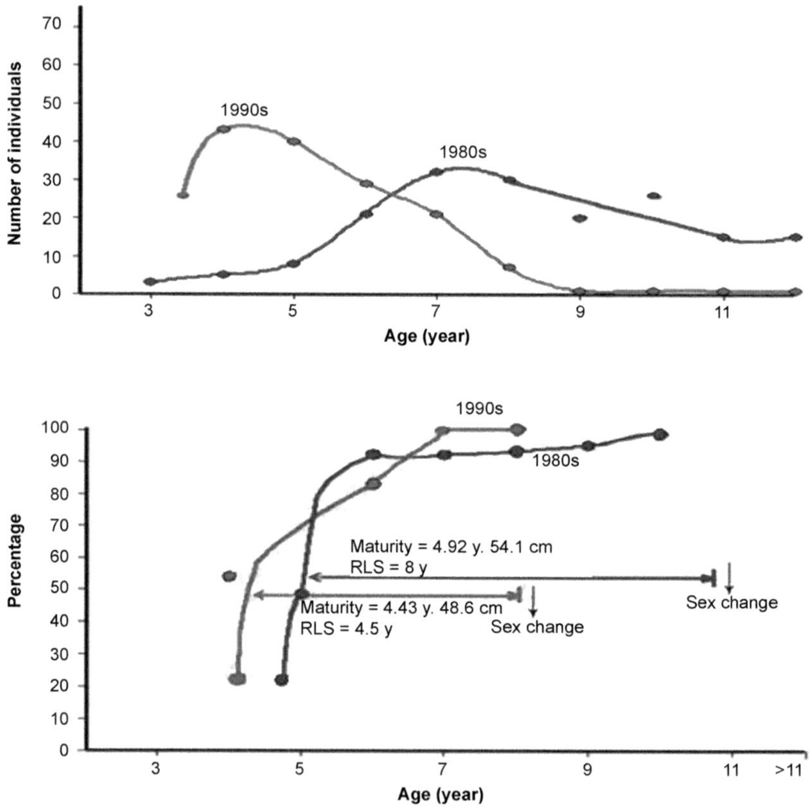

Fig. 2.17. Number (upper panel) and percentage (lower panel) of female snowy grouper *Epinephelus niveatus* collected off North Carolina during 1980s and 1990s. Note the precocious sexual maturity and sex change and decrease in Reproductive Life Span (RLS) in females of 1990s (compiled from Wyanski et al., 2000).

Color image of this figure appears in the color plate section at the end of the book.

et al., 2000) for the number of females collected during 1980s and 1990s are plotted against age. The number ranged from 15 to 59 for the age classes 3–11 years for females of 1980s, although one female continued to be present beyond 11 years. Since the sex change is known to occur upto age classes of 13 years, the effective RLS of the females collected during 1980s was 8 years, i.e., commencing from 5 years to 13 years. However, the RLS for the females of 1990s commenced at the age of 4.43 years is likely to have been terminated at the age of <9 years (Fig. 2.17). Hence the RLS of females of 1990s is 4.5 years only. Briefly, overexploitation of snowy grouper during the period from 1980s to 1990s resulted in 50% reduction in RLS. Hence the assumptions by authors like Alonzo and Mangel (2004) that the SSB of females and fecundity remain unaffected in overexploitated protogynics may have to be reconsidered.

A similar picture for the decrease in RLS of male red porgy *P. pagrus* also became apparent during the period from 1980s to 1990s, although a recovery became possible by 1993. The size range, during which males were found, was 30 cm, i.e., from 27.5 cm (3.25 years) to 57.5 cm (>14 years) during 1980s; it was just 20 cm during 1990s but ranged from 22.5 cm (2 years) to 42.5 cm (>13 years). Notably, the size at maturity is advanced to 22.5 cm during 1990s from 27.5 cm during 1080s (Fig. 2.18). Secondly, the number of males was 219 and 97 during 1980s and 1990s, respectively. Evidently, the size, at which the sex is changed, was advanced from 57.5 cm during 1980s to 42.5 cm during 1990s. The RLS, which continued from 27.5 cm body length to 57.5 cm, i.e., the period required to support a body growth of 30 cm, is reduced to a period during which the fish grows from

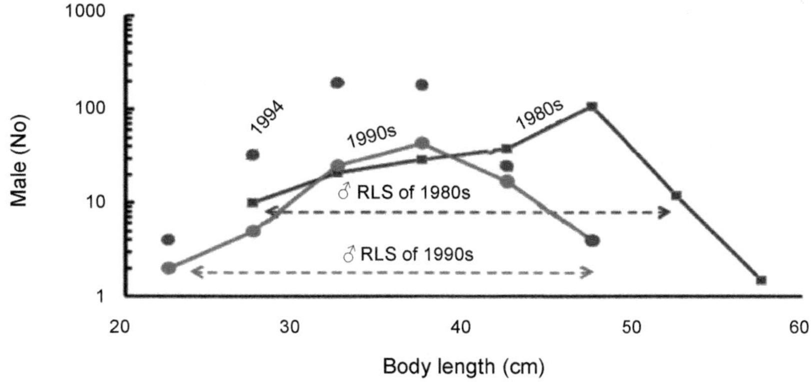

Fig. 2.18. Number of male red porgy *Pagrus pagrus* collected from North Carolina to Florida during 1980s and 1990s. Data are also shown for the number of males present after 1990s. Note the decrease in Reproductive Life Span (RLS) of males due to precocious sex change and capture of larger males (compiled from Harris and McGovern, 1997).

Color image of this figure appears in the color plate section at the end of the book.

22.5 cm to 42.5 cm, i.e., 20 cm only. The RLS of the 1990s porgy is reduced by 33%. Hence overexploitation of the porgy from 1980s to 1990s not only advanced the size at sexual maturity but also size at sex change and reduced the number of males as well. Clearly, these estimates suggest a significant decrease in SSB of males due to overexploitation. Consequently, SSB of males and possibly sperm availability may have also decreased considerably.

References

Ainley, D.G., N. Nur, J.T. Eastman et al. 2013. Decadal trend in abundance, size and condition of Antartic toothfish in McMurdo Sound, Antartica 1972–2011. Fish Fisher, 14: 343–363.

Aleja-Plata, C., P. Diaz-Jaimes and I.H. Salgado-Ugarte. 2011. Sex ratios, size at sexual maturity and spawning seasonality in dolphinfish (*Coryphaena hippurus*) captured in Gulf Tehuantepec, Mexico. Fisher Res, 110: 207–216.

Allsop, D.J. and S.A. West. 2003. Constant releative age and size at sex change in sequentially hermaphroditic fish. J Evol Biol, 16: 921–929.

Almodovar, A. and A. Nicola. 2004. Angling impact on conservation of Spanish stream-dwelling trout *Salmo trutta*. Fisher Mgmt Ecol, 11: 173–183.

Alonzo, S.H. and M. Mangel. 2004. The effect of size-selective fisheries on the stock dynamics and sperm limitation in sex changing fish. Fish Bull, 102: 1–13.

Anon. 2008. Teraglin (*Atractoscion aequidens*). Status of Fisheries Resource in NSW. 2008/09, pp. 341–343.

Arnason, T. and B. Bjornsson. 2012. Spawning-related mortality in captive Atlantic cod (*Gadus morhua* L.). Aquacult Res, 43: 292–296.

Ballon, M., C. Wosnitza-Mendo, R. Guevara-Carrasco and A. Bertrand. 2008. The impact of overfishing and El-Nino on the condition factor and reproductive success of Peruvian hake *Merluccius gayi*. Prog Ocenography, 79: 307.

Beal, R.E., J.C. Desfosse, J.D. Field et al. 1998. Review of interstate fishery management plans. Atlantic States Marine Fisheries Council, Washington, D.C.

Beverton, R.J.H. 2000. The R.J.H. Beverton Lectures at Woods Hole, Massachusetts, National Oceanic and Atmospheric Administration, Washington, DC, p. 16.

Black, M.P., M. Balthazat, M. Baillien and M.S. Grober. 2011. Rapid increase in aggressive behavior precedes the decrease in brain aromatase activity during socially mediated sex change in *Lythrypnus dalli*. Gen Comp Endocrinol, 170: 119–124.

Brouwer, S.L. and M.H. Griffiths. 2005. Reproductive biology of carpenter seabream (*Argyrozona argyrozona*) (Pisces: Sparidae) in a marine protected area. Fish Bull, 103: 258–269.

Bullough, W.S. 1947. Hermaphoditism in the lower vertebrates. Nature, 226: 828–830.

Cambray, J.A. 2003. Impact of indigenous species biodiversity caused by the globalization of alien recreational freshwater fisheries. Hydrobiologia, 500: 217–230.

Carter, A.B., A.J. Williams and G.R. Russ. 2009. Increased accuracy of batch fecundity estimates using oocyte stage ratios in *Plectropomus leopardus*. J Fish Biol, 75: 716–722.

Coleman, F.C., C.C. Koenig and L.A. Collins. 1996. Reproductive styles of shallow water groupers (Pisces: Serranidae) in the eastern Gulf of Mexico and the consequences of fishing and spawning aggregations. Env Biol Fish, 47: 129–141.

Cooke, S.J. and I.G. Cowx. 2006. Contrasting recreational and commercial fishing. Searching for common issues to promote unified conservation of fisheries resources and aquatic environments. Biol Conservation, 128: 93–108.

Cooper, D.W., K.P. Maslenikov and D.R. Gunderson. 2007. Natural mortality rate, annual fecundity and maturity at length for Greenland halibut (*Rheinhardtius hippoglossoides*) from the northeastern Pacific Ocean. Fish Bull, 105: 296–304.

Corten, A. 2013. Recuritment depressions in North Sea herring. ICES J Mar Sci, 70: 1–15.

DeVrries, D.A. 2007. No evidence of bias from fish behavior in the selectivity of size and sex of the protogynous red porgy (*Pagrus pagrus*, Sparidae) by hook and line-gear. Fish Bull, 105: 582–587.

Dwivedi, A.C., P. Nautiyal and K.D. Joshi. 2011. Sex ratio and structure of certain cyprinids of Vindhyan region of Central India. J Inland Fish Soc India, 43: 77–82.

Ebisawa, A. 1990. Reproductive biology of *Lethrinus nebulosus* (Pisces: Lethrinidae) around the Okinawan waters. Nippon Suisan Gakkaishi, 56: 1941–1952.

Ebisawa, A. 1997. Some aspects of reproduction and sexuality in the spotcheek emperor *Lethrinus rubrioperculatus* in waters off the Ryuku Islands. Ichthyol Res, 44: 201–212.

Emerson, L.S., M. Greer-Walker and P.R. Witthames. 1990. A sterological method for estimating fish fecundity. J Fish Biol, 36: 721–730.

Fallon, L.D. and E. Stratford. 2003. Issues of sustainability in Souther Ocean Fishes in the case of the Patagonian toothfish. University of Tasmania Publication.

FAO. 2010. General situation of world fish stocks. United Nation Food and Agriculture Organization, Rome, p. 113.

FAO. 2012. The State of World Fisheries and Aquaculture. FAO, Rome, p. 230.

Finn, R.N., H.J. Fyhn, B. Norberg et al. 2000. Oocyte hydration as a key feature in the adaptive evolution of teleost fishes to sea water. Proc Sixth Internatl Symp Reprod Physiol Fish. B. Norberg, O.S. Kjesbu, G.L. Taranger et al. (eds.). University of Bergen, Norway.

Fitzhugh, G.R., K.W. Shertzer, G.T. Kellison and D.M. Wyanski. 2012. Review of size and age dependence in batch spawning: Implications for stock assessment of fish species exhibiting indeterminate fecundity. Fish Bull, 110: 413–425.

Fricke, H.W. and S. Fricke. 1977. Social control of sex reversal in a coral reef fish. Nature, 266: 829–832.

Frisch, A. 2004. Sex change and gonadal steroids in sequentially hermaphroditic teleost fish. Rev Fish Biol Fisher, 14: 481–499.

Funge-Smith, S. and S. Kennely. 2013. Managing tropical fishes using trawl fisheries to look at issues and solutions. Proc Symp Mgmt Conservation Mar Fisher, Kochi. E. Vivekanandan (ed.). pp. 101–104.

Gaughan, D.J., T.I. Leary, R.W. Mitchell and I.W. Wright. 2004. A sudden collapse of in distribution of Pacific sardine (*Sardinops sagax*) of southwestern Australia enables an objective re-assessment of biomass estimates. Fish Bull, 102: 617–633.

Gessner, J., M. Tautenhahn, S. Spratte et al. 2011. Development of a German Action Plan for the restoration of the European sturgeon *Acipenser struio* L.—implementing international commitments on a national scale. J Appl Ichthyol, 27: 192–198.

Godwin, J., D. Crews and R.R. Warner. 1996. Behavioural sex change in the absence of gonads in a coral reef fish. Proc R Soc Lond, 263B: 1683–1688.

Gomathy, S. 2013. Trophodynamics of selected fishes along Chennai coast. Ph.D. Thesis, University of Madras.

Hampton, J., J.R. Sibert, P. Kleiber et al. 2005. Decline in Pacific tuna populations exaggerated. Nature, 434: E1–E2.

Hansen, M.J., A.H. Fayram and S.P. Newman. 2011. Natural mortality in relation to age and fishing mortality of walleye in Escanaba Lake, Wisconsin during 1956–2009. North Am J Fisher Mgmt, 31: 506–514.

Harris, P.J. and J.C. McGovern. 1997. Changes in the life history of red porgy *Pagrus pagrus* from the south eastern United States 1972–1994. Fish Bull, 95: 732–747.

He, W.-P., Y.-X. Li, M. Liu et al. 2011. Reproductive biology of *Coilia mystus* (Linnaeus) from the Yangtze Estuary, China responses to overexploitation. J Appl Ichthyol, 27: 1197–1202.

Heibo, E., C. Magnhagen and A. Vollestad. 2005. Latitudinal variation in life-history traits in Eurasian perch. Ecology, 86: 3377–3386.

Hunter, J.R., B.J. Macewicz, N.C.-H. Lo and C.A. Kimbell. 1992. Fecundity, spawning and maturity of Dover sole *Microstomus pacificus* with an evaluation of assumptions and precision. Fish Bull, 90: 101–128.

Hussy, K. 2011. Review of Western Baltic cod (*Gadus morhua*) recruitment dynamics. ICES J Mar Sci, 68: 1459–1471.
Hutchings, J.A. and J.D. Reynolds. 2004. Marine fish population collapses: Consequences for recovery and extinction risk. Bioscience, 54: 297–309.
Kazancioglu, E. and S.H. Alonzo. 2009. Costs of changing sex do not explain why sequential hermaphroditism is rare. Am Nat, 173: 327–336.
Kjesbu, O.S., H. Murua, F. Saborido-Rey et al. 2010. Method development and evaluation of stock reproductive potential of marine fish. Fisher Res, 104: 1–7.
Kokokris, L., S. Brusle, M. Kentouri and A. Fostier. 1999. Sexual maturity and hermaphroditism of the red porgy *Pagrus pagrus* (Teleostei: Sparidae). Mar Biol, 134: 621–629.
Koslow, J.A., J. Bell, P. Virtue and D.C. Smith. 1995. Fecundity and its variability in orange roughy: effects of population density, conditions, egg size and senescence. J Fish Biol, 47: 1063–1080.
Kraus, G., K. Muller, K. Trella and F.W. Koster. 2000. Fecundity of Baltic cod: temporal and spatial variation. J Fish Biol, 56: 1327–1341.
Kumar, G. 2012. Genetic variation in two tuna species *Euthynnus affinis* (Cantor) and *Auxis thozard* (Lacepede) along the Indian coast. Ph.D. Thesis, Goa University, Talaigoa, India.
Laroche, J.K. and S.L. Richardson. 1980. Reproduction of northern anchovy *Engraulis mordax* off Oregon and Washington. Fish Bull, 78: 603–618.
Lorenzi, V., R.L. Earley and M.S. Grober. 2006. Preventing behavioural interactions with a male facilitates sex change in female bluebanded gobies *Lythrypnus dalli*. Behav Ecol Sociobiol, 59: 715–722.
Lutnesky, M.M.F. 1994. Density-dependent protogynous sex change in territorial-haremic fishes: models and evidence. Behav Ecol, 5: 375–383.
Mangel, M. and P.E. Smith. 1990. Presence-absence sampling for fisheries management. Can J Fish Aquat Sci, 47: 1875–1887.
Marconato, A., V. Tessari and G. Martin. 1995. The mating system of *Xyrichthys novacula*: sperm economy and fertilization success. J Fish Biol, 47: 292–301.
Marconato, A., D.Y. Shapiro, C.W. Petersen et al. 1997. Methodological analysis of fertilization rate in the bluehead wrasse *Thalassoma bifasciatum*: pair versus group spawns. Mar Ecol Prog Ser, 161: 61–70.
McDermott, S.F., K.P. Masleniko and D.R. Gunderson. 2007. Annual fecundity, batch fecundity and oocyte atresia of Atka mackerel (*Pleurogrammus monopterygius*) in Alaska waters. Fish Bull, 105: 19–29.
McDonough, C.J., W.A. Roumiller and C.A. Wenner. 2005. Sexual differentiation and gonad development in striped mullet (*Mugil cephalus* L.) from South Carolina estuaries. Fish Bull, 103: 601–613.
McGovern, J.C., D.M. Wyanski, O. Pashuk et al. 1998. Changes in the sex ratio and size at maturity of gag *Mycteroperca microlepis* from the Atlantic coast of the southeastern United States during 1976–1995. Fish Bull, 96: 797–807.
Mike, A. and I.G. Cowx. 1996. The contribution of recreational fishing to the fisheries sector in northwest Trinidad. Fisher Mgmt Ecol, 3: 219–228.
Milton, D.A., S.J. Blaber and N.J.F. Rawlinson. 1995. Fecundity and egg production of four species of short-lived clupeoid from Solomon Islands, tropical South Pacific. ICES J Mar Sci, 52: 111–125.
Minto, C. and C.P. Nolan. 2006. Fecundity and maturity of orange roughy (*Hoplostethus atlanticus* Collet 1889) on Porcupine Bank, Northeast Atlantic. Env Biol Fish, 77: 39–50.
Mire, J.B. and L. Millet. 1994. Size of mother does not determine size of eggs or fry in the Owens pupfish, *Cyprinodon radiosus*. Copeia, 1994: 100–107.
Molloy, F.P., N.B. Godwin, I.M. Cole et al. 2007. Sperm competition and sex change: A comparative analysis across fishes. Evolution, 61: 640–652.
Morales-Nin, B., J. Moranda, C. Garcia et al. 2005. The recreational fishery off Majorca Island (western Mediterranean): some complications for coastal resource management. ICES J Mar Sci, 62: 727–739.

Morgan, M.J., C.E. Wilson and L.W. Crim. 1999. The effect of stress on reproduction in Atlantic cod. J Fish Biol, 54: 477–488.

Murua, H., L. Ibaibarriage, P. Alvarez et al. 2010. The daily egg production method: A valid tool for application of European hake in the Bay of Biscay. Fisher Res, 104: 100–110.

Myers, R.A. and B. Worms. 2003. Rapid worldwide depletion of predatory fish communities. Nature, 423: 280–283.

Nagashima, Y., Y. Sakai, K. Karino and T. Kuwamura. 2000. Female-female spawning and sex change in a haremic coral reef fish *Labroides dimidiatus*. Zool Sci, 17: 967–970.

Nash, R.D.M., G.M. Pilling, L.T. Kell et al. 2010. Investment in maturity at age and length in north east Atlantic cod stocks. Fisher Res, 104: 89–99.

Nissling, A. and L. Westin. 1997. Salinity requirement for successful spawning of Baltic and Belt Sea cod and the potential for cod stock interactions in the Baltic Sea. Mar Ecol Prog Ser, 152: 261–271.

Ojanguren, A.F. and R.-G.F. Brana. 1996. Effect of egg size on offspring development and fitness in brown trout *Salmo trutta* L. Aquaculture, 147: 9–20.

O'Leary, S.J., L.A. Hica, K.A. Feldherim et al. 2013. Severe inbreeding and small effective number of breeders in a formerly abundant marine fish. PLoS ONE, 8: e66126. Doi: 10.1371/journal.pone.0066126.

Olin, M., J. Jutila, H. Lehtonen et al. 2012. Importance of maternal size on the reproductive success of perch *Perca fluviatilis* in small forest lakes. Implications for fisheries management. Fisher Mgmt Ecol, 19: 363–374.

Pajuelo, J.C., M. Lorenzo, O.A. Bilbao and A.G. Ramos. 2006. Reproductive characteristics of the benthic coastal fish *Diplodus vulgaris* (Teleostei: Sparidae) in the Canarian Archipelago, Northwest Africa. J Appl Ichthyol, 22: 414–418.

Pandian, T.J. 2010. *Sexuality in Fishes*, Science Publishers, Enfield/CRC Press, Boca Raton, USA, p. 208.

Pandian, T.J. 2011. *Sex Determination in Fish*. Science Publishers, Enfield/CRC Press, Boca Raton, USA, p. 281.

Pandian, T.J. 2012. *Genetic Sex Differentiation in Fish*. CRC Press, Boca Raton, USA, p. 214.

Pandian, T.J. 2013. *Endocrine Sex Differentiation in Fish*. CRC Press, Boca Raton, USA, p. 302.

Pauley, D., V. Christensen, J. Dalsgaard et al. 1998. Fishing down marine food webs. Science, 279: 860–863.

Pepin, D., C. Orr and J.T. Anderson. 1997. Time to hatch and larval size in relation to temperature and egg size in Atlantic cod (*Gadus morhua*). Can J Fish Aquat Sci, 545: 2–10.

Post, J.R., M. Sullivan, S. Cox et al. 2002. Canada's recreational fishery: the invisible collapse? Fisheries, 27: 6–17.

Reznick, D., M. Bryant and D. Holmes. 2006. The evolution of senescence and post-reproductive life span in guppies (*Poecilia reticulata*). PLoS Biol, 4. Doi: 10.1371/journal.pbio.0010007.

Rhodes, K.L. and M.H. Tupper. 2008. The vulnerability of reproductively active squartail coral grouper (*Plectropomus areolatus*) to fishing. Fish Bull, 106: 194–203.

Rodgveller, C.J., C.R. Lunsford and J.T. Fujioka. 2012. Effect of maternal age and size on embryonic energy reserves, developmental timing and fecundity in quillback rockfish (*Sebastes maliger*). Fish Bull, 110: 35–45.

Rougier, T., P. Lambert, H. Drouineau et al. 2012. Collapse of allis shad *Alosa alosa* in the Gronde system (southwest France): environmental change, fishing mortality or Allee effect. ICES J Mar Sci, 69: 1802–1811.

Rurangwa, E., I. Roelants, G. Huyskens et al. 1998. The minimum effective spermatozoa: egg ratio for artificial insemination and the effects of mercury on sperm motility and fertilization ability in *Clarias gariepinus*. J Fish Biol, 53: 402–413.

Sadovy, Y.Y. 1996. Reproduction of reef fish species. pp. 15–59. *In*: N.V.C. Polunin and C.M. Roberts (eds.). *Reef Fishes*. Chapman and Hall, London.

Sadovy, Y. and D.Y. Shapiro. 1987. Criteria for diagnosis of hermaphroditism in fishes. Copeia, 1987: 136–157.

Sakai, Y., M. Kohda and M. Kuwamura. 2001. Effect of changing harem on timing of sex change in female cleaner fish *Labroides dimidiatus*. Anim Behav, 62: 251–257.
Schwenke, K.L. and J.A. Buckel. 2008. Age growth and reproduction of dolphinfish (*Coryphaena hippurus*) caught off the coast of North Carolina. Fish Bull, 106: 82–92.
Semsar, K. and J. Godwin. 2003. Social influences on the arginine vasotocin system are independent of gonads in a sex changing fish. J Neurosci, 23: 4386–4393.
Semsar, K. and J. Godwin. 2004. Multiple mechanisms of phenotypic development in the bluehead wrasse. Horm Behav, 45: 345–353.
Shapiro, D.Y., Y. Sadovy and M.A. McGehee. 1993. Size composition and age structure of the annual spawning aggregation of the red hind *Epinephelus guttatus* (Pisces: Serranidae). Copeia, 1993: 399–406.
Skjaeraasen, J.P., A.G. Salvanes, O. Karlson et al. 2004. The effect of photoperiod on sexual maturation, appetite and growth in wild Atlantic cod (*Gadus morhua* L.). Fish Physiol Biochem, 30: 163–174.
Spence, R. and C. Smith. 2005. Mate territoriality mediates density and sex ratio effects on oviposition in the zebrafish (*Danio rerio*). Anim Behav, 69: 1317–1323.
Stanley, R.D. and A.R. Kronlund. 2005. Life history characteristics for silvergray rockfish (*Sebastes brevispinis*) in British Columbia waters and the implications for stock assessment and management. Fish Bull, 103: 670–684.
Stergiou, K. 1999. Intraspecific variations in size and age-at maturity for red bandfish *Cepola macrophthalma*. Env Biol Fish, 54: 151–160.
Sun, C.L., S.Z. Yeh, Y.J. Chang et al. 2013. Reproductive biology of female bigeye *Thunnus obesus* in the western Pacific Ocean. J Fish Biol, 83: 250–271.
Takamoto, G., S. Seki, Y. Nagashima et al. 2003. Protogynous sex change in the haremic triggerfish *Sufflamen chrysopterus* (Tetradontiformes). Ichthyol Res, 50: 281–283.
Taranger, G.L., M. Carillo, M. Schulz et al. 2010. Control of puberty in farmed fish. Gen Comp Endocrinol, 165: 483–515.
Thorsen, A. and O.S. Kjesbu. 2001. A rapid method for estimation of oocyte size and potential fecundity in Atlantic cod using computer-aided particle analysis system. J Sea Res, 46: 245–308.
Thorsen, A., P.R. Witthames, G. Marteinsdotir et al. 2010. Fecundity and growth of Atlantic cod (*Gadus morhua*) along a latitudinal gradient. Fish Res, 104: 45–55.
Trippel, E.A. 1995. Age at maturity as a stress indicator in fisheries. Bioscience, 45: 759–791.
Trippel, E.A. 1999. Estimation of stock reproductive potential: history and challenges for Canadian Atlantic gadoid stock assessments. J Northwest Atlantic Fisher Sci, 25: 61–81.
Vaughan, D.S. and M.H. Prager. 2002. Severe decline in abundance of the red porgy (*Pagrus pagrus*) population off the southeastern United States. Fish Bull, 100: 351–375.
Villasante, S. 2010. Global assessment of the European Union fishing fleet: an update. Mar Policy, 34: 663–770.
Vivekanandan, E. 2013a. Precocious maturity and variations in reproductive potentials of threadfin bream *Nemipterus japonicus* in southwest Bay of Bengal. Sex Early Dev Aquat Organism (communicated).
Vivekanandan, E. 2013b. Seawater warming coincides with shift in the spawning season of a tropical bream in the southwest Bay of Bengal. Rev Fish Biol Fisher (in press).
Vivekanandan, E., M. Srinath and S. Kuriakose. 2005. Fishing the marine food web along the Indian coast. Fisher Res, 72: 241–252.
Vivekanadan, E., V.V. Singh and J.K. Kizhokudan. 2013. Carbon footprint by marine fishing boats of India. Curr Sci, 104: 361–366.
von Westernhagen, H. 1970. Erbluten der Eier von Dorsch (*Gadus morhua*), Flounder (*Pleuronectes flesus*) unter Kombinierten Temperatur- und Salzgehaltsbedingungen. Helgolander wissen Meeresunter, 21: 21–102.

Warner, R.R., D.Y. Shapiro, A. Marconato and C.W. Petersen. 1995. Sexual conflict: Males with highest mating success convey the lowest fertilization benefits to females. Proc R Soc Lond, 262B: 135–139.

Watanabe, Y., Y. Yamashita and Y. Oozeki. 1996. *Survival Strategics in Early Life Stages of Marine Resources.* A.-A. Balkema Uitegerers, b.v. Rotterdam.

Witthames, P.R., A. Thorsen, H. Murua et al. 2009. Advances in methods for determining fecundity: application of the new methods to some marine fishes. Fish Bull, 107: 148–164.

Wooton, R.J. 1992. Constraints in the evolution of fish life histories. Neth J Zool, 42: 291–303.

Wooton, R.J. 1998. *Ecology of Teleost Fishes.* Kluwer Acadamic Publications, Dordrecht.

Wu, G.C., S. Tomy and C.F. Chang. 2008. The *nrobl* and *nr* 504 during gonald development and sex change in protandrous black progy *Acanthopagrus schlegeli.* Biol Reprod, 78: 200–210.

Wyanski, D.M., D.B. White and C.A. Barans. 2000. Growth, population age structure and aspects of the reproductive biology of snowy grouper *Epinephelus niveatus,* off north Carolina and South Carolina. Fish Bull, 98: 199–218.

Xu, C., D.V. Schneider and C. Rideout. 2012. When reproductive value exceeds economic value: an example from the Newfoundland cod fishery. Fish Fisher, 14: 225–233.

Yamada, T., I. Aoki and I. Mitani. 1998. Spawning time, spawning frequency and fecundity of Japanese chub mackerel *Scomber japonicus* in the waters around the Izu Islands, Japan. Fisher Res, 38: 83–89.

Yoda, M. and M. Yoneda. 2009. Assessment of reproductive potential in multiple spawning fish with indeterminate fecundity: a case study of yellow sea bream *Dentex hypselosomus* in the East China Sea. J Fish Biol, 74: 2338–2354.

Zeller, D., P. Rossing, S. Harper et al. 2011. The Baltic Sea: Estimates of total fisheries removal 1950–2007. Fisher Res, 108: 356–363.

3

Aquaculture and Crowding

Introduction

Aquaculture is the fastest growing animal food-producing sector in the world (Nielsen, 2011). Fish and fishery products represent a very valuable source of protein and essential micronutrients for balanced nutrition and good health of humans all over the world. Globally, fishes provide about 7.5 billion people with 15–20% of their intake of animal protein. While capture fisheries from both marine and freshwater sectors are beginning to decline, global aquaculture production continues to grow in this new millennium. Since 1950s, the world aquaculture production is consistently growing and peaked to 60 mt in 2010 (Fig. 3.1). A dozen species/species groups contribute more than 50% of the 60 mt production (Table 3.1). China, India, Vietnam, Indonesia, Bangladesh, Thailand and Norway produce 49 mt. More than 87% of the world fishers and farmers are from Asia but they account for 67% of the world fish production (FAO, 2012), indicating the need for scientific and technological inputs to Asian aquaculture.

3.1 Human interventions

In the history of fishes, introduction of exotic fishes, hybridization, ploidy induction, hormonal manipulation and combinations of all the last three manipulations are important human interventions. Introduction of exotic fishes is not known to alter either the sex ratio or sex differentiation in fishes.

Hybridization, however, does alter sex ratios and cichlids represent good examples of it. Sex in tilapias is controlled by genetic determinants mostly located on sex chromosomes with XX and XY model (e.g., *Oreochromis*

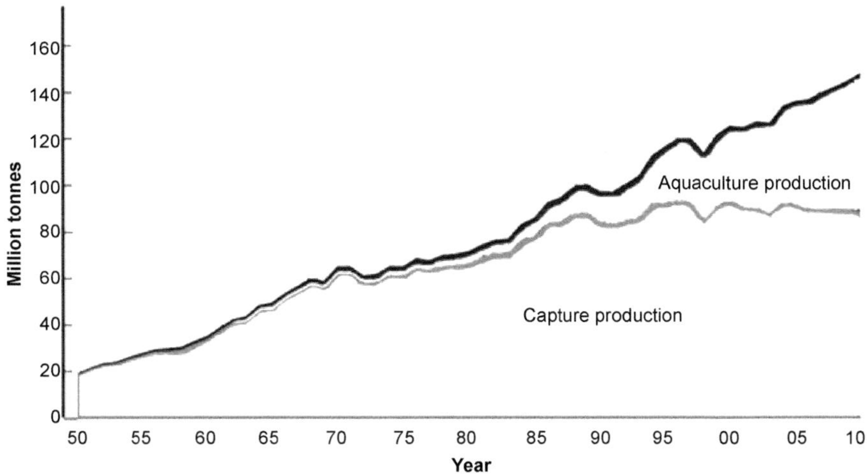

Fig. 3.1. Increasing trend in global aquaculture production from 1950 to 2010. The figure also shows the almost stabilized level of capture fish production during the same period (source: Food and Agricultural Organization of the United Nations, 2012. The State of World Fisheries and Aquaculture, Reproduced with permission).

Table 3.1. Top dozen species groups that are produced by aquaculture in 2010 (source: Food and Agricultural Organization of the United Nations, 2012. The State of World Fisheries and Aquaculture, Reproduced with permission).

Species group	Quantity (mt)
Silver bighead carps	6.8
Indian major carps	5.4
Grass carp	4.3
Common carp	3.4
Goldfish and crucian carp	2.3
Cichlids	3.5
Other cyprinids	2.2
Catfishes	1.8
Pangasius catfishes	1.6
Atlantic salmon	1.4
Milkfish	0.8
Rainbow trout	0.7

mossambicus, Pandian and Varadaraj, 1988). Table 3.2 lists the other models so far proposed for the cichlids. That most inter-specific crosses in *Oreochromis* result in altered sex ratio indicate that the autosomal genetic factors modify the primary effect of sex chromosomes. Incidentally, Δ supermales (YY) in *O. mossambicus* (Varadaraj and Pandian, 1989) and *O. niloticus* (Scott et al., 1989) have been generated. When *O. mossambicus* YY males were crossed with normal XX females, 2–12% unexpected females were produced (Pandian, 1993). However, 100% males were generated, when YY Δ

Table 3.2. Summary of crosses in *Oreochromis* with respective know/proposed genotypes and sex ratios of F_1 progenies (© 2011, Pandian, T.J., Sex Determination in Fish. Science Publishers/CRC Press, reproduced by Permission of Taylor and Francis Group, LLC, a division of Informa plc).

Female parent	Known/proposed genotype	Male parent	Known/proposed genotype	Male progeny (%)
O. mossambicus	XX	*O. hornorum*	YY	98–100
O. hornorum	XX	*O. mossambicus*	XY	75
O. niloticus	XX	*O. hornorum*	XY	98–100
O. hornorum	XX	*O. niloticus*	XY	75
O. niloticus	XX	*O. macrochir*	YY	100
O. macrochir	XY	*O. niloticus*	XY	75
O. niloticus	XX	*O. aureus*	ZW	50–100
O. aureus	ZW	*O. niloticus*	XY	75
O. niloticus	XX	*O. veriabilis*	XY ?	100
O. nigra	ZW ?	*O. hornorum*	YY	100
O. niloticus	XX	*O. nigra*	ZZ ?	85
O. nigra	ZW ?	*O. niloticus*	XY	43
O. nigra	ZW ?	*O. leucostica*	XY ?	96
O. niloticus	XX	*O. leucostica*	XY ?	94
O. aureus	ZW	*O. hornorum*	XY	90
O. vulcani	XX	*O. hornorum*	YY	98.5
O. niloticus	XX	*O. mossambicus*	XY	70
O. mossambicus	XX	*O. niloticus*	XY	70
O. vulcani	XX	*O. aureus*	ZW	93–98

supermale *O. niloticus* was crossed with YY Δ neo-superfemale (Muller-Belecke and Horstgen-Schwark, 2007), suggesting that two YY chromosomes are required to silence the expression of autosomal modifying genes.

Argue and Dunhan (1999) have listed a total of F_1 fertile 130 natural and 150 man-made F_1 hybrid fishes. Some 200 and odd hybrid records are published every year. Of more than 16,050 records on hybrids, fishes have a lion share of 21% (Schwenk et al., 2008). Human intervention rather than nature is responsible for the prevalence of hybridization events in fishes. However, survival of hybrids is low, fertility of surviving true hybrid is poor and is broken at F_3 in the ictalurid catfishes and F_8 in the hybrid salmonid 'spake'. Corresponding information is desired for hybrids among cichlids. For more information on hybridization in fishes, Pandian (2011, Chapter 3) may be consulted.

In many cichlids, frequent breeding leads to stunted growth. Gender-dependent growth to marketable size (e.g., *Dicentrarchus labrax*), meat taste and texture (e.g., salmonids) as well as body shape and color patterns, especially in ornamental fishes (e.g., *Betta splendens*) are all important for value addition and marketing of fish. Hence culturing of all-male or all-female, i.e., monosex culture is being widely practiced. Monosex culture may facilitate (i) reduction in number of broodstock required to produce a targeted number of progenies, (ii) selection of sex that grows faster and attains marketable size and (iii) prevention of undesired frequency of spawning. Table 3.3 lists some of these features that are relevant in monosex aquaculture of commercially important fishes.

Ploidy induction: Gynogenesis involves activation of development of egg by genome-eliminated/inactivated homospecific sperm or heterospecific sperm and retention of second polar body. Its induction in male heterogametic fishes usually results in all-female (XX) progenies (Pandian, 2011, Chapter 4). On the other hand, induction of androgenesis in male heterogametic fishes generates 50% normal XX females and 50% YY supermales (Pandian, 2011, Chapter 5). It involves activation of genome-inactivated/eliminated egg by a single haploid homospecific (e.g., *Puntius tetrazona*, Kirankumar and Pandian, 2003) or heterospecific fresh/cadarveric sperm (Kirankumar and Pandian, 2004a; David and Pandian, 2006b), or dispermic activation (Kirankumar and Pandian, 2004b; David and Pandian, 2008). Induction of maternal triploids involving retension of polar body has been widely reported and produces mostly sterile progenies. Rarely, paternal hybrid triploid, as in tetra, has been also generated by dispermic activation of egg but without retaining second polar body (David and Pandian, 2006a). For more information, Pandian (2011, Chapters 4, 5 and 6) may be consulted.

Table 3.3. Desired monosex culture in selected food- and -ornamental fishes (© 2013 from Pandian, T.J. Endocrine Sex Differentiation in Fish, CRC Press, Boca Raton, reproduced by permission of Tayler and Francis Group, LLC, a division of Informa plc).

Fishes	Remarks
	All–female progenies
Salmonids	Males mature earlier and have lower flesh quality than females
Common carp	At equal age females are 15% heavier than males
European sea bass	Females grow 20–50% faster than males
Atlantic halibut	Females reach marketable size earlier than males
Striped bass	At age 2+, females are heavier than males
Thai silver bass	Females grow faster and reach marketable size earlier than males
Czech tench	Females grow faster than male and attain the ultimate larger size
Acipenserids	Females alone produce expensive caviar
	All–male progenies
Tilapias	Males grow faster and monosex culture eliminates unwanted frequent spawning
Channel catfish	Male weighs 25% more than female at marketable size of 500 g
Fighter fish	Males are more attractive with more beautiful hood-like tail
	All sterile progenies
Grass carp	Introduction of exotics to control aquatic vegetation
Black carp	Introduction of exotics to control snails, intermediary hosts of human parasites

Hormonal manipulations: Administration of optimal dose of one or other natural or synthetic estrogen during the labile period produces all-female progenies. Similarly, that of an androgen results in the production of all-male progenies (see Pandian, 2013, Chapter 7). As freshly fertilized eggs are not accessible, viviparous fishes are not readily amenable to ploidy induction. In male heterogametic guppy *Poecilia reticulata*, Kavumpurath and Pandian, (1992, 1993) generated YY Δ supermales and YY Δ super-female by subjecting progenies of 3–5 generations to hormonal manipulations and progeny testing. In a female heterogametic molly *P. sphenops* too, production of ZZ Δ super-females has been achieved (George and Pandian, 1995). Different combinations of hormonal and ploidy manipulations along with progeny testing have generated Y^1Y^2 Δ supermales in male and female heterogametic fishes, and Z^1Z^2 Δ super-neofemales in female heterogametic fishes (Table 3.4). In *P. conchonius*, for example, both androgenic Y^2Y^2 Δ supermales and hormonally manipulated Y^1Y^2 Δ supermales have been produced. At the molecular level, no difference seems to exist between them (see Pandian and Kirankumar, 2003).

The Chinese creations: Between 2001 and 2007, Chinese scientists created three 'new species', synthesizing them by hybridization and tetraploidization. The first in the series is the bisexual tetraploid generated by Dr. S.J. Liu and his colleagues at the Hunan National University, Changsha. A hybrid line was established using 5% fertile males and 44% fertile females produced

Table 3.4. Generation of supermale and superfemales in fishes (© 2013 from Pandian, T.J. Endocrine Sex Differentiation in Fish, CRC Press, Boca Raton, reproduced by permission of Taylor and Francis Group, LLC, a division of Informa plc).

Species
Male heterogametics: Y^1Y^2 Δ supermale
Oryzias latipes
Carassius auratus
Oreochromis mossambicus
O. niloticus
Oncorhynchus tshawytscha
Poecilia reticulata
Ictalurus punctatus
Puntius conchonius
Male heterogametic: Y^1Y^2 Δ superfemale
P. reticulata
I. punctatus
O. niloticus
Female heterogametic: Z^1Z^2 Δ superfemale
O. niloticus
P. sphenops

by crossing *Carassius auratus gibelio* ♀ with *Cyprinus carpio* ♂. Strangely some F_3 hybrid females simultaneously produced haploid (1.3 mm), diploid (1.7 mm), and euploid 4n eggs (2.0 mm) and aneuploid eggs as well. Likewise some F_3 hybrid males also produced two-times larger volumes of milt consisting of 40% haploid, 49% diploid as well as 11% tetraploid, octoploid and aneuploid sperm. Fertilization of diploid eggs by diploid sperm produced bisexual (0.5♀ : 0.5♂) hybrid tetraploids (Fig. 3.2). Until

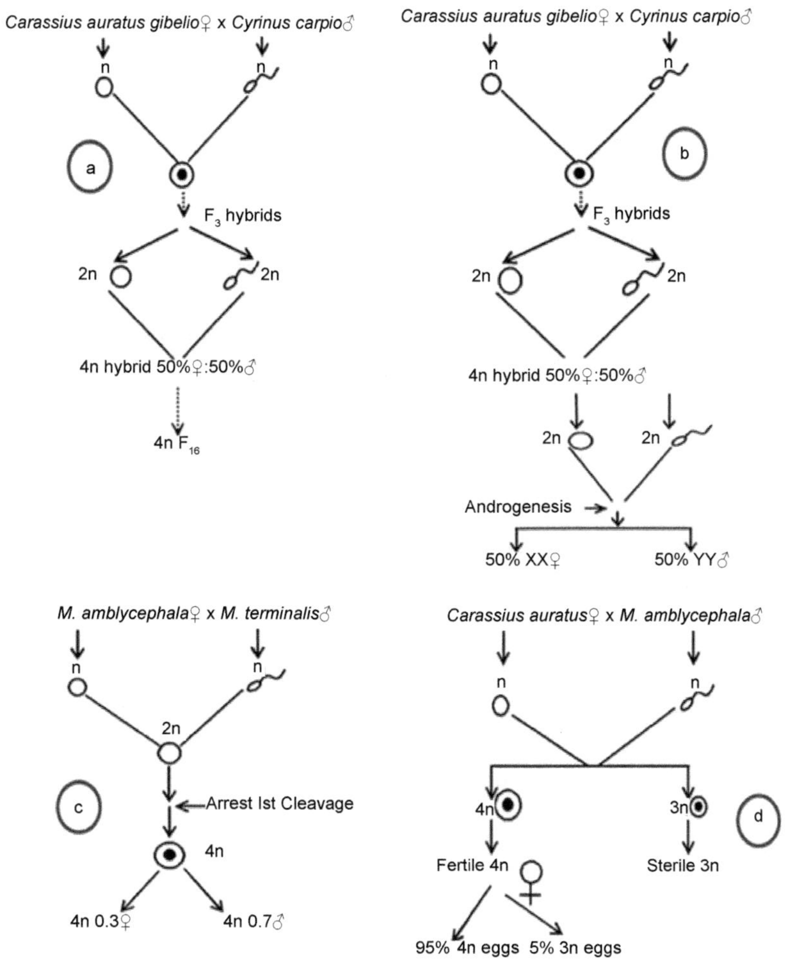

Fig. 3.2. The Chinese creation of new tetraploid 'species'. (a) Hybrid tetraploid between *Carassius auratus gibelio* ♀ and *Cyprinus carpio* ♂. (b) Production of androgenics of the hybrid tetraploid between the above said cross. (c) Production of mitotic tetraploid crossing *Megalobrama amblycephala* ♀ with *M. terminalis* ♂. (d) Production of hybrid tetraploid crossing *Carassius auratus* ♀ with *M. amblycephala* ♂ (Liu et al., 2001, 2007; Liu and Zhou, 1986; Sun et al., 2003; Duan et al., 2007; Zou et al., 2004).

now more than F_{16} tetraploid progenies were produced showing that their tetraploidy was stably inherited from one generation to another (Liu and Zhou, 1986; Liu et al., 2001; Sun et al., 2003). As a 'new species' with 200 chromosomes, 100 from each parent, the allotetraploid has become economically important. Their females, when crossed with the Japanese crucian carp 2n *C. auratus cuvieri*, produce triploids characterized by high survival and faster growth. In fact, these trigenomic triploids possess one genome each from *C. auratus gibelio*, *C. carpio* and *C. auratus cuvieri*. During recent years, >100,000 tetraploids and > 300 million triploids were annually produced in China for export to Japan. Interestingly, induction of gynogenesis in fertile allotetraploids produced all females and androgenesis 50% ($X^1x^1X^1x^1$) females and 50% ($Y^2y^2Y^2y^2$) Δ supermales. Crosses between these androgenics produced 85% 4n, 10% 3n and 5% 2n F_1 progenies; the tetraploids possess the desired smaller heads and shorter tails as well as resistance to diseases more vigorously (Duan et al., 2007).

The second in the series of the 'creation' is the bisexual allotetraploid generated by a simple but ingeniously designed distant cross between the omnivorous cyprininean red crucian carp (RCC) (2n = 100 chromosomes) *Carassius auratus* ♀ and the herbivorous cultinaen blunt snout bream (BSP) (2n = 48 chromosomes) *Megalobrama amblycephala* ♂ (Fig. 3.2). Fertilization and hatchling success were 60 and 50%, respectively. Among surviving F_1 progenies, 23 and 77% were 3n RB (124 chromosomes) and 4n RB (148 chromosomes); 4n RB hybrid females were fertile and produced 95% large (2.0 mm) unreduced tetraploid eggs and 5% smaller (1.7 mm) diploid eggs. However, the 4n RB males were only partially fertile and their 'watery milt' fertilized and produced just eight progenies (Liu et al., 2007; Yan et al., 2010).

The third in this series of success story is the 'synthesis' of a bisexual mitotic autotetraploid blunt snout bream *M. amblycephala* generated by S. Zou and his colleagues at the Shanghai Fisheries University. It involved crossing *M. amblycephala* ♀ with *M. terminalis* ♂ and production of 2n hybrid (Fig. 3.2). Using thermal shock to suppress the first mitotic cleavage, they have produced tetraploids. In 72% of F_1 progenies, tetraploidy was confirmed by the presence of 96 chromosomes. In them sex ratio is skewed 0.3♀: 0.7 ♂, the reason for which is not yet reported. Table 3.5 summarizes the available information on the reproductive performance of the tetraploid *M. amblycephala* (Zou et al., 2004).

3.2 Crowding and consequences

Realizing the costs of space and water, aquaculture operation is moving toward high-density and super high-density culture systems. Increasing density and consequent crowding may lead to increased mortality due to cannibalism and depletion of dissolved oxygen concentration. Size grading

Table 3.5. Reproductive performance of diploid and hybrid tetraploid blunt snout bream *Megalobrama amblycephala* (data as given to the author by Dr. S. Zou to review and improve the MS and with his permission).

Ploidy (n) and sex	Age (year)	Relative fecundity (no/g fish)	Egg size (mm)	Hatchability (%)	Net yield of fry (no/g fish)	Net yield of 4n fry as % of 2n cross
2♀ X 2♂	2	65.2	0.99	85	55.4	-
2♀ X 2♂	3	146.2	1.11	90	131.6	-
4♀ X 4♂	2	47.8	1.23	61	29.2	53
4♀ X 4♂	3	128.6	1.38	66	84.9	65

has been adopted to minimize cannibalism (e.g., *Dicentrarchus labrax*, Saillant et al., 2003). Many gill-breathing cichlids hold air-bubbles in the buccal cavity to aerate the water passing through the gills (e.g., *Pseudocrenilabrus multicolor victori*, Chi-Corri et al., 2008). Others like the hypoxic-tolerant cyprinids are periodic breathers, in contrast to salmonids, which continuously ventilate their gills (Nikinmaa and Rees, 2005). Understandably, the cyprinids and cichlids are amenable to crowding in aquaculture farms. Not surprisingly, the cyprinids and cichlids contribute maximum to aquaculture production (see Table 3.1). However, intense hypoxia may disrupt sex differentiation, which is discussed in Chapter 5.

High density culture systems are not known to disrupt sex differentiation in euteleosts (e.g., *D. labrax*, Saillant et al., 2003). However, crowding of eels in both aquaculture farms and natural habitats is reported to disrupt sex differentiation and alter sex ratio. A possible reason for this difference between euteleosts and eels may be traced to the ancient history of eels. The ancient history of fishes indicates that they underwent three genome-doubling events. The third genome-doubling event occurred after the eels split from euteleosts (Leggatt and Iwama, 2003). A good example of the genome-doubling is the presence of neural or brain aromatase gene *cyp19a1b* encoding p450 arom B highly expressed in brain of males and females (Patil and Gunasekara, 2008) and ovarian aromatase gene *cyp19a1a* encoding p450 arom A in gonads (Chiang et al., 2001) of euteleosts. Unlike euteleosts, eels possess a single *cyp19a1* alone and represent phylogenetically an ancient group of teleosts (see Pandian, 2013, p. 12). Hence eels perhaps respond to crowding differently from euteleosts. Contributing more than 0.3 mt (FAO, 2012) and being a delicacy, eel aquaculture plays an important role in the economy of many countries like Sweden (Andersson et al., 2012) and Taiwan. Crowding of eels in aquaculture farms and its role in disruption of sex differentiation is unique among the teleostean fishes. Hence it was chosen to provide an account of it.

During its complicated ontogeny, the European eel *Anguilla anguilla* passes through different stages from leptocephalus larva to glass eel and then to pigmented elver, immature yellow and mature silver eel (Beullens et al., 1997a). Being a catadromous fish, it enters into the rivers from the sea at the glass eel stage, grows through the said stages to metamorphose at the age of 3 to 9 years, and to the pubertal yellow stage, when sex differentiation occurs (Acou et al., 2003). After about 2+ years (Beullens et al., 1997b) or longer, it attains the adult silver stage and remains in the river for variable periods of time (10 to 40 years, Walsh et al., 2004). Males enter the migratory silver stage at an earlier age and smaller size than females (Vollestad, 1992). At equal age, silver females are larger than silver males (Wickstrom et al., 1996). The silver adults return to the spawning area in the Sargossa Sea, where they reproduce and die; the produced leptocedphali undertake long migration to the European and North African coasts, where they metamorphose into the glass eels, which enter river (Tesch, 2003).

"Sex differentiation relates to the events that occur during development and allows the expression of the genetic sex into appropriate phenotypic sex" (Piferrer, 2001). The gonadal differentiation process may be classified into three groups: 1. In the undifferentiated primary gonochore, differentiation proceeds from an undifferentiated gonad directly into ovary or testis (e.g., *Cyprinus carpio*, Komen et al., 1992). 2. In the differentiated secondary gonochores, the gonads first develop into non-functional ovary with oocytes. Later, the oocytes are degenerated by apopotosis in about 50% of presumptive male hatchlings and subsequently, the gonad differentiates into testis (e.g., *Gambusia affinis*, Koya et al., 2003). 3. In the tertiary gonochores, the indifferentiated gonad of some individuals directly differentiates into an ovary; however, the remaining individuals possess differentiated gonad; in them, a testicular differtentiation in juveniles proceeds through a transitional non-functional intersexual gonad consisting of single growing oocytes at the perinuclear stage (Beullens et al., 1997a); albeit evidence for degeneration of oogonia or oocytes is not yet available (e.g., *A. anguilla*, Beullens et al., 1997a).

In *A. anguilla*, one pair of heteromorphic chromosomes has been identified but the presence of these chromosomes could not be correlated with the male or female sex (Wiberg, 1983; Cau et al., 1992). In natural populations, the eel exhibits an unbalanced sex ratio. For example, females are predominantly present in the cooler north European waters, while male ratio ranges from 50 to almost 100% in the warmer southern European regimes (Tesch, 1977). Among various factors that influence sex differentiation, density (crowding), feeding conditions and temperature are considered important. Many authors noted that gonadal differentiation of eels depends more on size than age (e.g., Colombo et al., 1984) and it begins at 15–25 cm and at 10 g size (Holmgren and Mosegaard, 1996; Beullens et al., 1997a).

From their field observations, Klein Breteler et al. (1990) reported that higher the biomass of yellow eels, the lower was the percentage of females. The investigations of Beullens et al. (1997a) also suggested that female ratio increased from 16% at 600 t/ha to 36% at 20 t/ha. Krueger and Oliveira (1999) reported the effect of density on sex differentiation in the American eel *A. rostrata.* Their survey in the Annaquatucket River, Rhode Island lasted from 1977 to 1991. Among the 40% yellow eels, the sex ratio was 0.1♀ : 0.3♂ and the remaining 60% of the eels were undifferentiated. The ratio for the silver eel was 0.11♀: 0.89♂, indicating that majority of the undifferentiated yellow eels differentiated into males. Krueger and Oliveira (1999) also summarized available information on sex ratio of the eels both in natural habitats and aquaculture farms. For example, eel density was an order of magnitude higher in coastal waters than in inland waters. In the Esva estuary, Spain, where the density was as high as 3–5/m^2, 58% of the eels were sexually undifferentiated even at the sizes of 20–40 cm. This information together with that of Krueger and Oliveira shows that higher density delays sex differentiation in eels. At the densities of 0.018–0.023/m^3 in Georgia, females dominated the American eel population but the males dominated it at 0.18/m^3 in the Rhode Island. A reduction in the recruitment of the European eel from 1,100 to 250 per hectare in Comachio lagoon, Italy increased the number of females per male from three to nine. Reared communally in containers (0.8 m^3 each) at the density of 16.2/m^2, there were 27% males but no female. However, when reared individually, the observed sex ratio was 0.6♀ : 0.2♂ with 20% undifferentiated eels. Rearing 9,000 (1,540 g) American elvers at density of 2.25/m^2 produced 71% males but 27% males only, when 7,000 (1,170 g) elvers were reared at 1.75/m^2. These results suggest that it is the number and not the (size) biomass of elvers per unit area that determines sex.

If size is an important consideration, then it is necessary to know whether growth in body length or weight is more important in the differentiation process. Holmgren and Mosegaard (1996) reared the elver of *A. anguilla* for a period, during which the eel grew from 18–22 cm (10 g) to 38–45 cm (150 g) at 17,20 or 26°C and estimated sex ratio in different size groups. Interestingly, the undifferentiated gonad in the youngest glass eels consists of small numbers of Primordial Germ Cells (PGCs) characterized by a large size and centrally located round nucleus (Beullens et al., 1997a), which may be of interest to scientists working on xenogenesis. Growth pattern of the elvers did not overlap between temperatures within the actual ranges of length and weight (Fig. 3.3). Length increase of females seemed to be linear. Significant divergence between the growth of males and females occurred at about 35 cm length but that of weight initially at about 25 g and finally at 110 g, i.e., females grew faster in length until they reached 35 cm but males grew faster in weight until they attained a body weight of

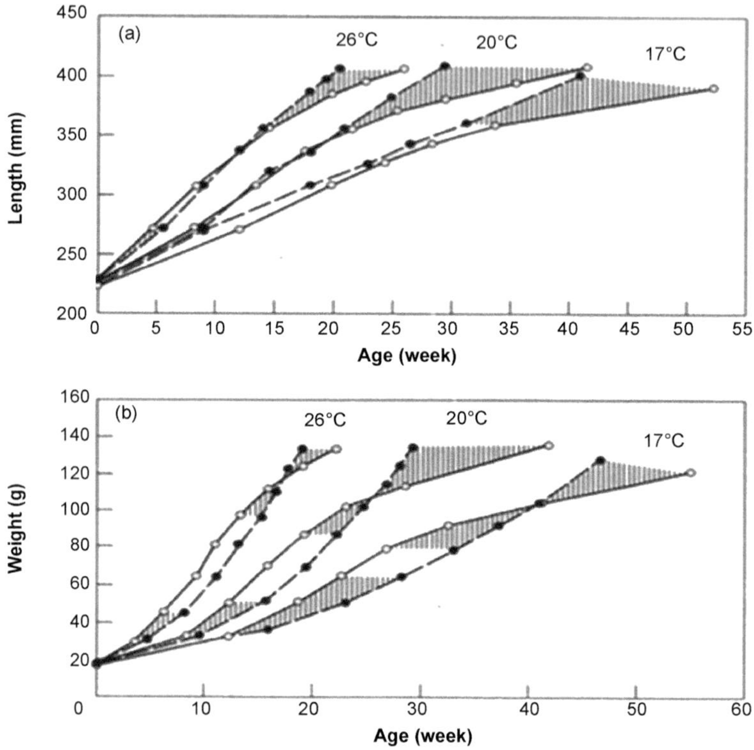

Fig. 3.3. Growth in length (upper panel, a) and weight (lower panel, b) during early stage of *Anguilla anguilla* as function of age at temperatures 17, 20 and 26°C. ● = female, ○ = male and |||||||| = intervals of length and weight, where absolute growth (slopes) differed significantly between sexes (permission by Wiley for Fisheries Society of the British Islets, from Holmgren, K. and H. Mosegaard. 1996. Implications of individual growth on the future sex of the European eel. J Fish Biol, 49: 910–925).

about 110 g. Figure 3.3 shows that females attain larger size in both length and weight at an early age. Hence, the investigation by Holmgren and Mosegaard (1996) indicates that rearing density may be a more important factor in influencing sex differentiation.

To study the effect of density on growth and food conversion efficiency, Roncarati et al. (1997) reared the elevers (0.45 g) of *A. anguilla* at densities of 800, 1,600 and 3,200 g/m^3. Male ratio increased from 69% at the lowest density of 800 g/m^3 to 78 and 96% in density groups of 1,600 g/m^3 and 3,200 g/m^3, respectively. Converting the food at higher efficiency (1.9 g food consumption to grow 1 g body weight), males at the lowest density attained 11 g body weight; males at 1,600 g/m^3 and 3,200 g/m^3 required 2.1 and 2.3 g food to grow 1 g body weight and grew to 10.2 g and 8.4 g, respectively (Roncarati et al., 1997). Clearly, male ratio is increased with increasing density and perhaps due to increased crowding stress (see Ramsay et al., 2006), males converted the food at reduced efficiency.

To study the effects of surface and volume density on growth and sex differentiation in *A. anguilla* at 26°C, Huertas and Cerda (2006) reared glass eels during period 1 lasting for 90 days and the same glass eels then metamorphosed into elvers during the period 2 from 90th day to the 720th day of their experiments. In period 1, the glass eels were stocked at the surface densities of 0.5 kg/m^2 and 2.0 kg/m^2 as well as at volume densities of 5 kg/m^3 and 10 kg/m^3. They were, however, designated as $S_{0.5}V_{0.5}$, $S_{0.5}V_{10}$, S_2V_5 and S_2V_{10} groups, respectively. During period 1, the increase in density in unit of number was 4, 10 and 30 times more in the last three groups (Table 3.6). Expectedly, mortality of glass eels in the group 4 (S_2 V_{10}) was as high as 83.3%, while that in other groups was < 42.0%. Specific Growth Rate in weight (SGR_W) of group 1 ($S_{0.5}$ V_5) was almost 50 to 100% higher than that in other groups during the first 30-days. Reversing the trend from the 30th day onwards, group 4 showed 1½ to 2 times higher SGR_W from the 30th to 90th days. Consequently, the final body weight of the group 4 was 2.6 g, in comparison to 1.2 g in the group 1. Clearly, high volume density facilitated faster growth in the weight of the glass eels. It may be interesting to know whether cortisol, an indicator of crowding stress (Ramsay et al., 2006), is secreted only after the glass eels metamorphosed into elvers. Crowding stress-induced cortisol effect can be mitigated, when feed is supplemented with 1.3% L-tryptonphan (Tejpal et al., 2009).

During period 2, Huertas and Cerda (2006) seemed to have opted to test effect of density on the reverse direction. Hence, the groups 1 ($S_{0.5}$ V_5), 3 (S_2 V_5) and 4 (S_2 V_{10}) were stocked at densities of 324, 281 and 60,

Table 3.6. Effect of surface and volume density on growth of glass eel (0.17 g) *Anguilla anguilla* for a period of 90 d (compiled from Huertas and Cerda, 2006).

Parameter	**Surface density (kg/m^2)**		**Volume density (kg/m^3)**	
	(0.5 kg/m^2)	**(2 kg/m^2)**	**(5 kg/m^3)**	**(10 kg/m^3)**
Designation	$S_{0.5}$ V_5	$S_{0.5}$ V_{10}	S_2 V_5	S_2 V_{10}
Density (no)	2941	11764	29412	58824
Mortality (%)	34.6	41.8	34.2	83.3
*Effective density (no)	2432	9305	24383	34294
SGR_W				
30 d	2.1	1.5	1.7	1.2
60 d	3.1	2.9	3.6	4.5
90 d	1.8	2.2	3.0	3.6
Final weight (g)	1.2	1.2	1.0	2.6
Metamorphosis (%)	26	24	27	43

*after deducting half of the respective mortality.

respectively (Table 3.7). At the lowest volume density, group 4 consisting of elvers sustained higher SGR_W, although it decreased from 1.5 on the 300th day to 0.6 on the 720th day of the experiment, in comparison with the drop from 1.5 to 0.2 in group 1. As a result, the elvers in group 4 attained a body weight of 609 g on the 720th day, in comparison to 114 g in group 1. Clearly, low volume density facilitated faster growth in weight from 90th day to 720th day. Incidentally, Holmgren and Mosegaard (1996) also reported faster growth in weight of glass eels-elvers *A. anguilla* reared at 26°C from 112th day (16th week).

Experimental studies of Geoffroy et al. (2013) suggest the existence of temporally separated two labile periods; accordingly, the undifferentiated gonad may differentiate directly into an ovary or an intersexual Syrski stage in 30–40% individuals (cf. Fig. 4.1, GST + A). The Syrski subsequently differentiate into either an ovary or testis. Huertas and Cerda (2006) also found that the gonads of all (100%) the silver eels in group 4 differentiated into ovaries. More than 60% elvers in groups 1 and 3 metamorphosed into silver eels (Table 3.7). However, 16 and 12% of the remaining silver eel in these groups possessed Syrski gonad and kept their option to differentiate into either testis or ovary, depending on the environmental cues. Whereas the Gonado Somatic Index (GSI) of testis remained at 0.06 at both stage I and II, the ovary developed faster, as its GSI increased from 0.03 in stage I to 1.19 in stage II. Notably, the values of SGR_W (Specific Growth Rate in weight) and SGR_L (Specific Growth Rate in length) for females after period 2 were 0.31 and 1.05, respectively and higher than those (0.96 and 0.28) of males. Interestingly, the ovarian differentiation induced growth

Table 3.7. Effect of surface and volume density on growth and sex differentiation in elvers of *Anguilla anguilla* for a period from 90th day to 720th day of the experiment (compiled from Huertas and Cerda, 2006).

Parameter	Surface density	Volume density	
Designation	$S_{0.5}\ V_5$	$S_2\ V_5$	$S_2\ V_{10}$
Density (no)	324	281	60
Initial weight (g)	1.2	2.0	2.6
SGR_W			
300th d	1.5	1.1	1.5
720th d	0.2	0.3	0.6
Final weight (g)	114	163	609
Silver eel (%)	60	70	100
Sex differentiation			
Undifferentiated (%)	16	12	0
Testis (%)	73	57	0
Ovary (%)	11	36	100

in body weight but that of the testis growth in body length during period 2. Huertas and Cerda (2006) provided food to all the 1, 3 and 4 groups at the equal ration of 4% body weight per day. Hence it was also not known whether SGR_W is controlled by density alone or other factors like ration (cf. Roncarati et al., 1997).

Sex ratio of the wild stocks of the sea bass (*Dicentrarchus labrux*) is female biased; the female ratio ranges from 0.59 (Vanderputte et al., 2007) to 0.66 (Baranabe, 1973). With increasing age from 4 to 7 years, it was 0.81–0.89 (Baranabe, 1973). However, many authors have repeatedly found that the ratio of the captive bass is around 0.25 only (Blazques et al., 1995, 1999, 2001; Barbaro et al., 1996; Saillant et al., 2001, 2002, 2003).

Saillant et al. (2003) tested the effects of (i) density, i.e., crowding leading to (ii) size hierarchy, the two major factors that masculinize the bass intensive culture. They generated 30 family lines by artificially fertilizing the eggs of three (A, B, C) dames with the sperm of 10 sires. Rearing procedures from larval rearing to nursery and subsequent pre-growing and on-growing stages in relation to both density and size gradient are summarized in Table 3.8. The results reported by Saillant et al. (2003) are briefly summarized: 1. Effect of rearing density is not detectable on sex ratio. As the treatment covered the entire stretch of the labile period, the extensive masculinization observed in intensive culture of the bass is not caused by the rearing density. 2. Size gradient at three different intervals to minimize size hierarchy, usually practiced in intensive culture of the bass, may not also affect sex ratio. However, the female ratio from the 84th dpf

Table 3.8. Experimental design to study the effect of density and size grading on growth and sex ratio of sea bass *Dicentrarchus labrax* (compiled from Saillant et al., 2003). Bold letters indicate commencement of feeding, age at size grading and sex ratio.

Rearing stage	High density	Low density
Egg	3,330	360
Larva (1–51 dpf)	200/1	20/1
Feeding commenced on 14 dpf	**82%**	**53%**
Larval growth	4 mm–21 mm	4 mm–25 mm
Nursery (51–84 dpf)	15/1	1/1
Size grading on 84 dpf	**0.65 g, 3.3 cm**	**0.65 g, 3.3 cm**
Pre-growing (84–128 dpf)	5.3/1	0.2/1
Size grading on 135 dpf	**5.6 g, 6.3 cm**	**5.6 g, 7.3 cm**
On-growing I (128–191 dpf)	2.7/1	0.2/1
Size grading (199 dpf)	**17.4 g, 9.3 cm**	**17.4 g, 10.4 cm**
On-growing II (191–419 dpf)	1.0/1	0.2/1
Pre-growing-on-growing	*54.0 g, 14 cm*	*48.0 g, 13 cm*
Female ratio	**0.24**	**0.17**

onwards remains higher (0.27) in the largest classes graded, against 0.07 and 0.04 in the middle and lower classes, respectively 3. Mortality does not directly modulate sex ratio but may indirectly alter it by inducing differential growth during larval stage, when most mortality occurs. 4. The overall effect of density indicates that the female ratio averages to 0.24 at low density but decreases to 0.17 at high density. 5. Parental effects, especially that of dames (cf. *Menidia menidia*, Conover and Heins, 1987) was very strong. An account of it is explained later.

In concluding remarks, Saillant et al. (2003) indicated that feeding rate and growth may account for the observed difference in the sex ratio at low and high densities. Huertas and Cerda (2006) also observed a shift in growth of glass eel of *A. anguilla* from the 60th day of rearing. Roncarati et al. (1997) noted that *A. anguilla* consumed less food but converted it with a higher efficiency at low density. Unlike freshwater fishes, marine fishes drink seawater to maintain osmo-concentration of their body fluids. Hence, the drunken seawater may neutralize the acidic stomach of the marine fishes like the bass. Decrease in pH was from 8.2 to 7.1 in the high density of the bass but was just from 8.1 to 7.8 in low density (Saillant et al., 2003). Secondly, IGF-1 is known to convey the somatic growth status of the reproductive endocrine system (Duan, 1998). Clearly, research on appetite, ration and conversion efficiency of the bass reared at high and low densities may resolve the riddle of excess masculinization of the bass in high density cultures.

To the question whether the SGR_w is controlled by density or other factors like ration, the findings of Paull et al. (2009) seem to provide an answer. To study the effect of ration on body growth in weight and subsequent differentiation, *Rutilus rutilus* were reared from 30 dpf (day post-fertilization), i.e., prior to gonadogenesis until the roaches sexually differentiated after 14 mpf (month post-fertilization). To simplify the results of Paull et al., high and low ration groups as well as low and high density groups alone are considered. The high ration was four times greater than that for low ration. In density series, there were 1.1 roaches/l and 8 roaches/l in low and high densities. In all, there were four groups: Low ration—low density, Low ration—high density, High ration—low density and High ration—high density. At the age of 14 mpf, the roaches were sacrificed to determine the sex. Figure 3.4a,b shows that differences in growth trends between densities were small but those between ration groups were very significant. Hence, it is the ration that determines body growth in weight rather than density level. Further the findings also showed that more and more roaches developed into females with increasing SGR_W (Fig. 3.4c). Therefore, food supply seems to affect body growth, which in turn, appears to influence sex differentiation.

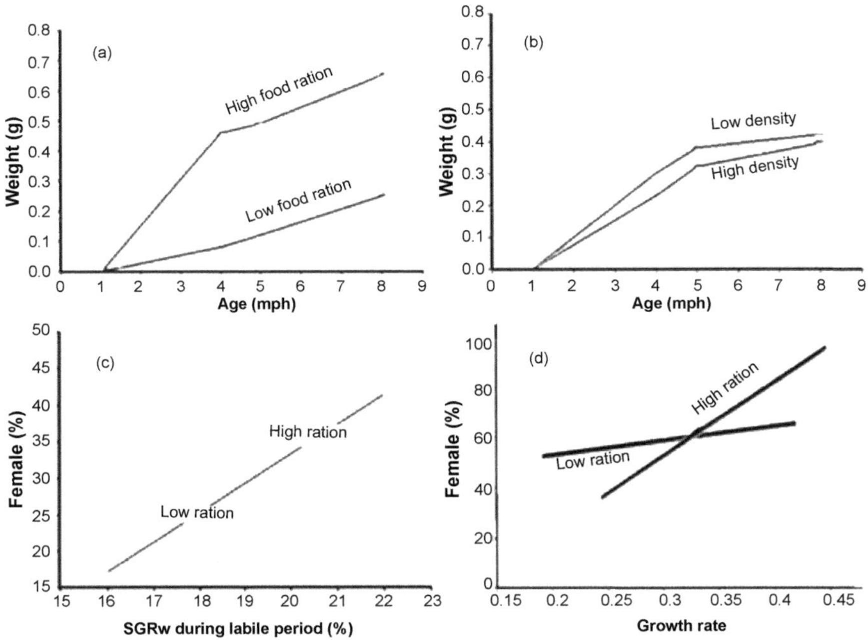

Fig. 3.4. Free hand drawings to show the growth of *Rutilus rutilus* as a function of age during the period of sex differentiation at (a) different ration and (b) density levels, (c) Percentage of females as a function of growth rate in *R. rutilus* (soruce: Paull et al., 2009). (d) Female ratio as a function of growth rate of *Danio rerio* fed low or high ration. Mph = month post-hatching (source: Lawrence et al., 2008).

Confirmation on high ration increasing female ratio also comes from a publication by Lawrence et al. (2008). They reared two genetically distinct zebrafish strains (i) Tubingen (TU) and (ii) wild caught from India (WK) in pairs separately at ≈28°C and pH 7.2–7.5 until their young ones were hatched and grown. In each of these strains, the high ration group received two-times more food than that of the low ration group. In these strains, adults are sexually dimorphic and dichromatic. The male has a more stream-lined in body shape than the female with its relatively more swollen belly. The non-black stripes are silver in color in females, while they are yellow in males. Hence, phenotypic sex can readily be identified. However, Lawrence et al. confirmed the identified sex by dissecting randomly selected progenies. Despite wide variations in female ratio ranging from 25 to 100%, the mean female ratio in both strains was 76% in the groups fed on high ration, in comparison to 51% in the low ratio-fed groups. The identified females in the high ration groups were also found to have attained a higher body length of 33 mm against 22 mm by the females of low ration groups. When the values for growth rate were plotted against sex ratio, simple linear relationships became apparent (Fig. 3.4d). Female ratio increased linearly with increasing growth rate. The increase in female ratio among high ration-fed, fast growing (in weight) fishes may due to the fact that the females have to store adequate reserve to meet the cost of vitellogenesis and provide adequate space for the ripening ovaries in the body cavity. However, males are able to reproduce at a significantly smaller body size than females (see Pandian, 2010, p. 174). As in *A. anguilla*, no sex specific chromosomes or sex specific molecular marker have so far been identified in the zebrafish *Danio rerio.* Being a secondary gonochore, all its offspring commence with ovaries but oocytes are degenerated by apoptosis 50% of the presumptive males and subsequently testicular differentiation occurs. It is to be known whether abundant food supply inhibits the process of apoptosis in the presumptive males.

Notably, density and food availability may alter sex differentiation process. For example, increased density resulting in crowding induces the appearance of more and more males in *Macropodus opercularis* (see Pandian, 2013, p. 138) and permeable social groups (e.g., *Anthias squamipinnis*, see Pandian, 2010, p. 128). In cyclic parthenogenetic crustaceans too, crowding (e.g., Fitzsimmons and Innes, 2006) and diminishing food availability (LaMontagne and McCauley, 2001) are known to induce the appearance of males. On the other hand, abundance of food induces C_4 (the one before the final molt) genetic male copepodid larvae to change sex to female phenotypes (Gusamao, 2009). Hence crowding and diminishing/abundant food supply altering sex ratio appear a common phenomenon among crustaceans and fishes. The medaka d-rR strain may prove an ideal fish to confirm whether the low ration but not high density disrupts the ovarian differentiation in XX female. The time point, at which the low ration (but not high density) disrupts the ovarian differentiation and produce recessive white colored males, can readily be observed (see p. 92). More ideal is the *GfpTr* transgenic medaka, in which differentiating ovary can readily be recognized by green fluorescence in transparent XX female and testis by its presence of leucophores (Selim et al., 2009).

Briefly, sex ratio in these tested fishes is altered indirectly through ration. However, the exact site in the endocrine cascade at which sex differentiation is derailed is not known. Insulin-like Growth Factor (IGF-1) is a key somatotropic hormone that conveys growth status to the reproductive endocrine system. IGF-1 is also known to prime pituitary gonadotropic cells of the pre-puberty GnRHR and FSH during the onset of puberty. However, the exact mechanism through which one or more specific site(s), in which sex differentiation are derailed remains to be known.

References

Acou, A., P.l. Leferbre, G. Contournet et al. 2003. Silvering of female eels (*Anguilla anguilla*) in two sub populations of the Rhode delta. Bull Fr Peche Piscic, 368: 55–68.

Andersson, J., A.-B. Florin and E. Petersson. 2012. Escapement of eel (*Anguilla anguilla*) in coastal areas in Sweden over a 50-year period. ICES J Mar Sci, 69: 991–999.

Argue, B.J. and R.A. Dunham. 1999. Hybrid fertility, introgression and back crossing in fish. Rev Fisher Sci, 7: 137–195.

Banabe, G. 1973. Contribution to the knowledge of growth and sexuality of bass (*Dicentrarchus labrax* L.) in Sete waters. Ann Inst Oceanogr, Paris, 49: 49–75.

Barbaro, A., P. Belvedere, P. Borgan et al. 1996. Chromosome set manipulation in the gilthead sea bream (*Sparus aurata* L.). pp. 227–230. *In*: B. Chatain, M. Saroglia, J. Sweetman and P. Lavens (eds.). Sea Bass and Sea Bream: Culture, Problems and Prospects. Internatl Workshop, Verona, Italy. European Aquaculture Society, Ostende, Belgium.

Beullens, K., E.H. Eding, P. Gilson et al. 1997a. Gonadal differentiation, intersexuality and sex ratios of European eel (*Anguilla anguilla* L.) maintained in captivity. Aquaculture, 153: 135–150.

Beullens, K., E.H. Eding, P. Olivier et al. 1997b. Sex differentiation changes in length, weight and eye-size before and after metamorphosis of European eel (*Anguilla anguilla* L.). Aquaculture, 153: 151–162.

Blazquez, M., F. Piferrer, S. Zanuy et al. 1995. Development of sex control techniques for European sea bass (*Dicentrarchus labrax* L.) aquaculture: effects of dietary 17α–methyltestosterone prior to sex differentiation. Aquaculture, 135: 329–342.

Blazquez, M., M. Carrillo, S. Zanuy and F. Piferrer. 1999. Sex ratios in offspring of sex reversal sea bass and the relationship between growth and phenotypic sex differentiation. J Fish Biol, 55: 916–930.

Blazquez, M., A. Felip, S. Zanuy et al. 2001. Critical period of androgen-inducible sex differentiation in a teleost fish, the European sea bass (*Dicentrarchus labrax*). J Fish Biol, 58: 342–358.

Conover, D.O. and S.W. Heins. 1987. Environmental and genetic components of sex ratio in *Menidia menidia* (Pisces: Atherinidae). Copeia, 1987: 732–745.

Cau, A., E. Coluccia, A.M. Deiana et al. 1992. Chormosomes and DNA of *Anguilla anguilla*: a study with restricted endonucleases. Genome, 35: 838–843.

Chiang, E.F.-L., Y.L. Yan, Y. Guiguen et al. 2001. Two *cyp19* (p450 aromatase) genes on duplicated zebrafish chromosomes are expressed in ovary or brain. Mol Biol Evol, 18: 542–540.

Chi-Corrie, L., L.J. Chapman and E.E. Reardon. 2008. Brood protection at a cost: mouth-brooding under hypoxia in an African cichlid. Evol Biol Fish, 82: 41–49.

Colombo, G., G. Grandi and R. Rossi. 1984. Gonad differentiation and body growth in *Anguilla anguilla* L. J Fish Biol, 24: 215–218.

David, C.J. and T.J. Pandian. 2006a. Maternal and paternal hybrid triploids of tetras. J Fish Biol, 69: 1–9.

David, C.J. and T.J. Pandian. 2006b. Cadaveric sperm induces intergeneic androgenesis in the fish *Hemigrammus caudovitattus*. Theriogenology, 65: 1048–1070.

David, C.J. and T.J. Pandian. 2008. Dispermic induction of interspecific andorgenesis in the fish Buenos Aeres tetra using surrogate eggs of widow tetra. Curr Sci, 95: 63–74.

Duan, C. 1998. Nutritional and developmental regulation in insulin-like growth factors in fish. J Nutri, 128: 3065–3145.

Duan, W., Q.-B. Qin, S. Chen et al. 2007. The formation of improved tetraploid population of red crucian carp X common carp hybrids by androgenesis. Sci China, Life Sci, 50C: 753–763.

FAO. 2012. The State of World Fisheries and Aquaculture. FAO, Rome, p. 230.

Fitzsimmons, J.M. and D.J. Innes. 2006. Inter-genotype variations in reproductive response to crowding among *Daphnia pulex*. Hydrobiologia, 568: 187–205.

Geoffroy, B., Y. Guiguen, A. Fostier and A. Bardonnet. 2013. New insights regarding gonad development in European eel: evidence for a direct ovarian differentiation. Fish Physiol Biochem, 39: 1129–1140.

George, T. and T.J. Pandian. 1995. Production of ZZ females in the female-heterogametic black molly *Poecilia sphenops* by endocrine sex reversal and progeny testing. Aquaculture, 136: 81–90.

Gusamao, L.F.M. 2009. Sex ratios, intersexuality and sex change in copepods. J Plankton Res, 31: 1101–1117.

Holmgren, K. and H. Moseguard. 1996. Implications of individual growth status on the future sex of the European eel. J Fish Biol, 49: 910–925.

Huertas, M. and J. Cerda. 2006. Stocking density at early developmental stages affects growth and sex ratio in the European eel (*Anguilla anguilla* L.). Biol Bull, 211: 286–296.

Kavumpurath, S. and T.J. Pandian. 1992. Production of YY male in the guppy *Poecilia reticulata* by endocrine sex reversal and progeny testing. Asian Fish Sci, 5: 265–276.

Kavumpurath, S. and T.J. Pandian. 1993. Production of YY female guppy *Poecilia reticulata* by endocrine sex reversal and progeny testing. Aquaculture, 116: 183–189.

Kirankumar, S. and T.J. Pandian. 2003. Production of androgenetic tiger barb *Puntius tetrazona*. Aquaculture, 228: 37–51.

Kirankumar, S. and T.J. Pandian. 2004a. Intraspecific androgenetic restoration of rosy barb using cadaveric sperm. Genome, 47: 66–73.

Kirankumar, S. and T.J. Pandian. 2004b. Use of heterologous sperm for the dispermic induction of androgenesis in barbs. J Fish Biol, 64: 1485–1497.

Klein Breteler, J.G.P., W. Dekker and L.H.R.R. Lammens. 1990. Growth and production of yellow eels and glass eels in ponds. Int Rev Ges Hydrobiol, 75: 189–205.

Komen, J., P. deBoer and C.J.J. Richter. 1992. Male sex reversed in gynogenetic XX females of common carp *Cyprinus carpio* L. by a recessive mutation in a sex determining gene. J Hered, 83: 431–434.

Koya, Y., A. Fujita, F. Niki and E. Ishihara. 2003. Sex differentiation and pubertal development of gonads in the viviparous mosquitofish *Gambusia affinis*. Zool Sci, 20: 1231–1242.

Kreuger, W.H. and K. Oliveira. 1999. Evidence for environmental sex determination in the American eel *Anguilla rostrata*. Env Biol Fish, 55: 381–389.

LaMontagne, J.M. and E. McCauley. 2001. Maternal effects in *Daphnia*: What mother are telling their offspring and do they listen? Ecol Lett, 4: 64–71.

Lawrence, C., J.P. Ebersole and R.V. Kesseli. 2008. Rapid growth and out-crossing promote female development in zebrafish (*Danio rerio*). Env Biol Fish, 81: 239–246.

Leggatt, R.A. and G.K. Iwama. 2003. Occurrence of polyploidy in fishes. Rev Fish Biol Fisher, 13: 237–246.

Liu, Y. and G. Zhou. 1986. Cytological study on the gonadal development of F_3 hybrids produced by crossing *Carassius auratus* (♀) with *Cyprinus carpio* (♂). Acta Hydrobiol Sinica, 10: 101–108.

Liu, S.J., Y. Liu, G.J. Zhou et al. 2001. The formation of tetraploid stocks of red crucian carp X common carp hybrids as an effect of interspecific hybridization. Aquaculture, 192: 172–186.

Liu, S.J., Q. Qin, Q. Ziao et al. 2007. The formation of the polyploidy hybrids from different subfamily fish crossings and evolutionary significance. Genetics, 176: 1023–1034.

Muller-Belecke, A. and G. Horstgen-Schwark. 2007. A YY male *Oreochromis niloticus* strain developed from an exceptional mitotic gynogenetic male and growth performance of genetically all male progenies. Auqacult Res, 38: 773–775.

Nielsen, R. 2011. Green and technical efficient growth in Danish freshwater aquaculture. Aquacult Econo Mgmt, 15: 262–277.

Nikinmaa, M. and B.B. Rees. 2005. Oxygen-dependent gene expression in fishes. Am J Physiol Regul Integ Comp Physiol, 288: R1079–R1090.

Pandian, T.J. 1993. Endocrine and chromosomal manipulation techniques for the production of all male and female population in food and ornamental fishes. Proc Ind Natl Sci Acad, 58B: 549–566.
Pandian, T.J. 2010. *Sexuality in Fishes.* Science Publishers, Enfield/CRC Press, Boca Raton, USA, p. 208.
Pandian, T.J. 2011. *Sex Determination in Fish.* Science Publishers, Enfield/CRC Press, Boca Raton, USA, p. 282.
Pandian, T.J. 2013. *Endocrine Sex Differentiation in Fish.* CRC Press, Boca Raton, USA, p. 303.
Pandian, T.J. and S. Kirankumar. 2003. Androgenesis and conservation of fishes. Curr Sci, 85: 917–931.
Pandian, T.J. and K. Varadaraj. 1988. Techniques for producing all male and allotriploid *Oreochromis mossambicus*. pp. 243–249. *In*: R.S. Pullin, T. Bhukuswan, K. Thungthai and J.L. MacClean (eds.). Tilapias in Aquaculture. ICLARM Conf Proc Bangkok.
Patil, J.G. and R.M. Gunasekara. 2008. Tissue and sexually dimorphic expression of ovarian and brain aromatase mRNA in the Japanese medaka (*Oryzias latipes*). Implications for their preferential roles in ovarian and neural differentiation and development. Gen Comp Endocrinol, 158: 131–137.
Paull, G.C., A.L. Filby and G.R. Tyler. 2009. Growth rate during early life affects sexual differentiation in roach (*Rutilus rutilus*). Env Biol Fish, 85: 277–284.
Piferrer, F. 2001. Endocrine sex control strategies for ferminization of teleost fish. Aquaculture, 197: 229–281.
Ramsay, J.M., G.W. Feist, Z.M. Varga et al. 2006. Whole body cortisol as an indicator of crowding stress in adult zebrafish *Danio rerio*. Aquaculture, 258: 565–574.
Roncarati, A., P. Melotti, O. Mordenti and L. Gennari. 1997. Influence of stocking density of European eel (*Anguilla anguilla* L.) elvers on sex differentiation and biotechnical performances. J Appl Ichthyol, 13: 131–136.
Saillant, E., B. Chatain, A. Fostier and C. Fauvel. 2001. Parental influence on early development in the European sea bass (*Dicentrarchus labrax*). J Fish Biol, 58: 1585–1600.
Saillant, E., A. Fostier, P. Haffray et al. 2002. Temperature effects and genotype-temperature interactions on sex determination in the European sea bass (*Dicentrarchus labrax* L.). J Exp Zool, 292: 494–505.
Saillant, E., A. Fostier, P. Haffray et al. 2003. Effect of rearing density, size, grading and parental factors on sex ratios of the sea bass (*Dicentrarchus labrax* L.) in intensive aquaculture. Aquaculture, 221: 183–206.
Selim, K.M., A. Shinomiya, H. Otake et al. 2009. Effects of higher temperature on sex differentiation and germ cell population in medaka *Oryzias latipes*. Aquaculture, 289: 340–349.
Schwenk, K., N. Breder and B. Streit. 2008. Introduction extent, processes and evolutionary impact of interspecific hybridization in animals. Phil Trans R Soc, 363B: 2805–2811.
Scott, A.G., D.J. Penman, J.A. Beardmore et al. 1989. The supermale in *Oreochromis niloticus* (L.) and its potential in aquaculture. Aquaculture, 78: 237–251.
Sun, Y.D., S.J. Liu, C. Zhang et al. 2003. Chromosome number and gonadal structure of F_0–F_{11} allotetraploid common carp x red crucian carp hybrids. Aquaculture, 264: 47–53.
Tejpal, C.S., A.K. Pall, N.P. Sahu et al. 2009. Dietary supplementation of L-tryptophan mitigates crowding stress and augments the growth in *Cirhinus mrigala* fingerlings. Aquaculture, 293: 272–277.
Tesch, F.W. 1977. *The Eel: Biology and Management of Anguilloid Eels.* Chapman and Hall, London, p. 434.
Tesch, F.W. 2003. *The Eel.* Blackwll Science, Oxford, UK.
Varadaraj, K. and T.J. Pandian. 1989. First report on production of supermale tilapia by integrating endocrine sex reversal with gynogenetic technique. Curr Sci, 58: 434–441.
Vanderputte, M., M. Duporit-Nivet, H. Chavanne and B. Chatain. 2007. A polygenic hypothesis for sex determination in the European sea bass *Dicentrarchus labrax*. Genetics, 176: 1049–1057.

Vollestad, L.A. 1992. Geographical variations in age and length at metamorphosis of maturing European eel: environmental effects and phenotypic plasticity. J Anim Ecol, 61: 41–48.
Walsh, C.T., B.C. Pease and D.J. Booth. 2004. Variations in the sex ratio, sex and age of long tinned eels within and among coastal catchments of south eastern Australia. J Fish Biol, 64: 1297–1312.
Wiberg, U.H. 1983. Sex determination in European eel (*Anguilla anguilla* L.). Cytogenet Cell Genet, 36: 589–598.
Wickstrom, H., L. Westin and P. Clevestam. 1996. The biological and economical yield from a long-term stocking experiment. Ecol Freshwater Fish, 5: 140–147.
Yan, J., L. Liu, S. Liu et al. 2010. Comperative analysis of mitochondrial control region in polyploid hybrids of red crucian carp (*Carassius auratus*) and blunt snout bream (*Megalobrama amblycephala*). Fish Physiol Biochem, 36: 263–272.
Zou, S., S. Li, W. Cai et al. 2004. Establishment of fertile tetraploid population of blunt snout bream (*Megalobrama amblycephala* ♀ X *Megalobrama terminalis* ♂) by heat shock. Aquacult Res, 39: 1322–1327.

4

Endocrine Disruption

Introduction

More than 60% of the 100,000 man-made chemicals are in routine use worldwide since 1990s. Every year 200 to 1,000 new synthetic chemicals enter the market (Shane, 1994). Over 900 of these chemicals are identified as established or potential endocrine disrupters (Soffker and Tyler, 2012). The chemicals that either mimic or antagonize the actions of endogenous hormones are known as endocrine disrupters or Endocrine Disrupting Chemicals (EDCs) (Hiramatsu et al., 2006). After use in domestic (into sewage), agricultural (e.g., pesticides, fungicides, herbicides) and industrial (e.g., 4-Nonylphenol) sectors, hundreds of estrogens and their mimics are discharged into aquatic habitats. These EDCs are broadly grouped as: 1. Natural and synthetic estrogens, 2. Phytoestrogens, 3. Mycoestrogens, 4. Phthalates, 5. Pesticides, 6. Herbicides and Fungicides, 7. Plasticizers, 8. Food antoxidants, 9. Organohalines and 10. Estrogen receptor modulators. Scholz and Kluver (2009), Iwanowicz and Blazer (2011) and Brander (2013) have listed some of them. For a list of robust estrogen-sensitive genes, Gunnarsson et al. (2007) may be consulted. As brain aromatase activity is up to 1,000 times higher in fishes than that of mammals, fishes are more susceptible to negative effects of estrogenic EDCs (Mills et al., 2014). Similarly, as many as 15 synthetic androgens (Soffker and Tyler, 2012) and antiestrogenics (e.g., industrial bi-products like β-sitosterol) are also released into waters; they are listed by Babin (1992), Pait and Nelson (2002) and Brander (2013). Briefly, the EDCs may be grouped under: (i) chemicals that mimic endogenous hormones (e.g., endosulfan, Balasubramani and Pandian, 2008a), (ii) chemicals that antagonize endogenous hormones (e.g., tributyltin [TBT], Balasubramani and Pandian, 2008b), (iii) chemicals that alter synthesis and metabolism of endogenous hormones (e.g., Barse et al., 2006) and (iv) chemicals that modify endogenous hormones (Sonnenschein and Soto 1998).

Publications on the negative effects of EDCs are pouring in. They are so diverse and rich that they have been reviewed from time to time (e.g., Jobling, 1998; Sonnenschein and Soto, 1998; Kime, 1999; Lister and Van der Kraak, 2001; Larkin et al., 2003; Hill et al., 2005; Mills and Chichester, 2005; Crain et al., 2007; Jezierska et al., 2009; Scholz and Kluver, 2009; Swedenberg et al., 2009; Leet et al., 2011). Availability of a large number (about 128) of publications on production of intersex in *Oryzias latipus* alone tempted Urushitani et al. (2007) to evaluate them in a review. The publications have also attracted many authors to contribute chapters to books; in 2007 alone, more than three chapters have been published (Denslow and Sepulveda, 2007; Finn, 2007; Jacobsen et al., 2007) and they have been preceded (e.g., Matthiessen and Sumpter, 1998; Kime, 1999) and followed by many (e.g., Thomas, 2008; Jezierska et al., 2009; Metcalfe et al., 2010). Many books have also been authored or edited (e.g., Kime, 1998; Naz, 2004; Di Giulio and Hinton, 2008; Leatherland and Woo, 2010).

Figure 4.1 shows the potential effects of EDCs on disrupted sex differentiation process reflected in alteration of structural, functional and behavioral components of reproductive system in fishes. EDCs can induce complete sex reversal in all individuals exposed to the threshold level during the labile stage (e.g., post-hatchlings of *Danio rerio* exposed to 15 ng 17α-Ethynylestradiol (EE_2)/l, Andersen et al., 2002). Typically, exogenous 17α-Methyltestosterone (MT) induction increases male ratio in medaka hatchlings with increasing dose up to a level, beyond which more and more intersexes and undifferentiated individuals are produced (Fig. 4.2). With increasing dose of EE_2, many more genetic males are feminized and consequently >90% females plus <10% intersexes are produced. Nevertheless, no EDC can induce sex reversal in adults of primary gonochores. On depuration, adults of secondary gonochores may undergo sex reversal from female to male but not from male to female (Pandian, 2013, p. 139). In adults of both of these gonochores, EDC exposure may significantly affect Secondary Sexual Characteristics (SSCs), gonadal structure, gametogenesis, vitellogenesis, spawning, spermiation, fertilization success, fecundity and/or hatchability. In rivulus, *Kryptolebias marmoratus*, exogenous 17β-estradiol (E_2) reverses sex from simultaneous hermaphrodite to female (Orlando et al., 2006). MT administration is reported to induce precocious sex reversal to male or direct differentiation to male in many groupers (Pandian, 2013, Table 31). Social induction can also reverse sex from male to female as well as from female to male in many protogynics and protandrics (Pandian, 2013, Table 35). However, publications on EDC induced changes in sexuality and

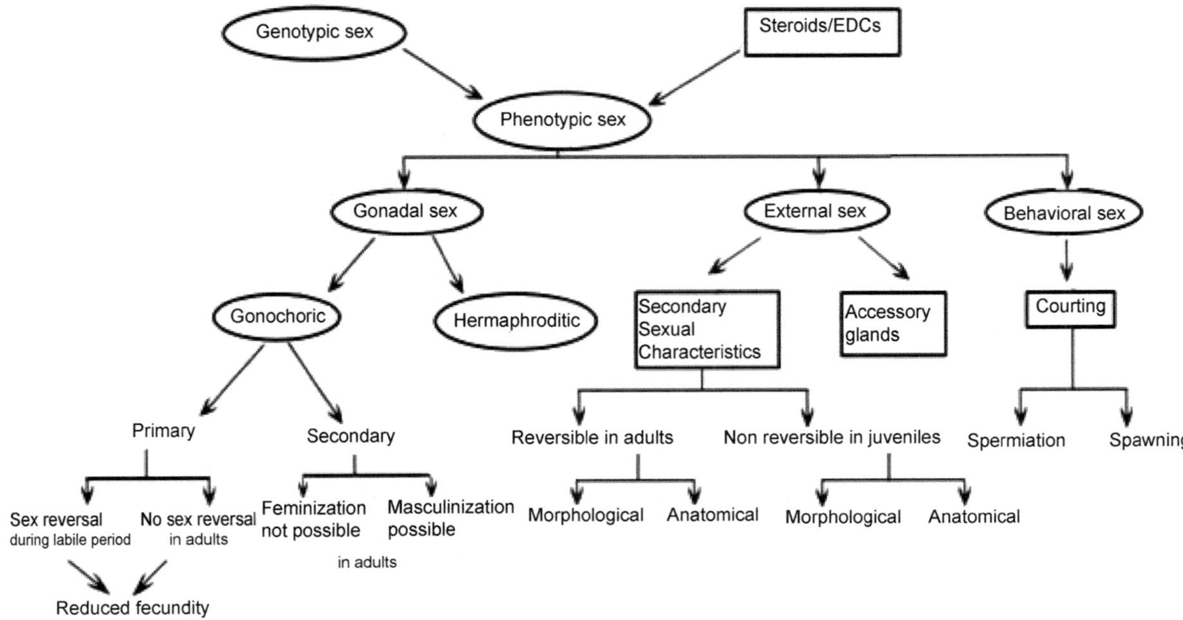

Fig. 4.1. EDC-induced disruption in sex differentiation process on the components of reproductive system of fishes.

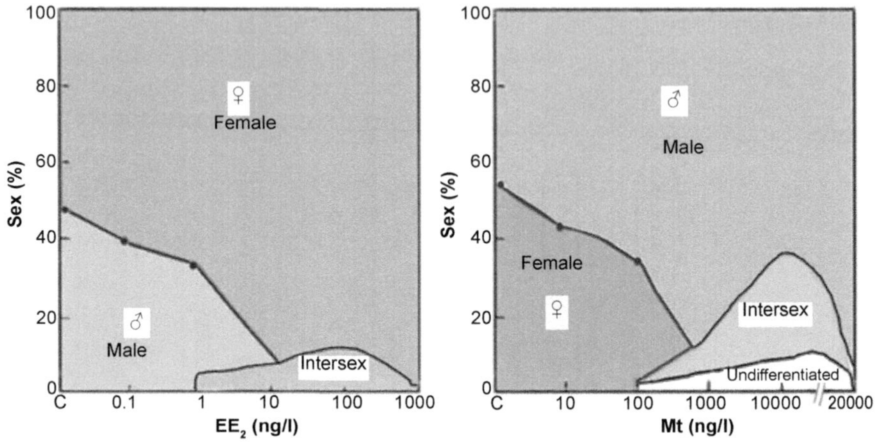

Fig. 4.2. The relative proportion of phenotypic male, female, intersex and undifferentiated *Oryzias latipes* exposed to different doses of EE_2 and MT (source: Metcalfe et al., 2010).

other components of reproductive system in sequential hermaphrodites are awaited. They are available on EDC-induced reversible and non-reversible effects on SSCs, although they are wanted for the EDC-induced alterations in accessory glands. Catfishes, gobies, poeciliids and syngnathids are ideal models to study the EDC effects on seminal vesicles (see Pandian, 2013, p. 92), Accessory Gland System (see Pandian, 2010, p. 120), spermatheca and brood pouch (see Sarria et al., 2013), respectively.

4.1 Labile and non-labile stages

The reviews and books cited above have considered endocrine disruption from the point of EDC groups listed above (e.g., Babin et al., 2007) or from species point of view (e.g., Mills and Chichester, 2005) or from the angle of the site of endocrine disruption in brain-pituitary-gonadal axis (e.g., Thomas, 2008), or from the angle of behavioral physiology (e.g., Soffker and Tyler, 2012), or from ecosystem point of view (e.g., Matthiessen, 2003; Sarkar et al., 2006). Long tables are available listing the effects of EDCs on one or more aspects of reproductive endocrine system of fishes (e.g., Pait and Nelson, 2002; Scholz and Kluver, 2009; Iwanowicz and Blazer, 2011; Brander, 2013). However, most publications on experimental and/or field observations have not clearly distinguished the disrupted reproductive performance by non-labile adults from the labile early stages. Briefly, the reported endocrine disruptions are mostly limited to induction of reversible vitellogenesis and non-reversible differentiation of oviducts in adult males. In the adult female, the disruptions induce development of gonopodium, nuptial tubercles and other SSCs of male. But the disruption is not capable of reversing original sex in adults of either sex. Hence, this chapter, which is by no means an exhaustive one, has chosen to approach endocrine disruption for the first time from the point of labile (covering egg, embryonic, alevin, fry and/or juvenile stage) and non-labile differentiated adult stages.

The labile period is defined as the period of time covering one or more life stage(s), when the sexually undifferentiated or differentiating gonad is responsive and amenable to the desired high percentage of sex reversal on exposure to an optimal dose of one or other EDC. In most fishes, ovarian differentiation precedes testicular differentiation (Piferrer, 2001), implying that the labile period for feminization is located earlier in development than that for masculinization. The periods through which a fish embryo passes from alevin to fry, juvenile, puberty and to adult stages, are marked by a specific event. Embryogenesis is terminated by hatching in oviparous and parturition in viviparous fishes. Alevins depend on yolk nutrition and do not feed. Opening of the mouth marks the entry of an alevin into the fry stage. The formation of morphogenic features like fin(s), eye, lateral line and so on indicates the entry into the juvenile stage. The appearance of SSCs, such as gonopodium marks the entry into puberty. Maturity and adult stage commence with the first wave of gametogenesis and release of gametes (Pandian, 2013, pp. 185–186).

In the fathead minnow *Pimephales promelas*, the labile period for ovarian differentiation is completed prior to 30 day post-hatching (dph), as indicated by the presence of differentiating ovaries. But the male gonad begins to differentiate after 63 dph (Bogers et al., 2006). In *Oryzias latipes* too, the ovarian differentiation is completed within the labile period from hatching to 14th dph. However, the testicular differentiation commences after 14th dph. In this context, the publication by Kuhl and Brouwer (2006) is relevant. In medaka strain d-rR, the male has a sex-linked orange/red phenotype with a genotype of X^rY^R, in which R stands for dominant xanthic pigmentation; female expresses white phenotype with a genotype of X^rX^r. Hence, the presence of white males and orange colored females is a definite and readily recognizable marker for successful induction of sex reversal. A male to female sex reversal in medaka on exposure to an EDC like o,p'-DDT is accompanied by an increase in brain aromatase expression and activity. Exposing d-rR strain of medaka hatchlings from 1st to 14th dph to o,p'-DDT and aromatase inhibitors (AI) like fadrozole (Fz) and/or TBT alone or in combination, Kuhl and Brouwer (2006) reported that Fz and TBT significantly reduced aromatase activity in d-rR medaka fry between 9 and 14 dph but it did not induce sex reversal in fry exposed to AI after 9 dph. Ovarian differentiation in medaka is almost irreversibly completed around 9 dph. Hence, once the induction process of feminization on exposure to o,p'-DDT from 1st to 14th dph is completed, then decrease in aromatase activity by Fz or TBT is unable to induce masculinization.

Based on the presence or absence of a non-functional intersexual phase during ontogenetic pathway of sexual differentiation, Yamamoto (1969) recognized two patterns of sexual differentiation: 1. Primary gonochores, in which sex differentiation proceeds from undifferentiated gonads according to genetic sex differentiation and develops ovary or testis. 2. Secondary gonochores, in which sex differentiation commences with differentiated ovary/ovotestis phase and proceeds with genotypic and phenotypic differentiation of sexes in parallel lines for a shorter or longer duration, but only one sex, as determined by genes and/or environment, becomes functional.

Primary gonochores are amenable to sex reversal by sex steroids or EDCs during the labile period, but are strictly not amenable during the non-labile adult stage. The following examples describe the unsuccessful attempts made to induce sex reversal in mature adults of primary gonochores. Liao et al. (2009) successfully induced feminization of fry and juveniles of *Gobiocypris rarus* immersed in 17β-estradiol (E_2) dissolved water for 21 days; by extending their experiment to adult males, they could induce only testis+ova with induction of vitellogenin (VTG). Swapna and Senthilkumaran (2009) also showed that exposure of *Clarias gariepinus* adults to MT and EE_2 affected gonadal recrudescence only; MT treatment led to precocious ovarian recrudescence, perhaps with availability more substrate for conversion into estrogen; in the EE_2-treated males, the male reproductive axis is more disturbed than that in MT-treated females. However, no sex reversal has been achieved. Ogawa et al. (2008) sensationally titled their publication as "Post-sex differentiational sex reversal…" but ended stating that "sexual bipotentiality of the ovary is maintained in the common carp, as in tilapia, at least before puberty" (see Pandian, 2013, pp. 186–187). In *Oncorhynchus mykiss*, Schwaiger et al. (2002) attempted to induce sex reversal by exposing 3-year old males and females intermittently (10 day/month) for a period of 40 days prior to spawning to 10 µl 4-*tert*-nonylphenol (NP)/l . They found no sex reversal in either the male or female. However, the offspring of the treated trout showed trans-generational hormonal imbalance with a two-folded E_2 increase in the plasma of male progenies and a 13-fold elevation of T in the plasma of female progenies. The objective of Mochida et al. (2000) was not to know whether a more potent anti-estrogenic TBT induces sex reversal in an adult. However, their data on sex ratio of *Fundulus heteroclitus* subjected to life long exposure to 1–2 µg TBT/l from early neurula to 200 day post-fertilization (dpf) provide an opportunity to understand whether a female sex reverses as an adult. Mochida et al. estimated the proportion of males as 63.4% on the 60th dpf and 62.3% on the 200th dpf. Understandably, once about 63% treated neurulae were differentiated into males, they continued as males, despite extended TBT exposure and no more males were added. Exposure of adult viviparous fishes to exogenous E_2 seems to throw more light on this subject. Exposure of pregnant adult female eelpout *Zoarces viviparus* (for 17 days prior to parturition, Rasmussen et al., 2002) and guppy *Poecilia reticulata* (26–30 days prior to parturition, Kinnberg et al., 2003) to environmentally relevant dose of 0.5 µg E_2/l did not induce sex reversal in either newly born offspring or mothers. A reason for the absence of any effect on sex ratio per se was that the environmentally relevant dose of 0.5 µg E_2/l was too low to cause such an effect. However, oral administration of a high dose of MT (400 µg/g diet) to pregnant guppy from 5 to 10 days prior to parturition resulted in the birth of 100% male guppies (Kavumpurath and Pandian, 1992, 1993) but the exposure failed to induce sex reversal in mother guppies (see also Pandian, 2013, p. 2019). Taken together, it is clear that sex differentiation is an irreversible process in primary gonochores.

However, the effects of exposure of differentiated non-labile adults of primary gonochoric fishes to an estrogen or anti-estrogenic EDC may reduce reproductive output only but not induce sex reversal. Incidentally, these observations bring additional support to the conclusion by Pandian (2012, pp. 100–102). Pandian brought evidence from seemingly unrelated aspects of sex differentiation to show that the ontogenetic pathway of differentiation leads to irreversible loss of bisexual potency by Germ Cells Supporting Somatic Cells (GCSSCs), while Primodial Germ Cells (PGCs) retain the potency. Once the GCSSCs lose bisexual potency and commit to pass through either the ovarian or testicular pathway, they cannot reverse sex during the non-labile adult stage of primary gonochores.

On the other hand, the threshold exposure of secondary gonochores to an estrogen or its mimics may reverse sex from adult female to adult male. For example, the exposure of zebrafish to 5 ng EE_2/l for four months from egg until sexual maturity produced all 100% females. Reared in clean water for 8 months, eight out of 25 EE_2-treated adult females reverted to males, indicating that estrogenic (protogynic) sex reversal was possible in zebrafish (Larsen et al., 2009). Figure 4.3 shows the effect of different depuration periods on male ratio as a result of protogynic 'sex reversal' in adult zebrafish that were exposed to 3 ng EE_2/l during early life (Fenske et al., 2005). These 'sex reversed' males contained spermatozoa in almost similar (49% occupancy) proportion in the testis, as that (58% occupancy) in the testis of unexposed males (Coe et al., 2010). However, none of the 17β-trenbolone (TB)-treated adult males, on being reared in clean water for 6 months, reverted to female suggesting that androgenic (protandric) sex reversal was not possible in zebrafish (Larsen and Baatrup, 2010). Being a secondary gonochore, the zebrafish initially develops ovary-like gonads leading to aromatase production and feminization (Fenske and Segner, 2004). However, about 20 days after hatching, the ovary along with PGCs suffers apoptosis (Uchida et al., 2002). Following the loss of ovary and aromatase, the

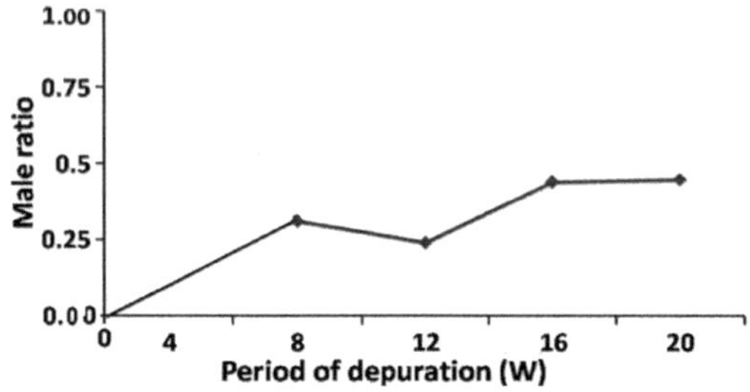

Fig. 4.3. Effect of depuration period on male ratio of *Danio rerio* exposed to 3 ng EE_2/l during early life stage (source: Fenske et al., 2005).

presumptive males develop testis. But with the loss of PGCs and ovary, the TB-treated male zebrafish was unable to regenerate the ovary (see Larsen et al., 2009). The castrated *Trichogaster trichopterus* (MT), *Macropodus opercularis* (Testosterone, T) and *Gasterosteus aculeatus* (11-Ketoandrostenedione, 11-KA) regenerated the testis, following administration of one or other steroid. Interestingly, the ovariectomized secondary gonochoric *Carassius auratus* female reverted to male on supplementation of 11-Ketotestosterone (KT) (Kobayashi et al., 1991) but failed to regenerate the ovary (Kobayashi and Stacey, 1993). The ovariectomized *Betta splendens*, presumably a seconday gonochore, regenerated testis only from the remaining oviduct, and ovarian wall (?). Adults of secondary gonochores belonging to families Poeciliidae (*Xiphophorus helleri*) and Cichlidae (*Crenicara punctulata*, *Matriaclima* cf. *livingstoni*) underwent social-mediated sex change from female to male. Briefly, all secondary gonochores are capable of protogynic sex change but not protandric sex change (see Pandian, 2013, pp. 139–140). Hence, threshold exposure to an estrogen from hatching to adult stage and subsequent depuration during adult stage restores the 'postponed sex change' and masculinizes 50% presumptive males. However, the exposure to an androgen (e.g., TB) or anti-estrogenic from hatchling to adult stage and subsequent depuration during adult stage does not feminize the presumptive 50% females.

Almost all publications that appeared prior to those of Larsen et al. (2009) and Larsen and Baatrup (2010) have focused on vitellogenin (VTG) induction rather than altered sex ratio in fishes including zebrafish that were exposed to one or another EDC. Despite carrying out the experiment, Brion et al. (2004) did not report sex ratio of E_1-treated adult zebrafish (Table 4.1). On depuration following a short period of exposure during labile period (2 to 69 dph) to EE_2, Hill and Janz (2003) and Coe et al. (2010) noted the appearance of 20 and 38% zebrafish males, respectively. Notably, Maack and Segner (2004) also reported the appearance of males after depuration for 190 days in adults, whose juveniles were exposed to 10 ng EE_2/1 from 70 to 99 dpf. However, the life long exposure of zebrafish commencing from labile

Table 4.1. Induction of sex reversal on estrogen exposure during larval, juvenile and adult stages of secondary gonochore *Danio rerio.*

Author(s)	Exposure duration	E_2/EE_2 Dose (ng/l)	♀ (%)	Remarks
Estrone (E_1)				
Holbech et al. (2006)	0–60 dph	98	85	15% males
E_2				
Brion et al. (2004)	0–21 dpf 21–42 dpf	5 25 100 5 25 100	55 56 59 28 27 35	For exposure of adults (42–200 dpf) at these doses, sex ratio is not reported
Drastichova et al. (2005)	20–65 dph	20 ng/g diet	100	
EE_2				
Hill and Janz (2003)	2–60 dph	10	75	20% males following depuration and intersex, increased follicular atresia
Andersen et al. (2004)	2–62 dpf	15	100	VTG induction and no information on males
Orn et al. (2003)	2–60 dpf	1–25		Feminization and 1 and 22% males at 1 and 25 µg/l, respectively
Maack and Segner (2004)	43–71 dpf	1.6 3 10	85 100 100	VTG induction and reduced fecundity Following exposure of hatchlings (15–42 dpf) and adults (72–99 dpf), males appear after 190 days of depuration
Coe et al. (2010)	2–60 dpf	2.8		62% females and 38% males
IDEA (2003)	Life long	10		Mating and spawning "virtually irreversibly inhibited". "Spawning" from aberrated gonad after 3 months inviable eggs. No data on restoration of males
Nash et al. (2004)	Life long	5		Functional ovary but non-functional testis
Bisphenol A (BPA)				
Drastichova et al. (2005)	20–65 dph	2000 ng/g diet	92	

period to 5 ng EE_2/1 (Nash et al., 2004) or 10 ng EE_2/1 (IDEA, 2003) resulted in the appearance of no males but intersex with non functional testis. Taken together, all these authors have noted that the threshold exposure to one or another estrogen skews the sex ratio in favor of females but certain proportions of functional males and/or non-functional intersexes were also produced. Even when 100% females were produced, a period of depuration induced 'the postponed masculinization' of presumptive males. Hence, the adults of secondary gonochoric fishes are able to reverse sex from females to males but not from males to females.

4.2 History and prevalence

A casual observation by Sweeting (1981) on 5% incidence of intersex roach *Rutilus rutilus* in the River Lea, Southeast England triggered the discovery of the occurrence of sexual disrupting estrogens in Sewage Treated Effluent (STE) water discharged into rivers. Subsequently, Jobling et al. (1998) reported widespread sexual disruption in wild fishes. Reports on similar sexual disruption in wild *Gobio gobio* (van Aerle et al., 2001), *Gasterosteus aculeatus* (see Matthiessen et al., 2003), *Esox lucius* and *Perca fluviatilis* (Vine et al., 2005) became frequent not only from the British waters but also from other European waters (e.g., *R. rutilus* in Denmark, Christiansen et al., 2000 and in France, Minier et al., 2000, *Abramis brama* in Germany, Hecker et al., 2002, 30–40% intersex of *Barbus plebejus* due to bisphenol-A (BPA) pollutant in River Po, Italy, Vigano et al., 2001, 2006). Reports from the USA are also not uncommon (e.g., *Pimephales promelas*, *Catostomus commersoni* in Boulder creek waters, Colorado, Vajda et al., 2011, Elkhorn River, Nebraska, Kolok et al., 2007). From a highly contaminated stretch of the Mississippi River, Harshbarger et al. (2000) captured two intersex sturgeon *Scaphirhynchus platyorhynchus* out of seven males. Analyzing more than 600 carps from 25 study sites of 11 major rivers of the United States, Kaiser (1996) found that estrogens of human origin present in treated effluents entering these rivers were adequate to induce VTG in males. Using Y-chromosome specific marker, Nagler et al. (2001) reported a high incidence (84%) of male specific marker in phenotypic female chinook salmon from the Columbia River. A survey of roach living in 51 British waters and study of a sample of 1,615 roaches revealed that the incidence of intersex ranged from 11% at the upstream to 44% at the sites, in which STE water was discharged (Jobling et al., 2005). An extensive survey of estrogens and their mimics in the Dutch waters revealed high plasma VTG level and prevalence of testis-ova in wild males of *A. brama* in small rivers receiving considerable load from STW; however, the survey did not indicate any significant estrogen effect in male flounder *Platichthys flesus* in the Dutch estuaries and open sea (Vethaak et al., 2005). Apparently, the open seawaters have not yet become estrogenic. Nevertheless, the deep sea fish *Mora mora* is reported to contain 7.4–12.6 ng DDT/g muscle (Sarkar et al., 2006). VTG induction is also reported in the flounder from the English and Danish coasts (for more examples, see Matthiessen et al., 2002). Evidence for inducing intersex or feminization in genotypic male/intersex roach was derived from the findings: 1. The number of roach with the normal testis in any population was inversely proportional to the number of intersex possessing testis-ova (Jobling et al., 1998). 2. The British sewage effluent waters are predominantly estrogenic (e.g., Rodgers-Gray et al., 2000) and anti-androgenic. 3. Unusually, elevated plasma VTG level was induced in males and intersexes (e.g., van Aerle et al., 2001). Table 4.2 lists environmentally relevant EDC concentrations reported from some developed countries.

Sources of EDCs: The primary source of natural steroidal estrogens is of human origin. A woman excretes between 10 and 100 µg estrogen daily. A pregnant woman excretes up to 30 µg estrogen daily (Aldercreutz et al., 1986). Another source is from the use of conjugated estrogens in treatment

Table 4.2. Reported median values for estrogenic chemicals present in sewage and industrial effluents of some developed countries.

Country	Level	Reference
	E_2 (ng/l)	
USA	160	Koplin et al. (2002)
UK	48	Desbrow et al. (1998)
Netherland	36.5	Vethaak et al. (2005)
Sweden	1.1	Larsson et al. (1999)
Italy	1.0	Baronti et al. (2000)
	E_3 (ng/l)	
UK	76	Desbrow et al. (1998)
Netherland	60.5	Vethaak et al. (2005)
Italy	9.3	Baronti et al. (2000)
Sweden	5.8	Larsson et al. (1999)
USA	2.7	Koplin et al. (2002)
	EE_2 (ng/l)	
UK	7	Desbrow et al. (1998)
Netherland	3.2	Vethaak et al. (2005)
USA	0.8	Snyder et al. (1999)
Italy	0.5	Baronti et al. (2000)
	BPA (ng/l)	
Sweden	490	Larsson et al. (1999)
Germany	258	Hansen et al. (1998)
	NP (µg/l)	
Netherland	333	Vethaak et al. (2005)
USA	332	Snyder et al. (1999)
UK	3	Lye et al. (1999)
Sweden	0.8	Larsson et al. (1999)
	OP (µg/l)	
USA	0.7	Snyder et al. (1999)
Netherland	0.7	Vethaak et al. (2005)

of cancer, osteoporosis, menopause and hypogonadalism with an estimated use of 1,690 kg/year in the USA alone (Arcand-Hoy et al., 1998). A major source of EE_2 in sewage effluent is derived from contraceptive pills. Lange et al. (2002) estimated that the US livestocks excrete annually 50 t natural and synthetic estrogens and androgens; synthetic hormones are excreted primarily from the approved implants of E_2, Testosterone propionate, TB and zeranol (Kolok and Sellin, 2008). Wastewater treatments plants efficiently remove of 80–90% of the hormones and alkylphenols (Vethaak et al., 2005). However, the remaining 10–20% of biologically inactive conjugated natural and synthetic estrogens eliminated from the body of women and animals are deconjugated during the sewage treatment process into biologically active parent compounds, which are then discharged into rivers. *Escherichia coli*, present in sewage sludge that produces β-glucuronidase, effectively transforms non-estrogenic glucuronid estradiol into estrogenic estradiol (Panter et al., 1998). Added to these estrogens of human and animal origin, agricultural and industrial sectors also generate hundreds of mimics of estrogens and anti-androgens.

Of thousands of EDCs produced by industries, some of them merit description. Organochlorine (OC) and alkylphenolic detergents are two classes of persistent estrogenics. OCs include industrial chemicals like polychlorinated bisphenols (PCBs) and polychlorinated dibenzo-*p*-dioxine (PCDOs) and insecticides. Organochlorine pesticides (OCPs) comprise of a large group of structurally diverse compounds used to control agricultural pests and vectors of diseases (Denslow and Serpulveda, 2007). The production and use of many OCPs have been banned in most developed countries but many of them continue to be used in developing countries. These industrial EDCs enter aquatic systems through several pathways including runoff, spray drift, contaminated groundwater, atmospheric deposition and sediment deposition. Not surprisingly, the persistence of EDCs ranges from Arctic to marine mammals (see Carlson et al., 2000).

Dichlorodiphenyl trichloethane (DDT): Using ^{14}C-labeled DDT, Pandian and Bhaskaran (1983) traced its uptake, elimination and accumulation in an air-breathing fish. Following 12-hours exposure to 480 μg DDT/l, *Channa striatus* accumulated 240 ng DDT from water but began to eliminate it from 60 hours after exposure. As a result, 6 ng DDT/g fish was retained. Then a fraction of it was metabolized into DDE (dichlorodiphenyl dichloethane) and DDD. DDE was more accumulated in the brain and muscle but DDD in intestinal adipose and liver. Despite diminishing use of DDT during the last few years, DDT and its daughter products continue to accumulate in organs and tissues of animals, because of the early widespread use and high chemical stability and lipophilicity of DDT.

Endosulfan: Being a broad spectrum pesticide, endosulfan is widely used in agriculture fields of more than 70 countries. Immediately following its applications, its concentration is increased to 44–56 µg/l in Asian rice fields (Suvaparp et al., 2001). It is then rapidly biocencentrated 2,682 times by freshwater algae (DeLorenzo et al., 2002).

Monomer bisphenyl A (BPA) is a key raw material for production of polychlorinated plastic and epoxy resins. In the United States alone, one billion kg of BPA is manufactured annually. On being hydrolyzed, BPA enters humans through food and liquids, and subsequently is released into sewage. Despite its environmental half-life of less than 5 days, it is present in concentration from 0.93 µg/l in groundwater to 21.0 µg/l in surface water. Amazingly, plastics floating in the surface of North Pacific Ocean 1,000 km off the United States are nearly six times more than that on land (Crain et al., 2007).

Alkylphenol polyethoxlates (APEOs) are non-ionic surfactants used primarily as industrial and agricultural cleaners and as detergents, emulsifiers as well as in formulations of pesticides, cosmetics and paints. Production of APEOs exceeds 200 million kg annually in the US alone, 60% of which are discharged into aquatic environment through sewage effluents (Naylor, 1992). Nonylphenol monoethoxylates (NP1EO) and octylphenol ethoxylates (OPEOs) are the most common APEOs with the market share of 82 and 15%, respectively. Over 83% of NP1EO production enters the sewage in Britain (Tyler et al., 2007). Hence, the river water concentrations of NP/OP are usually below 1 µg/l but can be as high as 4 µg/l for NP, 17 µg/l for NP1EO and 45 µg/l for NP1EC (nonylphenol monoethoxycarboxylates) (Bistodeau et al., 2006). Because of their lipophilicity, both NP and OP are capable of bioaccumulation in fish (Denslow and Sepulveda, 2007). As biodegradation of these compounds is usually very slow under anaerobic conditions, concentrations of alkylphenol tend to be higher in sediments; for example, 8 mg NP1EO/kg and 162 mg NP/kg sediment (Naylor, 1995).

Organotin compounds: In their inorganic form, most metals are insoluble in water. Hence, their bioavailability and toxicological effects are low. However, chemical modification of some metals greatly increases solubility and hence toxicity. Tributytin is a good example of it. TBTs are probably the most toxic compounds ever introduced deliberately in to marine environment. Organotin compounds (OTs) are used worldwide as biocides in anti-fouling paints. However, during the last two decades, many countries have strictly regulated the use of OTs, because of the persisting ecotoxicological impacts. In 1999, the International Maritime Organization adopted a resolution for global prohibition of the application of OTs as anti-fouling biocides in ships longer than 25 meter (Mochida et al., 2010). Nevertheless, there has been no regulation in Asian countries. Consequently, TBT concentrations are high in waters of these countries (Kannan et al., 1995).

Paper mill effluents are complex mixtures containing hundreds of compounds. Of them, plant sterols (e.g., β-sitosterol), lignin derivatives (e.g., polyphenols) and phytoestrogens (e.g., geistein, see Schiller et al., 2013) are important. Several publications implicate the plant-based β-sitosterol as responsible for a myriad of negative reproductive effects including pituitary function, reduced gonadal synthesis and peripheral metabolism of sex steroids, and derailed development of SSCs of fishes exposed to paper mill effluents (Denslow and Sepulveda, 2007).

Vinclozolin is a dicarboxiamide fungicide widely used to control fungal pathogens on vegetable and food crops. It is capable of disrupting sex differentiation in males in a manner similar to the therapeutic anti-androgen, flutamide. It has a demasculinizing effect on fishes (e.g., *Pimephales promelas*).

4.3 Routes of entry

EDCs are present in the aquatic habitats at fairly high concentrations and are persistent and bio-accumulating, although bisphenol A (BPA) and estradiol (E_2) may not bio-accumulate to greater levels (Tyler et al., 1998). However, these two are continuously entering the aquatic habitats. For fishes, water, food and parental transfer are the major sources of EDCs (Table 4.3). Water, which may carry EDCs, enters through body surface and gills; by drinking seawater to maintain osmo-concentration of their body, marine fishes may acquire a greater quantity of EDCs (cf. Matthiessen, 2003). In general, fish eggs imbibe water prior to hatching to gain buoyancy; however, pelagic spawners actively hydrate their oocytes; for example, an ovulated oocyte of cod imbibes sea water equivalent to eight times of its volume (see p. 44). Salmonid eggs are large (2–7 mm in diameter), have a low surface to volume ratio and are capable of buffering the interior from water quality perturbations (see Finn, 2007). Buccal brooders like cichlids actively irrigate the brooded eggs, which may gain additional dose of EDCs. Air-breathing fishes (400–500 species, Graham, 1997) may switch off gill-breathing and thereby avoid the entry of EDCs through gills. Gill cells of air-breathing fish *Channa striatus* are more sensitive to endosulfan than kidney cells (Pandey et al., 2006). During the period of its growth from 7 mg body weight to 750 mg, *C. striatus* develops an air-breathing organ and learns regular surfacing to breathe air. During this period, the fry intermittently 'hangs' above water surface projecting 3/4th of its body, avoids gill-breathing (Vivekanandan, 1977) and thereby may 'escape' from the EDC exposure during the sensitive labile period. Reared in 40 cm depth cylindrical aquarium (5 cm diameter), a fingerling (4.5 cm body length) surfaces to breathe air 1,879 times/day,

Table 4.3. Source and routes through which endocrine disrupting chemicals (EDCs) enter fishes during labile and non-labile stages.

Stage	Details	Remarks
	Labile stages	
Egg 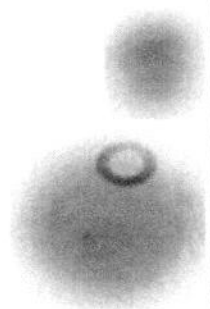	Non-hydrated demersal egg Hydrated pelagic eggs	EDCs maternally inherited and acquired via water EDCs maternally inherited and by 8–10 times to gain bouyancy
Embryo 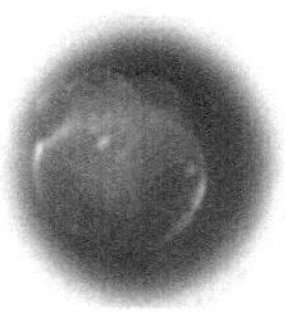	No parental care Maternal viviparity Paternal viviparity Buccal brooding	EDCs maternally inherited EDCs maternally inherited EDCs maternally and paternally inherited, and through water EDCs maternally inherited and water irrigating gills
Alevin 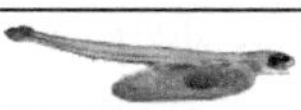	Non-feeding larva	EDCs are maternally inherited
Fry 	Feeding larva	Water and food
Juvenile	Feeding	Water and food
	Non-labile differentiated adult stage	
Freshwater fish	Feeding	Water and food
Marine fish	Feeding	Water irrigating gills and drinking seawater for osmo-regulation and food
'Sediment'eaters	Feeding	Water and food from highly EDCs concentrated sediment
Air-breather	Feeding	Water but can escape gill irrigation by air-breathing and food

travelling a distance of 1,503 m/day; it also 'hangs' to the surface for > 3 h/day (Pandian and Vivekanandan, 1976). There are many publications on the increased surfacing frequency of air-breathing fishes exposed to one or other insecticides (e.g., *Macropodus cupanus* exposed to carbaryl, Arunachalam and Palanichamy, 1982). The very sensitive (however, see Hamdoun and Epel, 2007) non-feeding eggs, embryos and alevins may acquire EDCs only through water. Alevin stage lasts for one or two days in *Pimephales promelas* but for a longer duration of > 19 dph in salmonids (e.g., *Oncorhynchus mykiss*, Andersen et al., 2000), in which the duration from egg to yolk absorption stage is the labile period (Fitzpatrick et al., 1993). Food is the second major source of entry for EDCs into the fish body. The threshold dose required to induce VTG mRNA expression in fathead minnow *P. promelas* is about 10 times greater, when NP is administered through diet (500 µg 4-NP/day) than through water (50 µg 4-NP/l) (Pickford et al., 2003). Sediments tend to accumulate persisting EDCs. For example, alkylphenol may be accumulated in sediments 2–3 orders of magnitudes higher than in water (Naylor, 1995). Estuarine core sediments from Newark, New Jersey contained 0.73 to 21 µg EDCs/kg and were reported to be toxic to medaka embryos (Cooper et al., 1993). Hence, sediment-feeders may acquire greater quantities of EDCs. The third source of EDCs is the parental contribution through eggs and embryos from the mother alone in oviparous and maternally viviparous fishes (e.g., *Zoarces viviparus,* Rasmussen et al., 2002) but from both mother and father in paternally viviparous fishes (e.g., *Syngnathus typhli*). Exposure of *S. abaster* to environmentally relevant 18 ng EE_2/l induced complete absence of male pregnancy (Sarria et al., 2013).

4.4 Mechanism of disruption

EDCs act on multiple sites in the reproductive endocrine system and by an array of mechanisms that disrupt endocrine function by (i) altering plasma levels of hormone through secretion and metabolic clearance and (ii) interfering hormone action at the receptor or other sites along the hormone signal transduction pathway (Thomas, 2008). Two different mechanisms of endocrine disruption on gonadal steroidogenesis are suggested: (ia) reduction in steroidogenic capacity of the gonad following the exposure of fish to β-sitosterol with associated decrease in mitochondrial cholesterol (Leusch and MacLatchy, 2003) and (ib) alterations in steroidogenic enzymes leading to reduced steroidogenesis. (ii) Andersen et al. (2002) reported that dieldrin, endosulfan, methiocarb and fenarinol act as both estrogen agonists and androgen antagonists but prochloraz acts as both estrogen and androgen antagonist. Fenarinol and prochloraz are potent aromatase inhibitors but endosulfan is a weak one.

Organochlorines affect neuroendocrine function and ultimately reproductive output. For example, methallibure is reported to inhibit gonadotropin secretion in male guppy *Poecilia reticulata* (Pandey, 1970). The PCB mixture of Anchor® markedly impairs LH secretion in *Micropogonias undulatus* (Thomas, 1989). The impairment inflicts a 30–35% decrease in secretion level in the hypothalmus, resulting in reduced LH secretion (Khan and Thomas, 1992).

A wide range of EDCs elicit estrogenic response in fish by competitively binding the nuclear estrogen receptor (*er*). EDCs like *orthopara* derivatives of DDT, nonylphenol, bisphenol A, hydroxylated metabolites of several PCB congenerer and kepone have relative binding affinity for fish *ers* 10^{-3} to 10^{-4} times lower than that of natural ligand E_2 (Loomis and Thomas, 1999, see also Table 3 of Crain et al., 2007). Nonylphenol and octylphenol show greater affinity for *erα* than *erβ* (Gale et al., 2004). In fact, alkylphenols (APs) are also capable of displacing E_2 from sex steroid binding site of fish in a dose dependent competitive manner (Tollefsen et al., 2004). Besides, APs may directly interact with cellular enzymes target resulting in derailed synthesis and metabolic clearance pathways of sex hormones in fish (Kazeto et al., 2004). The levels of steroids and their receptors as well as the potencies and action of steroid hormones change during reproductive cycle of fishes. Accordingly, both the pattern and degree of endocrine disruption caused by EDC also change. For example, estrogens and o,p′-DDT increase GnRH-induced LH secretion at the beginning of recrudescence but LH secretion level is reduced at the end of reproductive cycle (Khan et al., 1999).

The majority of EDCs that interact with nuclear androgen receptors antagonize the action of androgen (Gray et al., 2005). Conversely, metabolites of fungicides vinclozolin display anti-androgenic activity by competitively binding androgen receptors of fishes (see Thomas, 2008). The following examples show (i) how an antigen tamoxifen (Tx) masculinizes some fishes but not others and (ii) how xenobiotics interfere with steroid clearance pathways.

Molecular responses to E_2 are complex with most genes displaying differential responses between tissues and sexes (Filby et al., 2006). The estrogen E_2 is synthesized from the substrates T, KT and/or MT and its synthesis is catalyzed by the enzyme aromatase. Hence these precursor-androgens are classified together as aromatizable steroids. The other inducers are either steroids like 17α methyldihydrotestosterone (MDHT) or non-steroids like fadrozole (Fz), Tamoxifen (Tx) and so on (see Pandian, 2013, Table 41). Among the non-aromatizable steroids, Fz competes with other enzymes for the same sites in the substrate. Being a non-steroid type 1 anti-estrogen, Tx binds competitively to *er*. However, the receptor may still remain partially active, when the Tx-*er* complex is formed. Hence Tx is capable of acting both as estrogenic and anti-estrogenic inducer (MacGregor and Jordan, 1998). For example, Tx acts as anti-estrogenic at male-producing temperatures (MPT), but as estrogenic at female-producing temperatures (FPT) in alligators with temperature-sensitive sex differentiation (Lance and Bogart, 1991). In fishes too, Tx masculinizes sexually undifferentiated gonad by suppressing *cyp19a* expression in *Paralichthys olivaceus* (Kitano et al., 2007) and *Dicentrarchus labrax* (Navarro-Martin et al., 2009). But it fails to masculinize the all-female progenies of *Oncorhynchus mykiss* and *Oreochromis niloticus* (Guiguen et al., 1999). Arukwe et al. (1997) noted that NP regulation of reproductive hormones in juvenile *Salmo salar* by increasing the activity of steroid metabolizing enzymes at low concentrations but decreasing the activity of these enzymes at high concentration.

Sex steroids have two classical functions in differentiation: (i) to act as morphogenic factors during sex differentiation and (ii) to activate factors during sexual maturation. The interference of EDCs with synthesis and clearance of key sex hormones derails the bioavailability of the quantum of active hormones within a fish. Since a number of cytochrome p450 subfamilies (cyp2K, cyp3A) participate in metabolism of both steroid and xenobiotics in fishes (Buhler and Wang-Buhler, 1998; Thibaut et al., 2002), interactions between EDCs like alkylphenols and different p 450 isoforms involved in steroid metabolism have often been reported (e.g., Arukwe et al., 1997). Glucuronidation and sulfation are important routes of clearance of active hormones, resulting in a pool of hormone conjugates, which may be excreted or reconverted to active form in peripheral tissues. The inefficiency of most of the tested EDCs to inhibit testosterone (T) synthesis pathway indicates that 17β-HSDs are not the likely targets of EDCs. Measuring the effects of xenobiotic disruption on glucuronidation and sulfation, Thibaut and Porte (2004), however, estimated that T synthesis was significantly inhibited at 5 (TBT), 10 (fenarinol), 50 (NP and dicofol) and 100 μM 4-*tert*-pentylphenol (TPT), clearly indicating an alternate pathway, through which xenobiotics inhibit T synthesis. Among a dozen xenobiotics tested, NP, TPT and TBT were the only three compounds that inhibited sulfation. Certain compounds including phytoestrogens, xenoestrogens, fungicides and organotins may modulate aromatase cyp19 activity on the post-transcriptional level (Cheshenko et al., 2008).

4.5 Selected labile stages

A volume of literature is available on induction of sex reversal using one or other chemical. A detailed description on this aspect is provided elsewhere (Pandian, 2013, Chapter 7). Hence, this account is limited to eggs and embryos, and fishes exposed for life-long to EDCs.

Eggs and embryos: Unfertilized fish eggs contain a wide complement of maternally transferred hormones (Pandian, 2013, Section 2.5). The highest accumulation of orally administered o,p′-DDT is known to occur in gonads (e.g., *Micropogonias undulatus*, Ungerer and Thomas, 1996, see also Miller, 1993). *Labeo rohita* from the river Ganga contains 66 ng p′p′-DDT, 99 ng p′p′-DDE and 111 ng p′p′-DDD/ovary (Singh and Singh, 2008). The presence of 10 μg DDT/g eggs of *Salvelinus namayacush* is correlated with 50% alevin mortality (Burdick et al., 1964). Fish eggs also bioconcentrate many EDCs like NP and polyhybromediated diphenyl ether (PBDE), an EDC used in textile industry that displaces E_2 from estrogen receptor (*er*). For example, *Danio rerio* fry exposed to 100 μg 4-NP/l for 21 days accumulated 4–7 μg 4-NP/g eggs (Ishibashi et al., 2006). Eggs of *Acipenser sinensis* is reported to hold 178 pg PBDEs/g (Zhang et al., 2010).

Information available on accumulation, distribution and elimination of TBT in fish eggs is indeed rich. In natural waters, the detected content of triphenyltin (TpT), an estrogen antagonist, is 25.5 ng/g eggs of *A. sinensis*, which induced deformity in about 8% larvae (Hu et al., 2009). That for TBT is 150 and 190 ng/g eggs of *Aserina cernua* and *Clupea harengus*, respectively (Senthilkumar et al., 1999), and 311 and 431 ng/g eggs of *Sillago japonica* and *Trichurus lepturus*, respectively (Suzuki et al., 1992). Hano et al. (2007) reported abnormal development in embryos of *Oryzias latipes* nanoinjected with 160 µg TBT/g eggs. *O. latipes*, which was fed 1 µg TBT/g diet, contained > 20 ng TBT/g eggs (Nakayama et al., 2004, see also Nirmala et al., 1999)). From their study on the distribution of TBT in eggs of *S. japonica*, Shimasaki et al. (2008) reported the presence of 58, 29 and 13% of TBT in chorion, yolk and oil droplets of eggs. Interestingly, Shimasaki et al. (2008) also made an estimate of TBT eliminated through spawning by 100 g weighing whiting female. Fed on 20 µg TBT/g diet, the female accumulated 30,000 ng TBT /g. The TBT content of spawned eggs was 12.9 ng/g after a period from 5 to 30 days of feeding. The females spawned 38,300 eggs, equivalent to 11.3 g. Hence, the total amount of TBT eliminated through spawning is 141 ng, i.e., 0.5% of the TBT accumulated in its body.

Direct microinjection of an EDC into the embryonic yolk of fish has been chosen as a route of exposure to simulate the maternal transfer of lipophilic EDC. For example, injection of 5 ng EE_2/egg of *O. latipes* induced sex reversal in genetic males (Hano et al., 2005). However, higher doses of less potent E_2, DDT or DDE are required to achieve the same. From a preliminary study involving microinjection of 110 ng DDE/embryo drawn from monosex females of *Oncorhynchus mykiss*, Carlson et al. (2000) reported that the injected DDE did not significantly alter sex ratio. Explaining that the choice of rainbow trout was not ideal, Papoulias et al. (2003) landed on d-rR strain of *O. latipes*. However, Papoulias et al. also selected a low dose (0.5 µl/embryo) of relatively weak estrogenic DDE and could not induce sex reversal. Selecting a higher dose of stronger estrogenic (102 ng o,p'-DDT/ embryo) DDT, Metcalfe et al. (2000) too could not induce sex reversal in d-R medaka. However, the microinjection of 227 ng o,p'-DDT/embryo of medaka was strong enough to induce 83% sex reversal (Edmunds et al., 2000). Interestingly, microinjected o,p'-DDT did not undergo break down during embryonic development, as against DDT being broken to DDE and DDD in *Channa striatus* juvenile (Pandian and Baskaran, 1983) suggesting the mechanism for metabolization is not yet developed in embryos (see also p. 126). Fertilized eggs of the sex reversed X^rY^R orange females produced viable fry. Edmunds et al. (2000) also microinjected 20 ng E_2/embryo, which induced only 20% sex reversal of genetic males to females in medaka.

Very few studies are available on feminization by threshold exposure of embryos of fishes to estrone (E_1), EE_2 or diethylstilbestrol (DES) (Table 4.4). They indicate that the labile period is located during embryogenesis only in a few fish species. Exposure at nanogram levels of estrogens induced sex

Table 4.4. Effects of exogenous estrogen exposure during embryonic stage on feminization of primary gonochoric fishes.

Species and reference	Threshold level	Reported observations
	Estrone (E_1)	
Oryzias latipes Hagino et al. (2001) Hirai et al. (2006)	 32 ng/l 34 ng/l	 Feminization of genetic males Male to female sex reversal
	E_2	
Fundulus heteroclitus Urushitani et al. (2002)	10^{-8} M	Fertilization to 8 week post-hatching, 100% females
O. latipes, Kobayashi and Iwamatsu (2005)	1 µg/l	100% females following 20 hours exposure
	EE_2	
Pimephales promelas van Aerle et al. (2002)	10 ng/l	After 56 dph, feminization, intersex VTG induction
F. heteroclitus Peters et al. (2010)	100 ng/l	Increase in female ratio
Gasterosteus aculeatus Hahlbeck et al. (2004a)	10 µg/l	Exposure from fertilization to hatching failed to induce feminization
	Diethylstilbestrol (DES)	
Gobiocypris rarus Zhong et al. (2005)	500 ng/l	100% females

reversal from a slight increase in female ratio in *Fundulus heteroclitus* (Peters et al., 2010) to all females in *Gobiocypris rarus* (Zhong et al., 2005). However, exposure of *Gasterosteus aculeatus* to a dose of 10 µg EE_2/l failed to induce any signs of sex reversal, as its labile period occurs 2 weeks after hatching (Hahlbeck et al., 2004a). Due to low surface volume ratio, larger egg size may require stronger dose, as in *F. heteroclitus*. To induce reproductive effect, the E_2 dose required for rainbow trout is three orders of magnitude higher than that required for zebrafish (Thorpe et al., 2003).

Incidentally, discrete immersion of zebrafish fry to 350 ng endosulfan/l for a cumulative period of 14 hours (i.e., 2 hour/day each from 19th to 25th dph except on 21st dph) produced 72% females and 14% males plus 14% intersexes, despite rearing the treated individuals in clean water until 240 dph (Balasubramani and Pandian, 2008a). Interestingly, discrete immersion of zebrafish fry to TBT, an aromatase inhibitor, at the dose of 1µg/l for a cumulative period of 14 hours (i.e., 2 hour/day each from 18th to 24th dph except on 21st dph) produced 93% males and 7% females. Among the EDC-treated fishes, zebrafish is perhaps one of the few fishes, in which progeny testing was undertaken to bring an evidence for the EDC-induced sex reversal especially from female to male. In two out of four in the group exposed to 1 µg TBT/l were sex reversed females, i.e., there were 25% XY males and 25% sex reversed XX males (Balasubramani and Pandian, 2008b).

Life-long exposure: Experiments involving long term (generation time) or life-long exposure are expensive and time-consuming. Not surprisingly, only a few fishes characterized by a short life span have been tested. A few publications report on the effect of EDC exposure for a period of about 100 days, perhaps covering the length of a generation (Table 4.5). The 'claimed exposure' lasted for one to three generation time (see Andersen et al., 2000) or is limited to F_0 generation in some species and hardly in a few to F_1 generation (e.g., Yokota et al., 2001). The experimental design of a few authors is indeed very complicated. Despite the claim of transgenerational exposure, the durations of experimental exposure are limited to (i) embryo and (ii) hatchling of *G. rarus* (Zhong et al., 2005). Foran et al. (2002) too have exposed medaka hatchling for 2 weeks in an experiment; in the second, the treated hatchlings were reared to adults in clean water and subsequently they were exposed as adults to EDCs for 2 weeks to estimate the sex ratio of F_1. Life-long (e.g., Nash et al., 2004; Cripe et al., 2010) and multi-generational (e.g., Yokota et al., 2001) exposure of the fishes to selected EDC is limited to a few publications alone.

From the reported observations (Table 4.5), the following may be noted: 1. Unfortunately, many publications have not reported the EDC-induced change in sex ratio in F_o and/or F_1 (e.g., Diekmann et al., 2004; Nash et al., 2004). 2. However, from their investigations on exposure of medaka to E_2, EE_2, NP, OP, BPA, Di-*n*-butyl phthalate, p,p'o-DDE or o,p'-DDT, from 1–100 dph of F_o and 1–70 dph of F_1, the Ministry of Environment (MoE, 2004), Japan reported the presence of 95% females on exposure to EE_2 and 10–100% males with testis-ova on exposure to E_2, NP and OP (see Urushitani et al., 2007). Hence, exposure to estrogenic EDCs increases female ratio in a dose-dependent manner and the treated males possess fertile (e.g., Seki et al., 2002) or almost sterile testis-ova. 3a. The induced sex ratio remains the same in both F_o and F_1, as in *D. rerio* (Table 4.6), 3b. Decreases in female ratio of F_1 of *O. latipes* and 3c. Increases in male ratio of F_1 of *Cyprinus variegatus*. 4. Intersexes are produced, when *O. latipes* is exposed to a higher dose of 100 µg NP/l.

Fortunately, Orn et al. (2000) have reported sex ratios of zebrafish subjected to life-long exposure from F_o embryos to F_1 adults. They have tested at selected doses of E_2, EE_2 or MT (Table 4.6). With increasing dose, 92 and 100% females were produced at 10 and 25 ng EE_2/l, respectively. Presumably crossing F_o treated males at 0.01 ng EE_2/l with F_o females treated at 10 ng EE_2, Orn et al. found that the relation between female ratio and EE_2 dose remained more or less the same in F_o and F_1. Interestingly, the sex reversed genotypic (XY) F_o females, when crossed with the treated but sex reversed genetic (XY) F_o males, may have produced YY males. However, the fact that > 85% females were produced in F_1 indicates that EE_2 successfully induced feminization in both heterogametic XY and homogametic YY

Table 4.5. Effect of EDC exposure for a longer duration on sex ratio of F_0 and F_1 of some fishes. 0 = data not reported, NQO = 4-nitroquinoline-1-oxide, 4 TPP = 4-*tert*-pentylphenol, * = 1 3,4-dichloroaniline, ** = lindane, *** = intersex.

Species and Reference	EDC and dose	Exposure duration	Sex ratio		Remarks on observations
			F_0 ♀ : ♂	F_1 ♀ : ♂	
Danio rerio					
Nash et al. (2004)	5 ng EE_2/l	Hatching to 210 dph	0	0	Sex ratio not reported
Ensenbach and Nagel (1997)	200 µg/l*	Life long?	0		Sex ratio not reported
	40 µg/l**	Life long?	0		Sex ratio not reported
Diekmann et al. (2004)	2.9 µg NQO/l	1–96 dpf	0	0	Sex ratio not reported
Oryzias latipes					
Foran et al. (2002)	0.2–2000 ng EE_2/l	Hatching to 14 pdh	0.5:0.5	0.5:0.5	No change in sex ratio in F_0 and F_1
	0.8–8000 ng EE_2/l	Adult for 2 weeks	0.5:0.5	0.5:0.5	
Yokota et al. (2001)	52 µg NP/l	104 dph for F_0 1.3 dph for F_1	1.0:0.0	0.67:0.33	Sex ratio determination by appearance of SSCs
Balch and Metcalfe (2006)	100 µg NP/l	1–100 dph	0		Gonadal intersex in 80% ♂, suppression of anal papillae in 100% ♂, induction of mixed SSCs in 40% ♂
Cyprinus variegatus	27 ng TB/l		0.3:0.68		Sex ratio determination by appearance of SSC
Cripe et al. (2010)	27 ng TB/l			0.03:0.97	Histological sex ratio for 14 out of 15 was correct
Fundulus heteroclitus					
Mochida et al. (2010)	1 µg TBT/l	Life long	0.37:0.63		
Panter et al. (2006)	560 ng 4 TPP/l	1–107 dph	1.0:0.0		
Pimephales promelas					
Lange et al. (2001)	4 µg EE_2/l	Hatching to 306 dph	0		Sex ratio not reported
		Hatching to 56 dph	0.8:0.1+ 0.1***		
Bogers et al. (2006)	1 µg MDHT/l	Embryo to 114 dph	0.4:0.4		10% intersex and 10% with no gonad

Table 4.6. Effect of EE_2, E_2 and MT exposure for a longer duration on sex ratio of *Danio rerio* (source: Orn et al., 2000).

Dose	**F_0 sex ratio** ♀ : ♂	**F_1 sex ratio** ♀ : ♂	**F_1 sex ratio **** ♀ : ♂
	ng EE_2/l		
0.01	0.60 : 0.40	0.53:0.47	0.52:0.48
0.1	0.58 : 0.42	0.70 :0.30	0.40:0.60
1	0.60 : 0.40	0.55:0.45	0.45:0.55
10	0.92 : 0.08	0.90 :0.10	0.60:0.40
25	1.00:0.00	1.00:0.00	0.38:0.62
	µg E_2/l		
0.01	0.42:0.58		
0.1	0.45:0.55		
1.0	0.75: 0.25		
10	0.80:0.20		
	µg MT/l		
0.1	0.08:0.92		
1.0	0.13:0.73 + 0.04*		
10	0.68:0.18 + 0.14*		
50	1.00:0.00		

*intersex, ** not exposed but depurated.

males. On depuration, expectedly more F_1 males (40–62%) were produced. With increase of E_2 from 1 to 10 µg/l, 75–80% F_0 females were produced. On exposure to 10 and 50 µg MT/l, 68 and 100% females were produced due to paradoxical sex reversal.

4.6 Morphology and anatomy

An account on the EDC-induced structural alterations of morphological Secondary Sexual Characteristics (SSCs) and anatomical features is presented here. It should be noted that the induced alterations during the labile period are not reversible but those effected during the non-labile adult stage are reversible after a period of depuration.

SSCs: Many reviews and publications have not taken note of the fact that the morphological sex reversal involving SSCs may represent gonadal sex change, only when the fish is exposed to EDC during the labile period. But when the fish is exposed during the non-labile adult stage, sex reversal is limited to SSCs alone, while the original gonadal sex is retained but with altered structure and reduced egg/sperm production. A classical example is the development of pseudo-gonopodium without involving corresponding gonadal sex reversal in female mosquitofish exposed to paper mill effluent.

On exposure to the effluent in field (e.g., *Gambusia affinis holbrooki*, Howell et al., 1980; Parks et al., 2001) and laboratory (e.g., *G. affinis holbrooki*, Howell et al., 1994, *G. affinis affinis*, Sone et al., 2005), males precociously developed gonopodium and females an elongated anal fin resembling gonopodium. This gonopodium-like structure did not possess spines on the third ray and a gonopodial hook (Fig. 4.4). Regarding the manifestation of this gonopodium-like structure in *Gambussia* females exposed to androgenic chemicals during the non-labile adult stage, the report by Knapp et al. (2012) is relevant. Exposure of *G. affinis* females to 1 mg cortisol/l successfully induced the development of gonopodium-like structure and masculine copulatory behavior but no indication for the transdifferentiation from ovary to testis.

A second classical example is the development of nuptial tubercle in females exposed to MT. Fathead minnow *Pimephales promelas* is sexually dimorphic. The male has a short, blunt head with nuptial tubercles on the snout and dorsally located fatpad (Fig. 4.4). The female is lighter in color with a slender tapered head devoid of the tubercles. Exposure of males to 5.4 E_2/l for 10 days imparted a more feminine appearance to sexually mature males. The reported decreases were from 9.5 mm to 5.5 mm in diameter of the tubercle and from 1.2 mm to 0.9 mm in size of the fatpad (Miles-Richardson et al., 1999a). Incidentally, the exposure to NP also decreased the thickness of the pad from 3.5 mm in the untreated male to 0.5 mm in males exposed to 100 µg NP/l (Harries et al., 2000). Nevertheless, the E_2 induction in adults was limited to the structural degenerative changes in the testis but it did not reverse the gonadal sex (Miles-Richardson et al., 1999a; see also Miles-Richardson et al., 1999b). Hornung et al. (2004) showed that 7-day

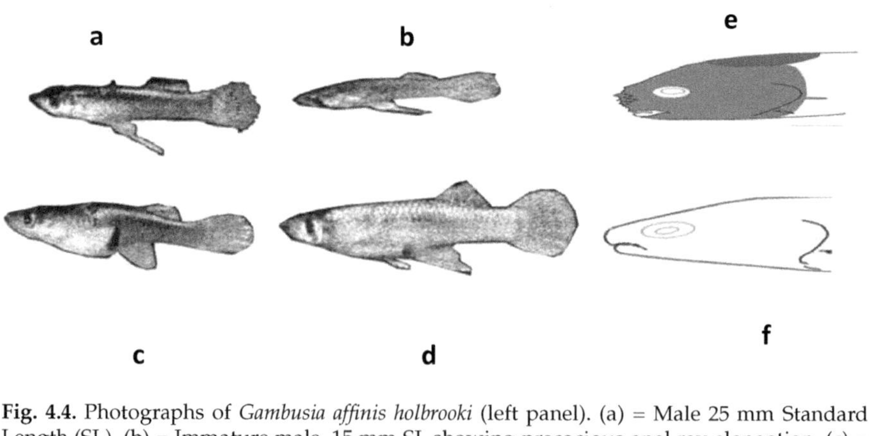

Fig. 4.4. Photographs of *Gambusia affinis holbrooki* (left panel). (a) = Male 25 mm Standard Length (SL), (b) = Immature male, 15 mm SL showing precocious anal ray elongation, (c) = Pregnant female, 31 mm SL (d) = Pregnant female, 34 mm SL with gonopodium (from Howell et al., 1980). Free hand drawings to show the dorsal fatpad and nuptial tubercles in *Pimephales promelas* male (e) and their absence in female (f) (source: Miles-Richardson et al., 1999a).

exposure of the minnow females to 20 µg MT/l induced androgenic effect with tubercle formation in adult females and VTG induction in both males and females. Further, they demonstrated the conversion of MT via aromatase activity to 17α-Methylestradiol (ME_2). However, the aromatization of MT to E_2 was strongly inhibited by fadrozole (Fz) (Zerula et al., 2002). From an interesting series of experiments, Panter et al. (2004) reported that the exposure of pre-spawning adult minnows to selected doses of androgen dihydrotestosterone (DHT), anti-androgen, flutamide and aromatase inhibitor, Fz for 4–21 days increased the number of tubercles from 6–10 to 12–18 in males but none in females (Table 4.7). Even a short term (13–21 days) exposure to MT or Fz or a combination of MT and Fz also induced the formation of tubercles and pigmentation in the dorsal fin in all the treated female minnows. Exposure of adult female minnows to trenbolone (TB) acetate for 21 days also induced the formation of nuptial tubercle and increased its number from 3 at 0.5 µg/l to 20 at 5.0 µg/l (Ankley et al., 2003).

An example for female specific SSCs is the Uro-Genital Papillae (UGP) in *Danio rerio*. The papillae are well developed in females but not in males. The exposure to E_2 during embryo-larval or juvenile stage did not produce persisting UGP in males but the exposure of adults led to the development of UGP in males in a dose-dependent manner (Brion et al., 2004). Regarding the report by Brion et al. (2004) on the appearance of visible (score 1) but not well developed (score 3) UGP in male zebrafish exposed to E_2 during the adult stage, an explanation is required. Notably, E_2 treated males developed visible but not well developed UGP. According to Brion et al. "none of the testis in E_2-treated adult males contained oocytes" and "active gametogenesis was seen in the testis with no obvious difference between control and E_2–treated adult males". Obviously, the presence of barely visible UPG in the E_2-treated adult zebrafish may not indicate gonadal sex reversal. NP-induction (100 µg NP/1) too suppressed the development of anal papillae in 100% males and also induced gonadal intersex in 80% males. However, no gonadal sex reversal was observed (Balch and

Table 4.7. Effects of dihydrotestosterone (DHT, androgen), flutamide (anti-androgen) and fadrozole (Fz, aromatase inhibitor) on nuptial tubercles, gonado somatic index (GSI) and vitellogenin induction in *Pimephales promelas* adults exposed to 14–21 d (source : Panter et al., 2004). ➞ Indicates the direction of decrease or increase.

EDC and dose	Tubercles (no)	GSI	Vitellogenin (µg/ml plasma)
DHT 8.6 µg/l ♂	8.8 ➞ 18	1.3 ➞ 0.9	0.11 ➞ 5,000
♀	0 ➞ 0	9.8 ➞ 9.0	10,000 ➞ 10,000
Flutamide 95 µg/l ♂	6 ➞ 12	1.1 ➞ 1.5	0.14 ➞ 0.16
♀	0 ➞ 0	9.0 ➞ 12.0	1,050 ➞ 11,000
Fz 24.8 µg/l ♂	10.5 ➞ 17	1.4 ➞ 1.5	1.1 ➞ 1.0
♀	0 ➞ 0	16 ➞ 10	2,000 ➞ 10

Metcalfe, 2006). Hence, reversal of SSC in fishes exposed to EDC during the non-labile adult stage may not indicate the gonadal sex.

On the other hand, exposure of males to estrogenic pollutants induces female specific SSCs. For example, 30–50% adult males of *Barbus plebejus* exposed to BPA-contaminated tributaries of the River Po, Italy exhibited barbels, the typical female SSC (Vigano et al., 2006). The presence of barbels in *Barbus* male exposed to treated estrogenic sewage effluent water may indicate the presence of testis-ova in intersex (Vigano et al., 2006). Interestingly, exposure of fertilized eggs of the secondary gonochoric zebrafish (cf. Fig. 4.3) to about 11.5 µg NP or OP/l from fertilization to 16 dph induced the formation of female SSC but a period of 2 month depuration reverted them to those of males (Seki et al., 2003). Notably, the EDC-induced male specific changes in SSCs are reversible on depuration, irrespective whether the induction was imparted during the labile and non-labile stages of secondary gonochoric males. Briefly, the presence of sex specific SSCs need not necessarily represent the gonadal sex in males and females. Further, hormones responsible for gonadal sex change and SSCs change are different.

However, the SSCs may also represent gonadal sex, when the sex reversal was induced during the labile period. The exposure to an EDC during the labile period not only reversed sex specific SSCs but also gonadal sex. For example, all female *G. affinis affinis* fry exposed to 1 or 10 µg TB/l during the labile period of 28 days developed testis with spermatozoa and true gonopodium (Sone et al., 2005). Turner (1947) showed that complete regeneration of true gonopodium in incompletely castrated male with testis as small as % of normal size, indicating the need for 11β-hydroxytestosterone for development of true gonopodium (see Pandian, 2012, p. 3). However, on administration of 1.0 µg EE_2/g diet for 150 days, male mosquitofish developed a shorter gonopodium (Angus et al., 2005).

A second example for SSCs truly representing gonadal sex is from *Oryzias latipes*. The life cyclic exposure of medaka to NP during the labile period, the sex reversed SSCs truly represented the gonadal sex at all doses of exposure, barring at one dose, in which the presence of testis-ova complicated the identification (Yokota et al., 2001). A third example is the exposure of fathead minnow during the labile period from embryo to 114 dph to 0.1 µg MDHT/l. Bogers et al. (2006) found that out of 98% males displaying nuptial tubercles, > 90% possessed testes, i.e., the development of masculine SSCs in females truly represented the gonadal sex of the phenotype (see Figs. 3 and 5 of Bogers et al.). Fathead minnow exposed to 100% mixtures of NP-OP during the labile period for 64 days permanently reduced the scores on tubercle formation in males (Bistodeau et al., 2006). A fourth example is from the tail of sheep minnow (*Cyprinus variegatus*); the tail of male minnow is decorated with black border, while that of female is

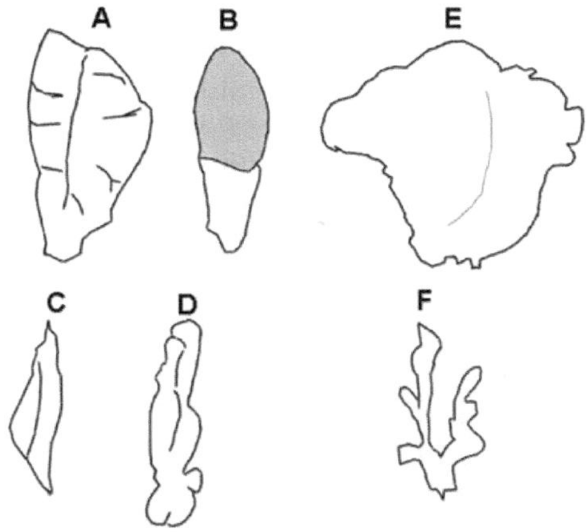

Fig. 4.5. Free hand drawings of the gonads of *Perca fluviatilis* (A to D) and *Danio rerio* (E and F). A = Paired ovary of XY male, B = Ovotestis, C = Paired testes of XY male and D = Single twisted testis of XX male (source: Rougeot et al., 2002), E = Fully developed ovary of normal female and F = Atretic ovary of female that was exposed to 25 ng EE_2/l for 3 w (source: Van den Belt et al., 2001).

not. Exposing the minnow to 27 ng TB/l from hatching to 294 dph, Cripe et al. (2010) noted that the sex ratio of the treated minnow from the appearance of SSCs was correct with histological sex ratio for 14 out of 15. Treating medaka from 3.3 to 11.6 µg NP/l or from 6.94 to 11.4 µg OP/l during the labile period from fertilization to 60 dph, Seki et al. (2003) also reported the sex distribution, as assessed by SSCs, agreed with the gonadal sex ratio. However, the sex of a certain percentage of OP-treated intersex medaka remained ambiquous between SSCs and testis-ova. On depuration in clean water for 2 months, the NP and OP-treated ambiguous intersex reverted from testis-ova to testis and from female specific to male specific SSCs. But gonadal sex reversal to ovary in a genetic female—once completed naturally or on EDC induction—remained irreversible. Notably, the recovery of male specific SSCs representing the recovery of testis from testis-ova indicates the gonadal recovery from intersex to male in fishes that were exposed to EDC like NP/OP during the labile period is still possible.

Anatomy: A brief account on the mesenterial gonadal attachment, gonadal structure and gonadal duct is provided below: In fathead minnow, all female gonads (ovary stage 2-cortical alveolus) have a double attachment to the mesentery, whereas that of males has a single attachment (Panter et al., 2006). The same holds true of zebrafish (Brion et al., 2004) and common carp (Gimeno et al., 1998a). On exposure to potent non-aromatizable androgen (0.1 µg MDHT/l) during the entire stretch of the labile period lasting from blastula to 114 dph, the minnows developed nuptial tubercles and testis but with double attachment to the mesentery (see Figs. 3, 5 and 6 of Bogers et al., 2006). This observation reveals that (i) the very presence of the tubercle represents also the EDC induced gonadal sex reversal and (ii) an exposure held during the labile period involves both morphological and gonadal sex reversal. The double attachment of the testis is, however, retained indicating that the presence of germ cells in the 'ovotestis' (see below) may also induce the double attachment (cf. single testis and double ovary in *Perca fluviatilis*, Rougeot et al., 2002).

The gonads, in which germ cells remain undifferentiated into spermatogenic or oogenetic lineages, are designated as 'sexually undifferentiated'. Gonadal sex of a fish is defined as 'male', when cells of spermatogenic lineage are histologically recognized or 'female', when cells of oogenic lineage are present. When both male and female germ cells are found within a gonad, it is called 'ovotestis', as in some protogynics (e.g., *Pagrus pagrus*), protandrics (e.g., *Acanthopagrus schlegeli*) and gonochorics (e.g., *P. fluviatilis*, see Fig. 4.5). However, the gonads containing male and female gamates in fishes exposed to androgens or anti-estrogenics are also called 'ovotestis'. Hence, there is a need to distinguish delimited ovotestis, in which testicular and ovarian zones are clearly separated within a gonad and undelimited ovotestis, in which both female and male germ cells are intermingled within a gonad (Fig. 2.12). The important differences and their implications between delimited and undelimited ovotestis are discussed elsewhere (see Pandian, 2013, p. 74, 230–232). Unfortunately, most publications reporting negative effects of one or another EDCs have not given consideration to the gonadal configuration. A comparative study on the EDC effects on *P. fluviatilis* with delimited ovotestis and *P. promelas* with undelimited ovotestis may prove rewarding.

The contribution by Rougeot et al. (2002) is unique, as it is perhaps the only publication, which has considered gonadal morphology, while reporting the negative effects of exogenous MT administration. At the threshold level of 30 mg MT/kg diet and treatment duration from 30 to 62 dph produced 100% males in Eurasian perch *P. fluviatilis*. Super-threshold dose and/or treatment duration induced the formation of ovotestis and/or underdeveloped gonad (Table 4.8). In fact, the described MT-induced the formation of four types of gonads (i) males possessing a single twisted testis, (ii) males possessing a pair of testes, (iii) females possessing a single ovary and (iv) intersexes possessing a single ovotestis (Fig. 4.5). Progeny testing revealed that the MT-treated males possessing single testis were indeed genetic XX females, as they produced 97–100% female F_1 progenies, when crossed with normal females. However, males possessing double testes were all males, which on crossing with normal females, produced progenies with balanced sex ratio. In perches, paired ovaries become a single unit during ontogenetic development, while testes remain paired.

Table 4.8. Effects of MT treatment on gonadal sex of Eurasian perch *Perca fluviatilis* (compiled from Rougeot et al., 2002).

Age (d)	Dose (mg/kg diet)	Duration (dph)	Testis (%)	Ovary (%)	Ovotestis (%)	Undeveloped (%)
32	30	32–62	100	0	0	0
42	60	42–122	78	3	19	0
45	60	45–75	70	29	1	0
45	40	45–125	80	0	0	21
45	60	45–125	72	0	1	27

Malison et al. (1986) identified sex reversed XX Δ neomales by the presence of single testis and sex reversed XY Δ neofemales with double ovaries. The presence of sex reversed XX Δ neomales with single testis has also been reported from pike perch *Stizostedion luciperca* (Zakes and Zakes, 1999) and *Dicentrarchus labrax* (Chatain et al., 1999). Most of the XX Δ neomales with single testis were unable to release sperm, probably due to abnormal gonadal morphology and/or the lack of sperm duct (cf. Rodgers-Gray et al., 2001). Unfortunately, Rougeot et al. (2002) remained silent on the structure and function of ovotestis.

Exposure of *D. labrax* to estrogens (E_2 and EE_2) during early and long period of ontogenesis (60–260 dpf) resulted in the development of two different types of gonads: (i) intersex and (ii) sterile. Intersex gonads contained germinal tissues of both sexes. A rudimentary central cavity was surrounded by groups of oogonia and pre-vitellogenic oocytes and peripherally-located testicular lobules (Blazquez et al., 1998). The sterile gonads of *D. labrax* are usually filiform shaped and consist of highly vacuolized fibrous connective tissues and are totally devoid of germ cells. A high proportion of connective tissues are also a common feature of intersex gonad (e.g., *Danio rerio*, Balasubramani and Pandian, 2008a,b). With hormonal incompatibility, a malformed non-elongated testis is formed in *Platichthys flesus* (Allen et al., 1999). Total absence of gonads in Zm, 184, 154-treated medaka has also been reported by Metcalfe et al. (2010).

To an EDC exposure, the progenitor cells of the presumptive reproductive tract are more sensitive than the germinal layer comprising of Primordial Germ Cells (PGCs). Super-threshold steroid doses are capable of disrupting multiplication of PGCs and/or development of reproductive duct in an opposite direction (see Jobling et al., 1998). Using RNA probes to *vasa* mRNA, Willey and Krone (2001) found that the exposure of 24-hours old zebrafish embryo to endosulfan or NP did not significantly alter the total number of PGCs; however, the exposure altered anterio-posterior axis of PGCs distribution following their migration to the genital ridge (cf. Pandian, 2011, p. 133). In *Cyprinus carpio*, the multiplication of PGCs is commenced at around 50 dph producing 150 cells on 130 dph. The number of PGCs in the testis of the carp exposed to TPP (4-*tert*-pentylnonylphenol)/l decreased dose-dependently. All the feminized testis of *C. carpio* developed oviduct following exposure to > 90 µg TPP/l for 60 or 90 days, when only a few PGCs were present (Gimeno et al., 1998a). In another study, Gimeno et al. (1997) found precocious development of oviduct on 51 dph in male carps that were previously exposed to TPP for 21 days prior to the commencement of the labile period. Further, the development of oviduct in male carp was an irreversible feature, as transfer of the carp to clean waters failed to restore vas defens. Confirming the experimental observation of Gimeno et al. (1998a), Rodgers-Gray et al. (2001) reported the development of female-like reproductive duct in male roach *Rutilus rutilus* exposed to estrogenic treated sewage effluent water during the labile period from 5–100 dph. Even a 100-days depuration did not restore vas defens in the treated males. On exposure to 10 ng EE_2/l, the proportion of *P. promelas* males possessing oviduct steeply increased from about 8% in those exposed from egg to 5 dph to 62% in those exposed from 15 to 20 dph: however, it is not clear why the appearance of oviduct was limited to 20% only in those exposed from egg to 20 dph (van Aerle et al., 2002). In the minnow too, feminization of reproductive duct occurred in the male reproductive system of all males that were exposed to 180 µg TPP/l. The alteration of vas defens to oviduct was irreversible, when the minnows were transferred to clean waters (Panter et al., 2006). Clearly, the EDC induced development of oviduct is a non-reversible feature, as the carp, roach and minnow were exposed to EDC during the labile period.

The majority of fathead minnow males exposed to 180 µg TPP/l during the labile period from embryo to 107 dph developed an ovarian like cavity. This histological change was not reversible even after prolonged depuration (Panter et al., 2006). Brion et al. (2004) too reported the persistence of female-like retrogonadal cavity developed on exposure of zebrafish embryo to E_2 even after 21 days depuration.

Conversely, the pregnant eelpout (*Zoarces viviparus*), on exposure to OP or E_2, developed abnormal male gonads with structures resembling the endo-ovarian cavity of female gonad (Rasmussen et al., 2002). Similarly, juvenile *Gasterosteus aculeatus* exposed to 1 µg MT/l from 26 to 44 dph is reported to develop large branched cavities in the gonad (Hahlbeck et al., 2004a). Unfortunately, eelpout and three spined stickleback were not transferred to clean waters to know whether the observed alterations are reversible. It is very likely that the eelpout recovers its normal gonads on transfer to clean waters.

4.7 Ovotestis and testis-ova

Ovotestis or testis-ova can be developed in sexually undifferentiated larva/juvenile and sexually differentiated adults by virtue of fact that PCGs and their derivatives Spermatogonial Stem Cells (SSCs) and Oogonial Stem Cells (OSCs) retain bisexual potency even after sexual maturity and spawning/spermiation (Pandian, 2012, p. 113, 2013, pp. 83–85). The term 'intersex' has been used to convey the presence of differentiated but non-functional ovarian stage in secondary gonochores (e.g., *Danio rerio*), ovarian (e.g., *Suezichthys gracilis*) or ovarian testicular (e.g., *Diplodus vulgaris*) stage in hermaphrodites (see Pandian, 2012, p. 7, 13). In an intersex, genotypic and phenotypic sexes differentiate in parallel lines for a short (e.g., *D. rerio*) or longer (e.g., *Dascyllus flavicaudus*) duration but only one sex, as genetically and/or environmentally determined, becomes functional and the other may persist as usual tissue (e.g., *Nemipterus bathybius*) or may not persist (Pandian, 2013, p. 21). Hence, an intersex is a natural developmental stage, which ultimately develops into 50% males and 50% females in secondary gonochores, but into 100% females in protogynics and almost 100% males in protandrics.

In the context of endocrine disruption, the term 'intersex' refers to an artificially induced gonadal stage. In general, estrogenic EDCs like EE_2 induce testicular feminization in a dose dependent manner; at the sub-threshold or super-threshold levels of exposure, intersexes are produced and the level of gonadal disruption may range from mild with functional gonads (e.g., *Oryzias latipes*, Seki et al., 2002) to complete atretic ovary with no yolky eggs, as in zebrafish (Fig. 4.5) exposed to super-threshold level of 10–25 ng EE_2/l. Hence, a distinction has to be made between 'natural' and 'induced' intersexes. Secondly, ovotestis and testis-ova are terms used to show the ambiguous sex differentiation status of gonad in an intersex (e.g., ovotestis in *Perca fluviatilis*, Rougeot et al. (2002) and testis-ova in *Oryzias latipes* (MoE, 2004). Shibata and Hamaguchi (1988) were perhaps the first to introduce the term testis-ova, which originated from type B spermatogonia of medaka hatchling exposed to 4 µg E_2/l for 48 hours. It is proposed to name ovarian masculinization as ovotestis and testicular feminization as testis-ova. Table 4.9 lists estrogenic induction of intersex with testis-ova in *O. latipes*, a model fish.

Structural alteration induced by EDCs ranges from extensive fibrosis to subtle disorganization of the spermatocytic cysts. The number, distribution and developmental stage of oocytes within testicular tissues vary greatly. The structural alterations depend on the potency and dose of EDC as well as exposure duration. Functional and behavior response of the testis-ova of cyprinid intersexes (Table 4.10) remarkably differs from that of *O. latipes* belonging to the family Adrianichthyidae. In *O. latipes*, fibrosis surrounding the testicular tubules (e.g., Metcalfe et al., 2001) and presence of oocytes within testicular tissues do not inhibit spermatogenesis and spermiogenesis. But the poor reproductive success of intersex is more due to altered mating behavior rather than due to reduced spermatogenic activity. Conversely,

Table 4.9. Effects of selected estrogenic EDCs on female ratio and males possessing testis-ova in *Oryzias latipes* (source: MoE 2004), Gray et al., 1999a,b*; Urushitani et al., 2007).

EDC	Exposure		Reported observations
	dose (µg /l)	duration (dph)	
E_1	100	100	*More males and all males with testis-ova
E_2	92.4 30 µg/g diet	100 70	F_0 Low fertility F_1 35% males with testis-ova
E_3	10	100	More females and all males with testis-ova
EE_2	93.4 100	F_0 100 F_1 70 21 d-adult	Low fertility in F_0, all males with testis-ova 95% F_1 females *Males with fertile testis-ova
NP	183	$F_0$100 F_1 70	20% males with testis-ova; females do not spawn
OP	100 70 100	$F_0$100 F_1 70 180	25% males with testis-ova 20% males with testis-ova Abnormalities, low incidence of testis-ova, altered mating behavior
BPA	1185 1820	$F_0$100 F_1 70 60	 10% males with testis-ova No SSCs in males
p′p′-DDE	31	$F_0$100 F_1 70	Low fertility All males with testis-ova
O, p′-DDE	1.94	$F_0$100 F_1 70	Low fertility, 30% males with testis-ova 10% males with testis-ova

the intersexes (with testis-ova) of many cyprinids like *Rutilus rutilus* and *D. rerio* suffer from suppressed gonadal development, reduced Gonado Somatic Index (GSI), sperm count and milt volume, and less motile sperm as well as inhibited spermatogenesis (Table 4.10). Evidently, EDCs disrupt the steroidogenesis in the testis-ova of cyprinids (cf. Sohoni et al., 2001) and others like *Pimephales promelas* and *Margaricus margarita*. However, the disruption of spermatogenesis by EDCs on the functions of testis-ova of *O. latipes* intersex is marginal. In other words, reproductive success of *O. latipes* intersex with testis-ova is disrupted at the courtship behavioral level but that in intersexes of cyprinids and others with testis-ova is inhibited at the functional steroidogenesis level.

Publications on ovarian masculinization of intersex with ovotestis are few. Exposure of medaka embryos to 100 µg T/l for one week induced intersex (Koger et al., 2000). In general, incidence of intersex in genetic male is low (e.g., 5% in *D. rerio* exposed to 4 µg TBT/l, Balasubramani and Pandian, 2008b). In *Gasterosteus aculeatus*, exposure to 1 µg MT/l from 0 to 39 dph and 0 to 51 dph produced intersexes with underdeveloped testis with abnormal cavity (Hahlbeck et al., 2004a,b; see also Rasmussen

Table 4.10. Functions and courtship behavior of *Oryzias latipes* and cyprinid intersex possessing testis-ova.

Authors	Reported observations
	Oryzias latipes
Gray et al. (1999a,b)	Intersex males treated at 100 µg NP/l for 6 months had testis-ova but still was able to fertilize eggs of untreated females
Kang et al. (2002)	Intersex males treated at < 463 ng E_2/l fertilized eggs, indicating the presence of testis-ova did not impair reproduction
Koger et al. (2000)	Treated for 6 days at 15 µg E_2/l, the intersex gonad was functionally a male and produced mature spermatozoa
Seki et al. (2002)	Treated at 64 ng EE_2/l, spermatogenesis was active in gonad
Balch et al. (2004)	Presence of oocytes in testis did not affect the ability of males to fertilize eggs. **Poor reproductive success was due to altered mating behavior**
	Pimephales promelas
Miles-Richardson et al. (1999b)	Exposed to 0.05–3.4 ng NP/l for 42 days induced increased necrotic spermatozoa and germ cell syncytia in a dose-dependent manner
Lange et al. (2001)	Treatment at 4 ng EE_2/l for 6 days produced intersexes but testis–ova became recognizable at age 3+ years
Sohoni et al. (2001)	Treated at 16 µg BPA/l induced 28% inhibition in sperm formation. But it did not reduce the ability of intersex to fertilize eggs
	Margaricus margarita
Palace et al. (2006)	More than 50% reduction in GSI in both males and females following the contamination by EE_2
	Rutilus rutilus
Jobling et al. (2002a,b)	Reduced sperm count and milt volume and less motile sperm
	Danio rerio
Andersen et al. (2003a)	Exposure to 15.4 ng EE_2/l from 20 to 40 dph produce 69% intersexes with testis-ova. Exposure of 0.1–1.0 ng EE_2/l suppressed gonadal development

et al., 2002). Interestingly, the exposure of *G. aculeatus* to 10 µg E_2/l or 0.05 µg EE_2/l from 0 to 13 dph, 0 to 39 dph, 0 to 42 dph and 0 to 51 induced testis-ova in genetic males but that to 1 µg MT/l from 0 to 42 dph and 0 to 51 dph produced paradoxically ovotestis mostly in genetic females (Hahlbeck et al., 2004a). Histological descriptions of ovotestis are not many (Van den Ven et al., 2003; Leino et al., 2005). Further, it is not known how feminizing and masculinizing steroids function within an ovotestis or testis-ova.

4.8 Vitellogenesis (VTGs)

Vitellogenin (VTG) is normally synthesized by the liver in response to endogenous estrogen (see Pandian, 2013, pp. 27–30) in oviparous and some viviparous (see Pandian, 2013, p. 39) females. Following its sequestration by developing oocytes, it is stored as yolk in the egg. For example, an oocyte weight of barbin flounder *Verasper moseri* is increased through deposition of yolk from < 1 mg to 3 mg during the course of seven days (Matsubara and Koyo, 1997). VTGs is usually restricted to females but males do have the *VTG* gene. Hence the exposure to natural and synthetic estrogens can induce its expression resulting in VTGs. VTG can be detected at µg level by immunohistochemistry (see Hiratmatzu et al., 2006) and ng level by Enzyme Linked Immunosorbent Assay (ELIZA) technique (e.g., Brion et al., 2002). For example, the threshold level of estrogenic chemicals for induction of VTGs in males ranges from a few ng/l in adults of *Danio rerio* (Andersen et al., 2003b) to 10 µg NP/l in *Pimephales promelas* (Pickford et al., 2003) and to 256 µg TPP/l in adult *Cyprinus carpio* (Gimeno et al., 1998b). Rightly, the zebrafish is selected by many authors as the model fish to detect the presence of estrogenicity of aquatic habitat (e.g., Andersen et al., 2000).

To detect the presence and quantify VTG, a selected EDC is administered through water, treated sewage effluent water (e.g., Diniz et al., 2005), diet (e.g., Pickford et al., 2003) or rarely by injection (e.g., Donohoe and Curtis, 1996). To quantify VTG, homogenate of larva/juvenile zebrafish is used and expressed in ng/g fish or ng/fish (Table 4.11). However, plasma is the usual source to quantify VTG, although, serum is also used (e.g., Hiramatsu et al., 2006). Rarely, the surface mucus has also been used to detect the presence of VTG in the South American *Cichlasoma dimerus* exposed to E_2 (Moncaut et al., 2003).

Understandably, VTG level in *Rutilus rutilus* from highly polluted estrogenic sewage effluent waters increased from 1.1 µg/ml in male to ≈9 and ≈150 µg/ml in intersex and female, respectively (Jobling et al., 1998). The presence of VTG in males of *R. rutilus*, *Cyprinus carpio*, *Gobio gobio*, *Oncorhynchus mykiss*, *Stizostedion vitreum*, *Platichthyes flesus* inhabiting treated estrogenic effluent waters indicates the prevalence of VTG induction (see Pait and Nelson, 2002). In general, it increases with increasing dose of estrogenic E_2, EE_2, E_3 (estriol) and aromatizable synthetic androgen MT in model fishes like *Danio rerio* (Table 4.11) and *P. promelas* (Table 4.12). Its level is higher at any tested dose in adults (Table 4.12) in comparison to larva and juvenile (Table 4.11). Exposure of zebrafish larva from 0 to 21 dpf and juvenile from 21 to 42 dpf to 5 ng E_2/l did not induce VTG; however, the same 5 ng E_2/l induced it in adults exposed to 21 days (Brion et al., 2004), indicating the absence of a fully developed liver, the site of VTG synthesis in larva and juveniles. In general, its levels are higher in females

Table 4.11. Dose-dependent changes in whole body VTG of *Danio rerio* exposed to different doses of estrogenic and androgenic chemicals.

Authors and duration	Doses	Changes in VTG level
Valle et al. (2001) Juvenile Adult		 150 ng/g fish 100 ng/g fish
Orn et al. (2003) 0–38 d post hatch 0–60 d post hatch	 1–25 ng EE_2/l 26–1000 ng MT/l	 26 → 1000 ng/g fish 40 → 10 ng/g at 500 ng MT/l, then to 70 ng/g at 1000 ng MT/l
Brion et al. (2004) 0–21 d larva 0–42 d juvenile Adult male Adult female	5–100 ng E_2/l	 60 → 200 ng/ml 110 → 1100 ng/ml 115 → 1×10^5 ng/ml 11×10^3 → 1×10^6 ng/l
Holbech et al. (2006) 0–38 d post hatch 20–38 d post hatch 20–30 d post hatch	 4–260 ng E_2/l 1.1 to 21.7 ng E_3/l 5–2779 ng TB/l	 5×10^2 → 9×10^7 ng/g 1×10^3 → 1×10^4 ng/g 60 → 25 ng/g
Van den Belt et al. (2001) Male Female	5, 10 ng E_2/l	 5 → 32 mg/ml 7 → 33 ng/ml

Table 4.12. Dose-dependent changes in plasma VTG of *Pimephales promelas* exposed to different doses of estrogenic and androgenic chemicals for a period of 3 weeks except Pickford et al. (2003), in which the duration was 2 weeks only.

Authors and sex	Dose	VTG Levels
Pawlowski et al. (2004) Adult female Adult male	 0.1–50 µg MT/l	 1.2×10^4 → 1.2×10^5 ng/ml 53 → 8×10^5 ng/ml
Hornung et al. (2004) Adult female Adult male	 20 or 200 µg MT/l	 25 → 140 µl/ml 55 → 112 µl/ml
Pickford et al. (2003) Adult male Adult male	 10 or 50 µg NP/l 100 to 1000 µg NP/day**	 100 → 8×10^6* 2×10^3 to 8×10^3*
Ankley et al. (2007) Adult male Adult female	6 to 40 µg KTC***/l	 3 → 12 µg/ml 12 → 13 mg/ml
Ankley et al. (2007) Adult female	0.005 to 5 µg TB/l	27 → 0 ng/ml

*VTG/g total RNA, **delivered as gel diet through water, ***a fungicide, Ketoconazole.

than in males but males do tend to catch up that level, when exposed to higher but environmentally irrelevant doses. Interestingly, the androgen DHT, flutamide and aromatase inhibitor Fz inhibited VTGs in males and maintained or reduced its levels in females (e.g., *P. promelas*, Table 4.7).

VTG is present in the blood long before it begins to be taken up by developing oocytes. "It may increase by up to a million fold from nanograms per milliliter to milligrams per milliliter" (Hutchinson et al., 2006). However, it is deposited in oocyte only after the stimulation by gonadotropin; however, it is not deposited in the sterile or ovarietomized (e.g., Ishibashi et al., 2004) ovary containing no oocyte to take up the circulating VTG. Its accumulation outside the gonad is known to induce kidney failure (Herman and Kincaid, 1988) and cause disruption in hemodynamics and functions (Scholz and Gutzeit, 2000). These negative effects of precocious and/or excess VTG may be a major reason for mortality. For example, 98% larvae of fathead minnow did not survive following their exposure to 200% NP-OP mixture (Bistodeau et al., 2006).

It is an established fact that estrogenic EDCs induce and elevate VTG in both males and females of many fishes. However, very few publications have demonstrated the reversibility of VTG on depuration in adult fishes that were exposed to exogenous estrogenics for a shorter or longer duration. Presumably, the elevated VTG induced by exposure during the labile period may not be reversible. Van den Belt et al. (2002) exposed adult zebrafish to 10 or 25 ng EE_2/l for 24 days and then transferred to EE_2-free water. The VTG levels in the 10-ng EE_2/l exposed males significantly decreased on 6th day of depuration (Fig. 4.6). But the same depuration period of 6 days failed to do it in the treated females, as zebrafish is capable of protogynic 'sex reversal' but not protrandric sex reversal (see p. 100).

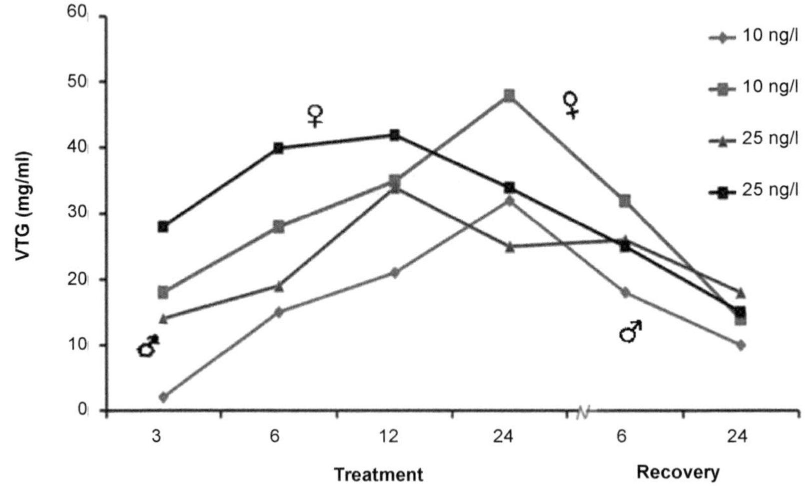

Fig. 4.6. Vitellogenesis following the exposure to 10 or 25 ng EE_2/l and subsequent depuration as a function of time in *Danio rerio* (compiled from Van den Belt et al., 2002).

Color image of this figure appears in the color plate section at the end of the book.

A second protein biomarker developed to detect VTG induction is the Vitelline Envelope Proteins (VEPs) (Arukwe et al., 1997). VEPs are usually synthesized in females only; however, as is true for VTG, male fish may also synthesize VEPs following exposure to exogenous estrogens (Arukwe et al., 2000). The genes encoding VTG and VEP are conserved and are products of several hundred million years of evolution (Arukwe and Goksoys, 2003). Being extremely hydrophobic, VEPs are difficult to measure, although VEP mRNA can be measured with relative ease.

4.9 Gametogenesis

Gametogenesis is the first gateway to the next generation. Table 4.13 provides an idea on the potential impacts of intersex and structurally altered gonad, delayed sexual maturity as well as reduced Testis Somatic Index (TSI) and Ovarian Somatic Index (OSI). Incidentally, the terms TSI and OSI are more meaningful than GSI. Age at sexual maturity of *Danio rerio* larva subjected to discrete immersion at 2 µg TBT/l for 14 hours was delayed by 33 and 66 days in male and female, respectively. Consequently, absolute fecundity was reduced from 1,813 eggs in control to 298 eggs in treated fish (Balasubramani and Pandian, 2008a). Similarly, a delay of 32 days in sexual maturity of zebrafish, which was exposed at hatchling stage to 350 µg endosulfan/l decreased the fecundity to 374 eggs (Balasubramani and Pandian, 2008b). The status of gametogenesis may range from almost fertile testis-ova in medaka male exposed to 64 ng EE_2/l for 21 days (Seki et al., 2002) to complete sterility, as in zebrafish exposed 100 µg NP/l or 1 ng EE_2/l from 2 to 60 dph (Table 4.16). Table 4.14 provides an idea of the effects of doses and durations of exposure to estrogenics and androgenics that resulted in reduced or no egg production in some fishes (see also Tables 4.22, 4.23).

Exogenous estrogen exposure induced VTG induction interfered with yolk deposition in oocytes leading to inhibition of spawning. For example, 19-day exposure of fathead minnow to 120 ng E_2/l inhibited egg production (Kramer et al., 1998). In zebrafish, the exposure to 10 or 25 ng EE_2/l for 3 weeks, despite elevated VTG in plasma, reduced yolk deposition and consequent reduction in OSI. Elevated plasma VTG is implicated for the decreased production of E_2, through a feed-back mechanism (Reis-Henriques et al., 1997). However, E_2 is required for VTGs and yolk deposition in oocytes (see Pandian, 2013, pp. 27–30). In the absence of yolky eggs, the EE_2-treated females did not spawn. Nevertheless, the elevated VTGs did not critically reduce TSI, which was around 0.7–0.8 in control and EE_2 or OP-treated males. Firstly, the EE_2 (25 ng/l) exposure did not affect 70–100% males that were ready to spermiate (see also Table 4.16). With its high potency, EE_2 (25 ng/l) exposure almost completely inhibited spawning in

Table 4.13. Delayed maturity of gonadal sex and alterations as well as decreased TSI and OSI in fishes exposed to exogenous EDCs (source: Mills and Chichester, 2005).

Fish	Hormone	Authors
	Intersex gonad	
Medaka	E_1	Metcalfe et al. (2001)
Medaka	E_2	Kang et al. (2002)
Medaka	E_3	Metcalfe et al. (2001)
Medaka	EE_2	Metcalfe et al. (2001)
Medaka	OP	Gray et al. (1999a)
Medaka	BPA	Metcalfe et al. (2001)
Carp	E_2	Gimeno et al. (1998b)
Medaka	T	Koger et al. (2000)
Zebrafish	MT	Orn et al. (2003)
Mullet	PCB	Ferreira et al. (2004)
	Delayed sexual maturity	
Zebrafish	EE_2	Segner et al. (2003)
Zebrafish	Endosulfan	Balasubramani and Pandian (2008a)
Sand goby	EE_2	Robinson et al. (2003)
Zebrafish	TBT	Balasubramani and Pandian (2008b)
Zebrafish	BPA	Segner et al. (2003)
Guppy	Vinclozolin	Bayley et al. (2002)
	Altered gonadal structure	
Zebrafish	EE_2	Segner et al. (2003)
Zebrafish	BPA	Segner et al. (2003)
Medaka	BPA	Metcalfe et al. (2001)
	Decreased TSI	
Fathead minnow	E_1	Panter et al. (1998)
Fathead minnow	E_2	Panter et al. (1998)
Medaka	E_2	Kang et al. (2002)
Goldfish	E_2	Bjerselius et al. (2001)
Carp	E_2	Gimeno et al. (1998b)
Rainbow trout	EE_2	Jobling et al. (1996)
Rainbow trout	OP	Jobling et al. (1996)
Carp	TPP	Gimeno et al. (1998b)
Fathead minnow	BPA	Sohoni et al. (2001)
	Decreased OSI	
Fathead minnow	Vinclozolin	Bayley et al. (2002)
Fathead minnow	TPP	Panter et al. (2006)
Zebrafish	E_2	Van den Belt et al. (2004)
	Decreased GSI	
Zebrafish	NP/OP	Bistodeau et al. (2006)
Nile tilapia	DDT	Marchand et al. (2008)
African catfish	Polluted water	Barnhoorn et al. (2004)
Perch	Polluted water	Hansson et al. (2006)

Table 4.14. Oogenesis in fishes exposed to EDCs. *dietary administration **3,4-dichloroaniline. *O. latipes* = *Oryzias latipes*, *P. promelas* = *Pimephales promelas*, *Po. minutus* = *Pomatoschistus minutus*, *D. rerio* = *Danio rerio*.

Species	Exposure		Reference
	Dose	Duration	
	Increased egg production		
O. latipes	0.54 µg NP/l	0–30 dph	Nimrod and Benson (1998)
P. promelas	0.1 ng EE_2/l	♀, 21 days	Jobling et al. (2006)
	Decreased egg production		
D. rerio	0.1 µg NQO/l	Life long	Diekmann et al. (2004)
O. latipes	83 µg OP/l	Hatch to adult	MoE (2004)
Po. minutus	6 ng EE_2/l	Juvenile to 7 months	Robinson et al. (2003)
O. latipes	3 µg E_2/g*	Adult, 21 days	Oshima et al. (2003)
O. latipes	350 µg endosulfan/l	Larva, 14 hours	Balasubramani and Pandian (2008a)
O. latipes	2 µg TPT/l	Larva, 14 hours	Balasubramani and Pandian (2008b)
O. latipes	20 µg OP/l	Adult, 21 days	Gronen et al. (1999)
O. latipes	2882 µg BPA/l	Adult, 14 days	Shioda and Wakabayashi (2000)
D. rerio	270 µg FTOH/l	Adult, 28 days	Liu et al. (2010)
P. promelas	280 µg BPA/l	Adult, 164 days	Sohoni et al. (2001)
P. promelas	100 µg NP/l	Adult, 21 days	Harries et al. (2000)
P. promelas	3.6 µg Methoxichlor/l	Adult, 21 days	Ankley et al. (2001)
P. promelas	10 ng EE_2/l	♀, 21 days	Pawlowski et al. (2004)
D. rerio	200** + 40 lindane/l	Adult, 60 days	Ensenbach and Nagel (1997)
	No egg production		
O. latipes	5 mg E_2/kg*	Hatch to adult	Patyna et al. (1999)
O. latipes	183 µg NP/l	Hatch to adult	MoE (2004)
P. promelas	120 µg T/l	Adult	Ankley et al. (2001)
P. promelas	100 ng EE_2/l	♀, 21 days	Pawlowski et al. (2004)

most females. However, the inhibition OP-treated females is not that high (Table 4.15). Secondly, the OSI of spawning female was smaller than that of non-spawning females, especially in females treated at 25 ng EE_2 and 12.5 µg OP/l (Van den Belt et al., 2001).

The OSI and TSI values reported by Brion et al. (2004) for zebrafish adults, which were exposed at larval and juvenile stages to E_2, are perhaps the only data available for OSI and TSI data on fish exposed to an estrogenic during the labile period. These values ranged from 12.2 to 17.1 (mean = 14.9) for OSI and from 1.0 to 1.6 (mean = 1.3) for TSI and were more comparable with those of the adults per se (cf. Coe et al., 2010). Hence, the impact of

E_2 exposure during the labile larval and juvenile stages is retained in the adults too. Interestingly, adult medaka exposed to 100 µg NP/l for 21 days produced eggs containing 4–7 µg NP/g eggs; however, the eggs of these medaka contained almost no NP following depuration for 25 days. Hence, the NP contained in eggs of the treated adult medaka is reversible on depuration (Ishibashi et al., 2006).

In a pioneering publication, Weber et al. (2003) reported relevant information on gametogenesis in *Danio rerio,* exposed to NP and EE_2 during the labile period from 2 to 60 dph (Table 4.16). The EE_2 induced alterations in gametogenesis were contrastingly differerent from those of NP at lower (10–30 µg/l) doses. A careful analysis of the data reported by Weber et al. suggests that exposure of zebrafish hatchlings to NP or EE_2 did not inhibit oogonial mitosis, as shown by increased occupancy of oogonial cells from 30% in control to 46 and 52% in those exposed to 100 µg NP/l and 10 ng EE_2/l, respectively (Table 4.16). On the other hand, these exposures completely inhibited vitellogenesis, as indicated by decreased occupancy of vitellogenic oocytes from 5–8 to 0%. Briefly, exposure of zebrafish hatchlings to EDCs inhibited vitellogenesis but allowed oogonial mitosis to proceed. Conversely, adult exposure to EDC allowed both oogonial mitosis and vitellogenesis to proceed more or less at the same level. For example, exposure of 4 months old adult zebrafish for 4 weeks to 8.2 fluorotelomer (FTOH), a precursor compound in manufacture of fluorinated surfactant, allowed both oogonial mitosis and vitellogenesis albeit with a progressive dose dependent decreases (Liu et al., 2010). Pawlowski et al. (2004) reported the effects of aromatizable MT on oogenesis of *Pimephales promelas* females exposed for 3 weeks. Classifying oogenetic stages, they estimated the spatial

Table 4.15. Effect of EE_2 or NP on OSI and spawning in female zebrafish treated for 3 w. Approximate values are given (compiled from Van den Belt et al., 2001).

Dose	OSI Spawning ♀	 Non-spawning♀	Non-spawning ♀ (%)
		EE_2 (ng/l)	
0	7	10	55
5	8	8	65
10	6	4	95
25	0	2	100
		OP (µg/l)	
0	6.5	11	55
12.5	7	14.5	25
25	8	7	30
50	7.5	7	60
100	7	5	45

Table 4.16. Effect of early exposure from 2 to 60 dph to NP or EE_2 and adult exposure to FTOH on gametogenesis in *Danio rerio*. Stage-wise occupany values are given in % (compiled from Weber et al., 2003; Liu et al., 2010).

Stage	Control (C)	NP (µg/l)			EE_2 (ng/l)	
		10	30	100	1	10
Early exposure: Oogenesis						
Oogonia	30	26	18	46	40	52
Previtellogenic	56	62	68	54	52	43
Vitellogenic	5	5	8	0	0	0
Post-ovulatory	5	3	2	0	1	0
Atretic	4	4	2	0	6	5
Early exposure: Spermatogenesis						
Spermatogonia	60	68	63	88	94	-
Spermatocyte	26	16	16	12	6	-
Mature	14	16	21	0	0	-
FTOH (µg/l)						
	C	10	30	90	270	
Adult exposure: Oogenesis						
Oogonia	28	27	25	24	23	
Previtellogenic	25	25	20	20	18	
Vitellogenic	25	25	23	19	17	
Post-ovulatory	22	23	22	36	42	
Adult exposure: Spermatogenesis						
Spermatogonia	10	10	12	12	12	
Spermatocyte	60	62	61	78	80	
Spermatid	30	28	27	10	8	

occupancy of primary (previtellogenic), secondary (vitellogenic), tertiary (post-ovulatory) and atretic follicles in the ovary. Unfortunately, Pawlowski et al. did not report the occupancy of oogonia. Nevertheless, the occupancy of vitellogenic follicles decreased from 12% in control to 6–7% at 0.1–1.0 µg MT/l reveals that MT did allow vitellogenesis albeit at reduced level (Fig. 4.7). Consequently, very high EDC doses reduced egg production to almost zero in some fishes (Table 4.14).

Exposure of zebrafish hatchling to 10 ng EE_2/l completely eliminated spermatogonia, resulting in sterile testis (Table 4.16), which may be useful for transfer of Spermatogonial Stem Cells or Oogonial Stem Cells to produce allogenics and xenogenics (cf. Pandian, 2011, pp. 149–150, see also Lacerda et al., 2006). The doses of 1 ng EE_2 and 100 µg NP inhibited spermatogenesis beyond spermatocyte stage. However, males treated at lower doses of NP produced mature sperm. This may explain the presence of mature sperm in intersex gonads of fishes like medaka (Seki et al., 2002, see also Table 4.10, p. 129).

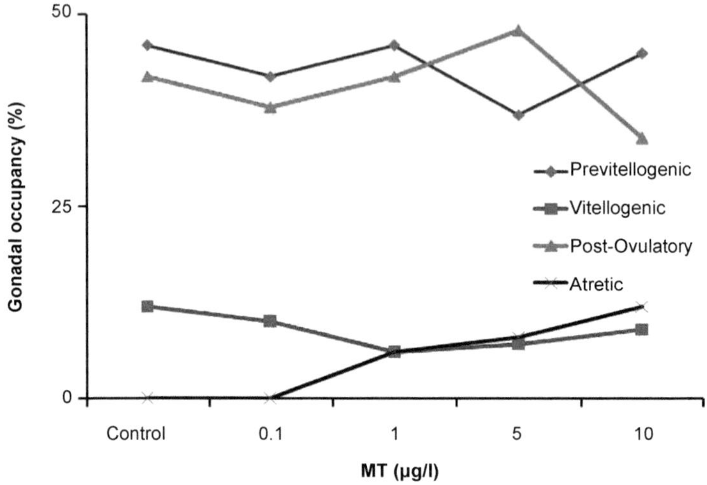

Fig. 4.7. Occupancy by oogenetic stages in the ovary of *Pimephales promelas* females exposed to different doses of MT for 21 days (compiled from Pawloski et al., 2004).

Color image of this figure appears in the color plate section at the end of the book.

Similarly, the exposure of zebrafish hatchlings to NP or EE_2 allowed spermatogonial mitosis, as shown by increased spermatogonial occupancy from 60 to 88 and 94% in those exposed to 100 μg NP/l and 1 ng EE_2/l, respectively (Table 4.16). However, the occupancy of mature sperm was zero at these doses of NP and EE_2. Hence, the exposure did not inhibit spermatogonial mitosis but progressively and dose dependently reduced the transformation of spermatocytes to spermatids. Conversely, exposure of adult (4 months old) zebrafish for 4 weeks to 270 μg FTOH/l allowed spermatogonial mitosis and spermatogenesis to proceed; hence, spermatogonia increased from 10% in control to 12% in treated males, spermatocytes from 60 to 80% but spermatocytes to spermatids and to spermatozoa were drastically reduced (Table 4.16). Hence, EDC exposure during labile period allowed increased spermatogonial mitosis but not the transformation of spermatogonia to spermatocyte and sperm. However, the adult exposure allowed spermatogonial mitosis as well as the transformation of spermatocyte to spermatid. Interestingly, the lowest doses that inhibited spermatogonial differentiation and spermatogenesis in adult fathead minnow were 640 and 16 μg BPA/l, respectively (Sohoni et al., 2001).

Conversly, spermatogenesis in adult *Poecilia reticulata* exposed to NP, E_2 or BPA for 21–30 days proceeded uninterruptedly from spermatocyte → spermatid → spermatozoa and to spermatozeugmata, albeit at reduced sperm count. But the exposure inhibited spermatogonial mitosis (Kinnberg and Toft, 2003). Briefly, exposure of adults to EDC inhibited spermatogonial mitosis but allowed the onward transformation of the existing spermatocytes to spermatozeugmata. For example, the sperm count was lower (7.01 x 10^5/ml) in guppy male exposed to more pesticide contaminated water from Lake Apopka than that (8.41 x 10^5/ml) in relatively less contaminated water from Boulware Spring (Toft and Guillette, 2005). Decreases in the count have also been reported in *Danio rerio* males exposed to DDT for one and two months (Njiwa and Muller, 2002) and DDT plus Arochlor 1254 (Njiwa et al., 2004a) as well as in *Salmo trulta f. fario* (Lahnsteiner et al., 2005).

Unlike that of mammals, fish sperm are immotile on ejacuation and attain motility only on coming in contact with water, which is followed by an extremely short duration of motility for < 1 minute. Hence, sperm count, motility and swimming velocity are correlated with fertilizability, an index of the male's success. Rurangwa et al. (1998) reported that operational ratio of 1,500 sperm to 1 egg in artificial insemination of *Clarias gariepinus* resulted in 80% fertilization success. A critical sperm to egg ratio to monitor the effects of EDCs is 1,500:1. In *Oreochromis mossambicus*, over 70% sperm were motile at the velocity of 88 μm/sec in the reference site Albasini dam but only 54% sperm were motile in DDT-sprayed Xikunda weir (Marchand et al., 2008). Exposure of *D. rerio* to 1 ng TBT/l from 0 to 70 dph produced abnormal flagellum-less spermatozoa in 50% males (McAllister and Kime, 2003). Balasubramani and Pandian (2008b) also reported the presence of spermatozoa with broken and without flagellum in 4 to 7% zebrafish males that were exposed to TBT at hatching (Fig. 4.8). The motility

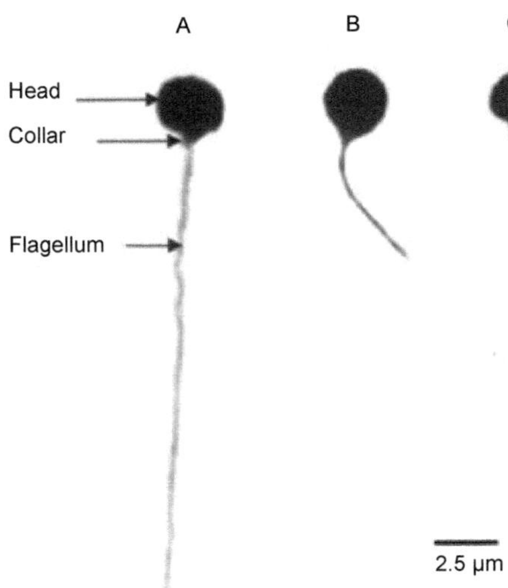

Fig. 4.8. Representative photomicrograph of (A) sperm from control, (B) sperm with broken flagellum and (C) sperm with collar alone in *Danio rerio* exposed to 4 μg TBT/l during early life stage (from: Balasubramani, 2006).

duration decreased from 105 seconds in control to 83 seconds in zebrafish, which were exposed as hatchlings to 350 µg endosulfan/l. Consequently, fertilizability also decreased from 91 to 28% (Balasubramani and Pandian, 2008a). Corresponding decreases reported for zebrafish, which were exposed as hatchlings to 2 µg TBT/l, were from 100 to 78 seconds for sperm motility duration and from 88 to 69% for fertilizability (Balasubramani and Pandian, 2008b). Incidentally, being an estrogen antagonist, TBT increased milt volume from 0.6 ml in control to 2.0 ml in males that were exposed to TBT as hatchlings (McAllister and Kime, 2003). The exposure of brown trout males to BPA during the pre-spawning season also reduced semen mass from 0.5 g to 0.1 g at the end of spawning (Lahnsteiner et al., 2005).

To minimize sperm loss and to extend life span of sperm, some fishes like grass goby *Zosterisessor ophiocephalus* adopt a sperm trail technique (Mazzolidi et al., 2000). Mucus surrounding the sperm requires some time before it is dissolved and the sperm is activated on coming in contact with water. In the sperm trail of *D. rerio*, the sperm was active for 29 minutes and its life span was 138 minutes. Sperm drawn from younger fish had longer life span than those drawn from older ones (Njiwa et al., 2004b). On exposure of the trail to 5.0 µg DDT/l, these values were decreased to 26 and 125 minutes, respectively (Njiwa and Muller, 2002). Spermatozoa packed in spermatozuegma or spermatophore (see Pandian, 2013, pp. 89–90) can be stored for 2–4 months in reproductive system of many viviparous fishes (see Pandian, 2011, pp. 14–15). The impact of EDCs on the life span of spermatozuegma or spermatophore in viviparous fishes remains to be investigated, in spite of *P. reticulata* being an ideal model for the EDC study.

4.10 Courting and mating

Courtship behavior plays a pivotal role in reproductive success. It facilitates recognition and selection of mating partners. Despite its importance, literature available on the effects of EDC on courtship behavior and mating is limited (e.g., Sloman and Wilson, 2006). Providing choices for the untreated zebrafish female to select between (i) untreated male and TBT-treated male, (ii) untreated male and TBT-treated sex reversed XX Δ neomale and (iii) TBT-treated XY male and TBT-treated sex reversed XX Δ neomale, Balasubramani and Pandian (2008b) made an interesting experiment in an all glass 'Y'-maze (see Yambe et al., 1999) to know the selection of these males by female. They found that the female preferred to court with an untreated male, against the treated XY male, an untreated male against XX Δ neomale and a treated XY male, against the treated sex reversed XX Δ neomale (Table 4.17). In a mate choice experiment, Coe et al. (2010) found that the normal female zebrafish attracted the attention of the male more effectively than EE_2-treated female.

Typically, a responsive zebrafish female, sexually aroused at a specific concentration of pheromones produced by its mating partner (see Pandian, 2013, pp. 80–81), is chased (stage 1a) and later nosed (stage 1b). It is followed by the male encircling the female (stage 2); if found not compatible, the partners may separate from each other. But on entering into stage 3, the female closely leads the male. Subsequently they move parallel to each other; at this stage, the partners reach a point of 'no return'. Then, they enter into 4th stage, when they tilt against each other to juxtapose the openings of their gonadal ducts. Finally, the gametes are released. The 4th stage of the courtship is repeated 15–20 times, releasing 6–20 eggs/pulse (see Darrow and Harris, 2004). Many times, the partners may skip one or other stage and pass into final 4th stage. Table 4.18 shows the dose-dependent suppression of the described stages in zebrafish, who were exposed to TBT as hatchling. Table 4.19 provides a comprehensive picture of MT effects on SSCs, and reproductive behavior and performance of *Betta splendens*, which were exposed to MT as fry (Kirankumar and Pandian, 2002). Exposure of zebrafish males to 1 ng EE_2 from hatching to sexual maturity did not suppress courtship behavior but those at 5 ng EE_2/l were no longer able to induce spawning (Larsen et al., 2008).

Poecilia reticulata seems to have been an ideal model to study the effect of EDC on courtship behavior. The exposure to OP (Bayley et al., 1999; Gray et al., 1999b) or administration of EDC supplemented diet during the labile period from birth to sexual maturity or longer suppressed courtship (e.g., Soffker and Tyler, 2012) behavior. Conversely, the dietary administration of p,p'-DDE, flutamide or vinclozolin to adult guppies for 30 days disrupted courtship behavior and reduced the quantum of ejaculated

Table 4.17. Mate choice between untreated male, treated XY male and treated XX neomale by a normal zebrafish female (from Balsubramani, 2006).

Entry of female into the chamber of male (number of times)		
Control XY male	**Treated XY male**	**Treated XX neomale**
7.3	2.7	-
8.3	-	1.7
-	6.3	3.7

Table 4.18. Effect of tributyltin exposure on the frequency of events in the courtship behavior of zebrafish during the observation period of 30 minutes (from Balsubramani, 2006).

Dose (µg/l)	Chasing	Nosing	Encircling	Leading
0	11	14	12	10
1	6	6	3	3
2	6	4	4	4
8	3	3	2	2

Table 4.19. Courtship behavior and reproductive performance of normal male and MT-treated sex reversed male of *Betta splendens* (© 2013 from Pandian, T.J. Endocrine Sex Differentiation in Fish, CRC Press, Boca Raton, reproduced by permission of Tayler and Francis Group, LLC, a division of Informa plc).

Parameter	**Courtship and reproduction**	
	Control male	**MT-treated male**
Age at maturity (day)	110	145
Milting cycle (day/cycle)	4	13
Body length (cm)	6	3.5
Dorsal fin (cm	4	2.8
Caudal fin (cm)	3.5	2.3
Anal fin (cm)	4.2	2.9
Pairing duration stage 4 (seconds)	6	3
Completion stage 5 (seconds)	19	8
Sperm count (x 10^5/ml)	7.7	5.9
Motility duration (sec)	100	78
Male induced spawning pulse (no)	19	16
Spawned eggs (no)	130	78

sperm (Baatrup and Junge, 2001). For example, control male performed 0.96 display/10-minutes toward the female, whereas the males that received 10 µg vinclozolin/mg dry feed performed only 0.35 display/10-minutes. The quantum of sperm ejaculated also significantly decreased from 3.0 x 10^5 to 1.1 x 10^5 sperm/male. Consequently, the number of progenies born of the first clutch significantly decreased. Interestingly, the mating between normal males and virgin or pregnant female(s) that were previously fed on 10 µg vinclozolin/mg feed for 10–14 weeks did not significantly reduce the number of offspring (Bayley et al., 2003). In brown trout *Salmo. trutta f. fario* too, exposure of females to 1.75 or 2.40 µg BPA/l did not affect quality of eggs in terms of egg mass and fertility (Lahnsteiner et al., 2005). Investigating the EDC effect on interaction between individuals in a colony of zebrafish, Coe et al. (2008) exposed 2 male + 2 female groups (cf. Pandian, 2010, pp. 52–54) to 2 and 10 ng EE_2/l for 17 days and estimated the daily egg output. They found that the eggs sired by the more successful (dominant) male was 67% of the total egg output, while that of more dominant female was 74%. In the dominant males, the plasma KT level was almost two times higher than in the submissive males. Behaviorally dominant males were characterized by their activities in defending spawning territories (see also Pandian, 2013, p. 19) and aggressive against submissive males. The number of eggs sired by dominant male was significantly low following EE_2 exposure, especially during the last few days of observation. Conversely, the output by the dominant females remained unaffected by EE_2 exposure. EE_2 seems to suppress KT levels in the dominant males. Evidently, EDCs affected the male's courtship behavior but not the reproductive output of females.

In nest-holding fishes, nest building and defending are important criteria for selection of a male mating partner (see Pandian, 2010, pp. 25–27, 2011, pp. 13–15, 2013, pp. 120–121). Bubble nests built by male fighting fish exposed to E_2, equol, β-sitosterol and geistein are 84, 48, 45 and 32% as large, respectively, as nest built by control male (Clotfelter and Rodriquez, 2006). Conversely, the bubble nest built by *B. splendens* males, which were exposed to MT as fry, was smaller (Kirankumar and Pandian, 2002).

Other examples include disruption of reduced gluing frequency (e.g., Brian et al., 2006), delayed onset of nest building and/or reduced care for the nest (e.g., Wibe et al., 2002). Spiggin is the glue protein that male nest holders like *Gasterosteus aculeatus* use to build their nest and is directly controlled by androgen. Fenitrothion (FN) is a widely used organophosphorous pesticide that has structural similarity to the anti-androgen flutamide. Exposure of stickleback males to 50 μg FN/l for 26 days significantly reduced spiggin production as well as nest building activity. It also adversely disrupted the zig-zag dance and biting behavior of males (Sebire et al., 2009). Exposure to 100 μg flutamide/l for 21 days completely inhibited nest building by the stickleback (Sebire and Katsiadaki, 2008). Decreases in holding and defending nests were 70 and 30% by territorial male fathead minnows, who were exposed to 50 and 100% NP/OP mixtures as fry, respectively (Bistodeau et al., 2006). Hence, EDC exposure not only affects nest building but also its defense. Conversely, spiggin level elevated in juvenile stickleback after treatment with MT during labile period (Hahlbeck et al., 2004b).

4.11 Spawning and spermiation

Irrespective of exposure to EDC during the labile or non-labile period fertilization success (FS) and/or fecundity was reduced in many fishes. A few examples are briefed to show how EDCs disrupt egg production at spawning and sperm production at spermiation. In an interesting experimental design, Coe et al. (2010) investigated the effects of male to female ratio and two levels of EE_2 (2.76 ng/l, low concentration [LC] scenario and 9.86 ng/l, high concentration [HC] scenario) on reproductive performance of zebrafish, which were previously exposed to EE_2 from 20 to 60 dph. In sex ratio series, either two males competing for one female or two females competing for one male were placed in a series of tanks. In the competitive exposure pairings of either two males or two females one of the following possible three combinations were tested: (i) control (C) fish vs. C fish, (ii) C fish vs. LC scenario and (iii) C fish vs. H fish scenario were tested. Coe et al. found that the control female spawn more eggs than HC scenario female but the control male sired less number of eggs than LC scenario male.

To evaluate the apparently anti-estrogenic activity of polyaromatic hydrocarbon (PAHs), Cheek et al. (2001) exposed medaka hatchlings to anthracene (ANT) alone or in combination with E_2 for 2 or 8 weeks. ANT did not alter sex ratio regardless of dose (12 or 20 µg/l) or duration. Neither did it affect the reproductive success, as measured by the number of eggs spawned and fertilized. However, a combination of ANT and E_2 reduced egg output at spawning and FS. A cross between E_2-treated male with normal female produced about 4% fertilized eggs of the normal cross indicating that E_2 but not ANT reduced sperm production in treated males. Females treated with E_2 and ANT produced the same number of eggs, as did the controls, suggesting that E_2 prevented the ANT-induced reduction in the number of eggs spawned. When both ANT-treated males and females were crossed, FS significantly decreased. However, males treated with ANT alone showed no change in its fertilizability. FS of eggs produced by F_1 female was not affected by ANT exposure of F_0 medaka. Clearly, ANT affected egg output but E_2 the sperm production.

Lahnesteiner et al. (2005) investigated the effects of environmentally relevant concentrations (1.75, 2.40 or 5.00 µg/l) of BPA during different periods of spawning season in brown trout *Salmo trutta f fario*. In terms of quality of sperm, as assessed by (i) sperm count, (ii) proportion of motile sperm, (iii) swimming velocity and (iv) swimming pattern, spermiated by the end of spawning season in November was far superior to that during the commencement of spawning season in October. Exposure to BPA affected both quantity and quality of sperm. Semen mass was reduced from 0.5 g in control to 0.1 g in a male exposed to 5.0 µg BPA/l, motile sperm from 86 to about 3%, swimming velocity from 97 to 41 µm/seconds and linear swimming pattern from 25 to almost 0% (Table 4.20). Consequently, the brown trout inhabiting the BPA contaminated waters may spawn eggs but there may be no sperm to fertilize them.

Table 4.20. Effects of different spawning periods and BPA on quantity and quality of a sperm of *Salmo trutta f. fario* (compiled from Lahnsteiner et al., 2005).

Sperm: quantity and quality	Early spawning (µg BPA/l)			Late spawning (µg BPA/l)		
	0.00	**2.40**	**5.00**	**0.00**	**2.40**	**5.00**
Sperm mass (g)	0.5	0.6	0.05	0.46	0.65	0.09
Sperm count (10^{10} x ml)	2.91	2.68	2.62	2.33	2.42	2.54
Motile sperm (%)	85.5	30.8	4.0	85.8	80.7	3.5
Swimming velocity (µm/sec)	96.7	74.9	32.5	106.2	111.1	41.0
Circular (%)	55.5	46.7	-	19.8	39.5	-
Non-linear (%)	20.2	27.3	-	14.9	19.8	-
Linear (%)	24.6	26.0	-	65.3	40.7	-

4.12 Fertilization and fecundity

Fertilization is the most crucial event in the reproductive endeavor and fecundity is a decisively important feature in sustaining a population. The sensitivity and response to EE_2 exposure on induction of spawning seem to differ considerably. For example, life long exposure of zebrafish to 5 ng EE_2/l arrested VTGs but not spawning (Table 4.21). However, exposure of adult zebrafish to 5 ng EE_2/l for 21 days elicited a different response. It inhibited vitellogenesis and thereby produced non-spawning females (Table 4.15) but in the spawning females, fecundity was reduced (Coe et al., 2008, see Table 4.21). On exposure to EDCs, the hatchlings of *Oryzias latipes* and adults of *Pimephales promelas* also did not spawn (Table 4.14). Life long exposure of *Danio rerio* to an estrogen EE_2 and *Cyprinodon variegatus* to an androgen TB (Table 4.21) and *Fundulus heteroclitus* (Table 4.22) to an aromatase inhibitor TBT reduced FS and thereby fecundity. Life long exposure of *D. rerio* to 50 ng EE_2/l or *F. heteroclitus* to 0.1 µg TBT/l inhibited or significantly reduced VTG induction, respectively. Evidently, the inhibition or reduction of VTGs in fishes, which underwent exposure during early life stage, does not inhibit spawning. Conversely, exposure of *D. rerio* adult to 10 ng EE_2/l induced VTG but prevented yolk deposition in oocytes and inhibited spawning (Van den Belt et al., 2001).

Irrespective of life long exposure in *F. heteroclitus* (1 µg TBT/l) and *C. variegatus*, early exposure of *D. rerio* and *Oryzias latipes* and adult exposure of *D. rerio, O. latipes, F. heteroclitus* and *Gobiocypris rarus* to androgenic or estrogenic EDC, fertilization success (FS) decreased resulting in reduced fecundity to different levels (Tables 4.21, 4.22). Notably, reduction was from 120 to 0 eggs in adults of *F. heteroclitus* exposed to 120 ng E_2/l for 19 days (Kramer et al., 1998). Inviability of zebrafish eggs exposed to carbaryl increased with an extended duration of exposure, despite the eggs being fertilized (Todd and Van Leeuwen, 2002). Nevertheless, failure of FS may be an important cause for inviability of eggs. Many authors have not reported whether the eggs from treated females were fertilized with sperm from normal or treated males. In zebrafish males that were exposed to 270 µg/l FTOH for 4 months, sperm count decreased from 9.96×10^{10}/ml in control to 2.35×10^{10}/ml in males; the corresponding decrease in sperm production was 8.43×10^7/day to 3.69×10^7/day (Liu et al., 2010). Fertilizability of *Rutilus rutilus* intersex with 50% occupancy of testis in testis-ova is 10% only (Jobling et al., 2002a). Besides, the treated males may also suffer from possessing oviduct. Consequently, spermiation may be inhibited.

Fecundity: Spawning periodicity ranges from 19 minutes in Owens pupfish *Cyprinodon radiosus* to 1–2 days in *D. rerio*, 3–5 days in *F. heteroclitus*, 15 days in *Betta splendens*, 19–60 days in *Tilapia zilli* or annually *Oncorhynchus mykiss* (Mills and Chichester, 2005; Pandian, 2013, pp. 33–36). The periodicity in

Table 4.21. Fertilization success (FS) and viability of spawned eggs following life long, early life or adult exposure to EDCs.

Species and reference	Reported observations
Danio rerio Nash et al. (2004)	Life long exposure to EE_2 inhibited VTG induction but no spawning. With no fertilization, eggs were inviable
D. rerio Diekmann et al. (2004)	Life long exposure to 0.1 µg NQO/l induced 0.6% inviable eggs. FS reduced dose dependently from 71% to 63% at 2.9 µg NQO/l
Fundulus heteroclitus Mochida et al. (2010)	Life long exposure reduced FS dose dependently from 95% to 75% at 1.0 µg TBT/l, VTG induction significantly reduced in females
Cyprinodon variegatus Cripe et al. (2010)	Life long exposure to TB increased dose dependently infertile eggs from 5 to 22% at 4.1 µg/l in F_0. No spawning of eggs at 0.13 µg/l in F_2 and at 0.87 and 4.1 µg/l in F_1 and F_2
D. rerio, Todd and van Leeuwen (2002)	Eggs exposed to carbaryl reduced viability to 80%
D. rerio, Balasubramani and Pandian (2008a)	Exposure of hatchlings to endosulfan reduced fertilizability dose dependently to 28% at 350 ng/l
D. rerio, Hill and Janz (2003)	Over 90% viable eggs after larval exposure to NP. But exposure to 10 ng EE_2/l reduced egg viability to 85%
Oryzias latipes, Nimrod and Benson (1998)	Egg exposure to environmentally relevant doses of NP or methoxychlor (MXC) produced about 855 viable eggs
D. rerio Maack and Segner (2004)	FS reduced from 85 to 60% and to 70% in those exposed to 10 ng EE_2/l from 43 to 71 dpf and from 72 to 99 pdf, respectively. FS reduced from 85 to 50% with increasing EE_2 dose from 1.67 to 10.0 ng/l in juveniles exposed from 43 to 71 dpf
D. rerio, Balasubramani and Pandian (2008b)	Exposure of hatchlings to TBT reduced dose dependently fertilizability to 43% at 8 µg TBT/l
O. latipes, Ishibashi et al. (2006)	FS decreased from 100% in control to 80 and 8% in adults exposed to 100 µg NP/l during the 1st and 3rd weeks after exposure for 21 days
D. rerio, Coe et al. (2008), Santos et al. (2007)	Exposure of adults to 5 ng EE_2/l for 21 days reduced egg spawned especially after 15–20 days exposure
F. heteroclitus, Peters et al. (2007)	Exposure of adults to EE_2 for 21 days reduced FS to 55%
F. heteroclitus, Pawlowski et al. (2004)	In females exposed to 5 µg MT/l, FS decreased to 30%
Gobiocypris rarus, Zhong et al. (2005)	Adult exposure reduced FS from 79 to 71% at 1 µg DES/l
Rutilus rutilus (see Tyler et al., 2007)	Fertilizability (up to hatching) decreased from ≈ 78% for normal male to ≈10% for an intersex with 50% occupancy of testis in testis-ova

Table 4.22. Effects of life long and early life exposure to estrogenics and androgenics on fecundity of fishes.

Species and Reference	Reported observations
Danio rerio Diekmann et al. (2004)	Life long exposure to 0.1 µg NQO/l reduced fecundity from 2,154 to 793 eggs/spawning and from 45 to 17 eggs/♀/day
D. rerio, Ensenback and Nagel (1997)	Life long exposure to 200 µg DCA+ 40 µg lindane/l reduced fecundity from 426 to 174 eggs/group (4♀ + 8 ♂)/days
D. rerio Maack and Segner (2004)	Fecundity was reduced from 12 to 4 and to 7.5 eggs/♀/day in those exposed to 10 ng EE_2/l from 43 to 71 dpf and 72 to 99 dpf, respectively. It also decreased from 17.5 to 12.5 eggs/♀/day with increasing dose from 1.65 to 10 ng EE_2/l in juveniles exposed from 43 to 71 dpf
D. rerio, Balasubramani and Pandian (2008a)	Early life exposure to 350 µg endosulfan/1 reduced fecundity from 89 to 74 eggs and cumulative fecundity from 1,424 to 370 eggs/♀
Oryzias latipes, Gormley and Teather (2003)	Exposure of 4–6 dpf eggs to 0.1 µg endosulfan/1 reduced fecundity from ≈4 to ≈3 eggs/♀/day but it increased to 5 eggs /♀/day in 4–6 day-old hatchling exposed to 1 µg endosulfan/l
Pimophales promelas Makynan et al. (2000)	Exposure of 6 hour–old embryos to 100 µg vinclozolin/l for 34 days reduced fecundity from 121 to 49 eggs
D. rerio, Balasubramani and Pandian (2008b)	Early life exposure to 2 µg TBT/l reduced fecundity from 89 to 60 eggs and cumulative fecundity from 1632 to 45 eggs/♀
Fundulus heteroclitus, Mochida et al. (2010)	Life long exposure to 1 µg TBT/l reduced fecundity from 60 to 18 eggs/spawning
Cyprinodon variegatus Cripe et al. (2010)	Life long exposure to 27 ng TB/l reduced fecundity from 40 to 2 eggs/spawning in F_0, 80 to 0 eggs/spawning in F_1 and 10 to 0 eggs/spawning in F_2. Corresponding values were from 280 to 5 eggs/♀/day in F_0, 550 to 0 eggs/♀/day in F_1 and 750 to 0 eggs/♀/ day in F_2

most fishes, in which EDC effect on reduced fecundity is reported, is not longer than 15 days. Some authors have reported the EDC effect on reduced batch fecundity (e.g., Mochida et al., 2010) and cumulative (e.g., Peters et al., 2007)/total (e.g., Brion et al., 2004) fecundity, while others reported fecundity expressed in number of eggs spawned/♀/day (e.g., Maack and Segner, 2004; Ankley et al., 2007). The last one is more appropriate, as it provides cross comparison between species characterized by different durations of spawning periodicity and between EDCs doses and durations of exposure. Secondly, the spawning frequency is also reduced in some fishes, implying that fecundity in number of eggs spawned/♀/day is a preferred parameter to monitor EDC effect on fecundity. For example, the spawning frequency is reduced from 16 to 5 during 240 days period in *D. rerio* exposed to endosulfan (Balasubramani and Pandian, 2008a), from 22 to 5 during 300 days period in *D. rerio* exposed to endosulfan (Balasubramani and Pandian, 2008b), from 14 to 11 in zebrafish exposed to 100 ng E_2/l (Brion et al., 2004) and from ≈3 to 0.5 during 21 days period in fathead minnow exposed to KTC (Ankley et al., 2007). Thirdly, authors like Foran et al. reported fecundity of 4 ♀ + 8 ♂ exposed to EE_2 as a group.

Summaries of the scattered information on reduced fecundity in fishes exposed to EDC during labile embryonic, hatchling and juvenile stages (Table 4.22) and non-labile adult stage (Table 4.23) are listed. Irrespective of exposure to estrogenic or androgenic EDC and the labile or non-labile stage of exposure, fecundity is reduced to different levels, the reduction being more significant, when an androgen or anti aromatase was used. For example, during life long exposure to an androgen TB, egg production was totally arrested in F_1 and F_2 females of *C. variegatus* (Cripe et al., 2010). Early life exposure of *D. rerio* to TBT reduced cumulative fecundity from

Table 4.23. Effects of adult exposure to estrogenics and androgenics fecundity of fishes.

Species and Reference	Reported observations
Oryzias latipes Foran et al. (2002)	Adult exposure reduced cumulative fecundity from 200 to 5 eggs at 2 µg EE_2/l but egg exposure only to 150 eggs
Danio rerio Orn et al. (2000)	Adult exposure reduced fecundity from 120 to 60 but increased to 200 eggs/♀/day at 0.1 µg E_2/l and 10 ng E_2/l, respectively
D. rerio Brion et al. (2004)	Adult exposure reduced fecundity from 120 to 80 eggs at 100 ng E_2/l
D. rerio Liu et al. (2010)	Adult exposure reduced fecundity from 37 to 20 eggs /♀/d at 270 µg FTOH/l
O. latipes Ishibashi et al. (2006)	Adult exposure reduced fecundity from 70 to 20 eggs/pair for 3 weeks exposure to 100 µg NP/l
Gobiocypris rarus Zhong et al. (2005)	Adult exposure reduced fecundity from 425 to 275 eggs/ spawning and 115 to 40 eggs/♀/d at 5 µg DES/l, respectively
Fundulus heteroclitus Peters et al. (2007)	Adult exposure reduced cumulative fecundity from 300 to 100 eggs at 100 ng EE_2/l on the 5th day after full noon
Pimophales promelas Pawlowski et al. (2004)	Adult exposure reduced cumulative fecundity from 600 to 180 eggs at 5 µg NP/l
P. promelas Ankley et al. (2003)	Adult exposure reduced cumulative fecundity from 300 to 0 eggs at 500 µg TB/l
P. promelas Ankley et al. (2007)	Adult exposure to fungicide ketoconazole (KTC) reduced fecundity from 3 to 0.5 eggs/spawning and 15 to 2.5 eggs/ female/day at 400 µg KTC/l during 21 days observation

1,632 to 45 eggs (Balasubramani and Pandian, 2008b). In fathead minnow, adult exposure to TB reduced cumulative fecundity from 300 to 0 eggs at 50 µg/l (Ankley et al., 2003).

Among the authors referred in Tables 4.21 and 4.22, a few have reported data on levels of VTG changes following EDC exposure. Remarkably, egg production was arrested, once VTG induction was completely inhibited on exposure of fathead minnow females at 5.0 µgTB/l (Ankley et al., 2003). In *O. latipes*, *F. heteroclitus* and *P. promelas* exposed to EE_2 or MT, VTG levels increased to different levels; even on adult exposure to 400 µg KTC/l, VTG level was doubled in female minnow. However, it remained at equal levels in control and E_2-treated early hatchlings of zebrafish, although it increased in E_2-treated late hatchlings, presumably with development of the liver. With limited information, it may be too early to generalize a relation between VTG and fecundity in EDC-exposed fishes. Still, VTG level in treated female may be an important factor in deciding fecundity size.

4.13 Hatching and survival

Notably, publications reporting EDC effects on hatchability and survival of F_1 progeny of F_0-treated parents began to appear only from the beginning of this millennium. Irrespective of the labile or non-labile stage, in which F_0 parents were treated, incubation period of their fertilized eggs was prolonged from 1 to ≈10 days in almost all fishes (Table 4.25). Critically, adult zebrafish

Table 4.24. Effect of EDC exposure on VTG levels in female fishes.

Species and Reference	Reported observations
Danio rerio Brion et al. (2004)	Whole body VTG remained ≈ 90 ng/ml in control and hatchlings treated from 0 to 21 dpf at 100 ng E_2/l. But it increased from 110 to ≈ 1, 200 ng/ml in hatchlings treated at 25 ng E_2/l from 21 to 42 dpf
Gobiocypris rarus Zhong et al. (2005)	VTG increased from 20 µg/g fish exposed as larva-juvenile at 1 µg DES/l
Oryzias latipes Foran et al. (2002)	Hepatic VTG remained ≈ 0.5 level on egg exposure but increased to 3.25 level on adult exposure to 2 µg EE_2/l
Fundulus heteroclitus Peters et al. (2007)	Plasma VTG increased (after decreases at lower doses) from 2,000 to 2,500 µg/ml after exposure to 100 µg EE_2/l for 21 days
Pimophales promelas Pawlowski et al. (2004)	Plasma VTG increased from ≈20 to 100 mg/ml at 50 µg MT/l
P. promelas Ankley et al. (2007)	VTG increased from 6 to 12 µg/ml at 400 µg KTC/l
P. promelas Ankley et al. (2003)	VTG decreased from 27 to 0 mg/ml at 5.0 µg TB/l
Cyprinodon variegatus Cripe et al. (2010)	VTG decreased from 4.4 µg/ml on exposure to 0.9 µg TB/l to ≈ 0.55 µg/ml on exposure to 4.1 µg TB/l in F_0 ♀

Table 4.25. Effect of EDC exposure on incubation, hatchability and survival of EDC-treated fishes.

Species and Reference	Reported observations
	Incubation period
Oryzias latipes Gormley and Teather (2003)	Remains equal at ≈ 11–12 days in control and hatchlings exposed to 1 µg endosulfan/l
O. latipes Yokota et al. (2001)	Prolonged from 9.4 to 10.2 days in control and in F_1 eggs, whose F_0 parents were life long exposed at 51.5 µg NP/l
Fundulus heteroclitus Mochida et al. (2010)	Prolonged from 22 to 32 days after F_1 eggs whose F_0 parents were life long exposed to 1 µg TBT/l
O. latipes Ishibashi et al. (2006)	Prolonged from 9 to 12.3 days in F_1 eggs of F_0 (females exposed to NP for 3 weeks) fertilized by sperm of NP-treated F_0 male
Danio rerio Kime and Nash (1999)	Adult exposure to 5 µg EE_2/l for 12 days prior to spawning arrested development at early blastula
	Hatchability
F. heteroclitus Mochida et al. (2010)	Decreased from 78 to ≈ 40% eggs after the life long exposure of F_0 to 0.5 µg TBT/l
D. rerio, Balasubramani and Pandian (2008b)	Decreased from 87 to 15% in hatchlings, whose F_0 parents were exposed to 2 µg TBT/l
D. rerio, Balasubramani and Pandian (2008a)	Decreased from 87 to 76% in F_1 hatchlings, whose F_0 parents were exposed to 350 µg endosulfan/l
F. heteroclitus Urushitani et al. (2002)	Decreased from 55 to 40% in embryos exposed to 10^{-8} M E
O. latipes Cheek et al. (2001)	Decreased from 10 to < 5% in F_1 hatchlings, whose mother was exposed to 12–20 µg E_2/l
D. rerio Liu et al. (2010)	Decreased from 79 to 38% in adults exposed to 270 µg FTOH/l
D. rerio Brion et al. (2004)	Remained ≈ 95 to 98% hatchlings, whose parents were treated at 100 ng E_2/l
	Survival
D. rerio Hill and Janz (2003)	Cumulative survival of treated zebrafish decreased to 5% following early life exposure of hatchlings at 100 µg EE_2/l
F. heteroclitus Urushitani et al. (2002)	Survival of fry, whose parents were treated as embryos at 10^{-8} M E_2, decreased from 30 to 12%
O. latipes Cheek et al. (2001)	Survival decreased from 95 to 50% in 8 weeks old fry, whose parents were treated as post hatch at 20 µg E_2/l
O. latipes Yokota et al. (2001)	Cumulative survival of hatchlings decreased from 98 to 32% in 60 dph old hatchlings following life long exposure of F_0 at 51.5 µg NP/l. But it increased 5.5 to 18.7% in 'swim-up' stage

exposure to 5 µg EE_2/l prior to spawning arrested development at early blastula (Kime and Nash, 1999). For more information on developmental abnormalities owing to parental exposure to EDC, Boudreau et al. (2005), Denslow and Sepulveda (2007) may be consulted. The prolonged incubation exposes the embryos to greater risk of predation.

Hatchability: Irrespective of the labile or non-labile stage, in which F_0 parents were exposed to EDC, hatchability of F_1 progeny decreased in all species (Table 4.25). In general, decreases in hatchability of androgen treated fishes were more severe than in those treated with estrogenic EDC. For example, the decrease was greater from 87 to 15% in *Danio rerio* exposed to TBT (Balasubramani and Pandian, 2008b; see also Strmack and Brauenbeck, 1999) than that (87 to 76%) exposed to endosulfan (Balasubramani and Pandian, 2008a). Causes for the decrease in hatchability can be traced to (i) developmental abnormality and (ii) incompatability of available yolk to meet the cost of delayed hatching. EDC exposure not only reduced fecundity but also the quality of eggs. For example, egg production decreased from 37 to 20 eggs/pair/day in zebrafish exposed to 270 µg FTOH/l. Further, the egg quality was also decreased; the decreases were from 1.2 to 1.00 mm in egg diameter, from 3.26 to 2.65 µg protein/egg and from 0.84 to 0.68 µm for egg shell thickness (Liu et al., 2010). With reduced yolk protein, the delayed hatching compounded the risk of embryonic/larval mortality.

Survival: A good amount of literature is available on survival of EDC treated fishes. For example, cumulative survival of zebrafish exposed to 100 µg EE_2/l was reduced to 5% (Table 4.25). Survival of hatchlings and fry arising from EDC-treated F_0 parents is significantly reduced. However, if the larvae survives up to 'swim-up' stage, then survival is indeed increased (Yokota et al., 2001). On a long term basis, EDC exposure seems to reduce life span of fishes. The life span of *Microgadus tomcod* was reduced to from 6 to 4 years, when males suffer from exposure to pulp mill effluent (Couillard et al., 1999). Cosidering 19 biochemical physiological and histological variables in *Perca fluviatilis* from the polluted waters of Baltic Sea, Hansson et al. (2006) reported that the life span of the perch decreased from 5.7 years in 1988 to 4.5 years in 1993.

4.14 New from old

Hitherto endocrine disruption in fishes has been considered from the point of environmental pollution. However, this problem of pollution remains without a solution. A possibility that has to be explored is appropriate mixing of wastewater effluent discharged from cattle industry, in which approved estrogen E_2 and androgen TB are implanted. Another solution can be the establishment of paper industries in the vicinity of sewage treatment plants to neutralize estrogenic sewage mixed with β-estradiol.

But keeping with the objective of this book on sex differentiation, endocrine disruption has been viewed for the first time from the point of labile early life stages and the non-labile adult stage. In this process, relevant information from widely scattered publications has been identified, collated, sequenced and edited to synthesize a new concept from earlier literature (Table 4.26). Single starred reproductive characteristics are more or less confirmed from available literature. Others carrying two stars mark require confirmation from future research. Being an ideal model, zebrafish

Table 4.26. Summary of EDC effects on fishes that were exposed during labile early life stages and non-labile adult stage. *confirmed observations, **confirmation is awaited.

Characteristics	Early life labile stage	Non-labile adult stage
	Sex reversal	
Primary gonochore	*Possible	*not possible
Secondary gonochore	*Possible	*Possible from female to male but not male to female
Intersex	**Not reversible from ovotestis to testis or ovary *Ovotestis developed in genetic males and females	*Recovery from testis-ova to testis possible on depuration *Testicular recovery possible but not ovarian recovery **Testis-ova developed in genetic male
	Other characteristics	
Secondary sexual characteristics	*Not reversible *SSCs represent gonadal sex	*Reversible after depuration *SSCs do not represent gonadal sex
Ovarian like cavity	*Irreversible	*Reversible on depuration
Gonadal duct	*Irreversible	**Reversible on depuration
Vitellogenesis	*Absence or reduction of VTG induction inhibits spawning	*Elevated VTG induction prevents yolk deposition and inhibits spawning
OSI and TSI	*Almost equal in control and treated	*Decrease to different levels and to sterility in some fishes
Maternally inherited EDC in eggs	**Irreversible	*Reversible on depuration
Gametogenesis	*Oogonial and spermatogonial mitosis allowed even at a faster rate. But the remaining gametogenitic stages are suppressed	*Entire process of gametogenesis including oogonial and spermatogonial mitosis is suppressed
Courting	*Suppressed behavior	*Disrupted behavior
Nest building	*Precocious and elevated synthesis of spiggin glue	*Decreased spiggin glue, reduced nest building and defending

is perhaps one fish that has attracted more attention than any other. Many authors, who remained confused to interpret their data, may now find an explanation for their original contribution. Being a secondary gonochore, the ability of zebrafish to reverse adult sex from female to male but not from male to female may represent a good example.

References

Adlercreutz, H., T. Fotsis, C. Bannwart et al. 1986. Urinary estrogen profile determination in young Finnish vegetarian and omnivorous women. J Steroid Biochem, 24: 289–296.

Allen, Y., S.P. Scott, P. Matthiessen et al. 1999. Survey of estrogenic activity in United Kingdom estuaries and coastal waters and its effect on gonadal development of the flounder *Paralichthys flesus*. Environ Toxicol Chem, 18: 1791–1800.

Andersen, L., G.I. Peterson, K. Hylland and L. Novegren. 2000. Suitability of using zebrafish (*Danio rerio*) in the detection of endocrine disrupting effects. A literature survey. Tema Nord, 2000: 555 pp. 14–34.

Andersen, L., H. Holbech, Å. Gessbo et al. 2003a. Effects of exposure to 17α-ethinylestradiol during early development on sexual differentiation and induction of vitellogenin in zebrafish (*Danio rerio*). Comp Biochem Physiol Part C: Toxicol Pharmacol, 134: 365–374.

Andersen, L., P. Bjerregaard and B. Korsgaard. 2003b. Vitellogenin induction and brain aromatase activity in adult male and female zebrafish exposed to endocrine disrupters. Fish Physiol Biochem, 28: 319–321.

Andersen, L., K. Kinnberg, H. Holbech et al. 2004. Evaluation of a 40 day assay for testing endocrine disrupters: effects of an anti-estrogen and an aromatase inhibitor on sex ratio and vitellogenin concentrations in juvenile zebrafish (*Danio rerio*). Fish Physiol Biochem, 30: 257–266.

Andersen, R.H., A.M. Vinggaard, T.H. Rasmussen et al. 2002. Effects of currently used pesticides in assays for estrogenicity, androgenicity, and aromatase activity in *in vitro*. Toxicol Appl Pharmacol, 179: 1–12.

Angus, R.A., J. Stanko, R.L. Jenkins and R.D. Watson. 2005. Effects of 17α-ethynylestradiol on sexual development of male western mosquitofish (*Gambusia affinis*). Comp Biochem Physiol Part C: Toxicol Pharmacol, 140: 330–339.

Ankley, G.T., K.M. Jensen, M.D. Kahl et al. 2001. Description and evaluation of a short-term reproduction test with the fathead minnow (*Pimephales promelas*). Environ Toxicol Chem, 20: 1276–1290.

Ankley, G.T., K.M. Jensen, E.A. Makynan et al. 2003. Effects of the androgenic growth promoter 17-β-trenbolone on fecundity and reproductive endocrinology of the fathead minnow. Environ Toxicol Chem, 22: 1350–1360.

Ankley, G.T., K.M. Jensen, M.D. Kahl et al. 2007. Ketoconazole in the fathead minnow (*Pimephales promelas*): reproductive toxicity and biological compensation. Environ Toxicol Chem, 26: 1214–1223.

Arcand-Hoy, L.D., A.C. Nimrod and W.H. Benson. 1998. Endocrine-modulating substances in the environment: estrogenic effects of pharmaceutical products. Internatl J Toxicol, 17: 139–158.

Arukwe, A. and A. Goksøyr. 2003. Egg shell and egg yolk proteins in fish: hepatic proteins for the next generation: oogenetic, population, and evolutionary implications of endocrine disruption. Comp Hepetol, 2: 4.

Arukwe, A., L. Forlin and A. Goksoyr. 1997. Xenobiotic and steroid biotransformation enzymes in Atlantic salmon (*Salmo salar*) liver treated with an estrogenic compound, 4-nonylphenol. Environ Toxicol Chem, 16: 2576–2583.

Arukwe, A., R. Thibaut, K. Ingebrigtsen et al. 2000. *In vivo* and *in vitro* metabolism and organ distribution of nonylphenol in Atlantic salmon (*Salmo salar*). Aquat Toxicol, 49: 289–304.

Arunachalam, S. and S. Palanichamy. 1982. Sublethal effects of carbaryl on surfacing behaviour and food utilization in the air-breathing fish, *Macropodus cupanus*. Physiol Behav, 29: 23–27.
Baatrup, E. and M. Junge. 2001. Anti-androgenic pesticides disrupt sexual characteristics in adult male guppy (*Poecilia reticulata*). Environ Health Prospect, 109: 1063–1070.
Babin, P.J. 1992. Research needs for the risk assessment of health and environmental effects of endocrine disrupters: a report of the US-EPA sponsored workshop. Environ Health Perspect, 104: 715–740.
Babin, P.J., O. Carnevali, E. Lubzens and W.J. Schneider. 2007. Molecular aspects of oocyte vitellogenesis in fish. pp. 39–76. *In*: P.J. Babin, J. Ceard and E. Lubzens (eds.). *The Fish Oocyte: From Basic Studies to Biotechnological Applications*, Springer, Netherlands.
Balasubramani, A. 2006. Endocrine and genetic studies in selected ornamental fish. Ph.D. Thesis, Madurai Kamaraj University, India.
Balasubramani, A. and T.J. Pandian. 2008a. Endosulfan suppresses growth and reproduction in zebrafish. Curr Sci, 94: 883–890.
Balasubramani, A. and T.J. Pandian. 2008b. Effect of tributyltin chloride on survival, growth and reproduction in zebrafish (*Danio rerio*). J Aquacult, 21: 146–156.
Balch, G. and C.D. Metcalfe. 2006. Developmental effects in Japanese medaka (*Oryzias latipes*) exposed to nonylphenolethoxylates and their degradation products. Chemosphere, 62: 1214–1223.
Balch, G.C., C.A. Mackenzie and C.D. Metcalfe. 2004. Alterations to gonadal development and reproductive success in Japanese medaka (*Oryzias latipes*) exposed to 17α-ethinylestradiol. Environ Toxicol Chem, 23: 782–791.
Baronti, C., R. Curini, G. D'Ascenzo et al. 2000. Monitoring natural and synthetic estrogens at activated sludge sewage treatment plants and in a receiving river water. Environ Sci Technol, 34: 5059–5066.
Barnhoorn, I.E.J., M.S. Bomman, G.M. Pieterse and H.J. Van Vuran. 2004. Histological evidence of intersex in feral sharptooth catfish (*Clarias gariepinus*) from an estrogen-polluted water source in Gauteng, South Africa. Environ Toxicol, 19: 603–608.
Barse, A.V., T. Chakrabarti, T.K. Ghosh et al. 2006. One-tenth dose of LC_{50} of 4-*tert*-butylphenol causes endocrine disruption and metabolic changes in *Cyprinus carpio*. Pesticide Biochem Physiol, 86: 172–179.
Bayley, M., J.R. Nielsen and E. Baatrup. 1999. Guppy sexual behavior as an effect biomarker of estrogen mimics. Ecotoxicol Environ Saf, 43: 68–73.
Bayley, M., M. Junge and E. Baatrup. 2002. Exposure of juvenile guppies to three antiandrogens causes demasculinization and a reduced sperm count in adult males. Aquat Toxicol, 56: 227–239.
Bayley, M., P.F. Larsen, H. Backgaard and E. Baatrup. 2003. The effects of vinclozolin, an anti-androgenic fungicide on male guppy secondary sexual character and reproductive success. Biol Reprod, 69: 1951–1956.
Bistodeau, T.J., L.B. Barber, S.E. Bartell et al. 2006. Larval exposure to environmentally relevant mixture of alkylphenolethoxylates reduces reproductive competence in male fathead minnows. Aquat Toxicol, 79: 268–277.
Bjerselius, R., K. Lundstedt-Enkel, H. Olsén et al. 2001. Male goldfish reproductive behaviour and physiology are severely affected by exogenous exposure to 17β-estradiol. Aquat Toxicol, 53: 139–152.
Blázquez, M., S. Zanuy, M. Carrillo and F. Piferrer. 1998. Structural and functional effects of early exposure to estradiol-17β and 17α-ethynylestradiol on the gonads of the gonochoristic teleost *Dicentrarchus labrax*. Fish Physiol Biochem, 18: 37–47.
Bogers, R., S. DeVries-Buitenweg, M.V. Gils et al. 2006. Development of chronic tests for endocrine active chemicals: Part 2: An extended fish early-life stage test with an androgenic chemical in the fathead minnow (*Pimephales promelas*). Aquat Toxicol, 80: 119–130.

Boudreau, M., S.C. Courtenay, D.L. MacLatchy et al. 2005. Morphological abnormalities during early-life development of the estuarine mummichog, *Fundulus heteroclitus*, as an indicator of androgenic and anti-androgenic endocrine disruption. Aquat Toxicol, 71: 357–369.

Brander, S.M. 2013. Thinking outside the box: Assessing endocrine disruption in aquatic life. Monitoring Water Quality: Pollution Assessment, Analysis and Remediation: Elsevier. Amsterdam, Netherlands, pp. 103–145.

Brian, J.V., J.J. Angley and V.A. Braithwaite. 2006. Endocrine disrupting effect on nesting behavior of male three-spine stickleback *Gasterosteus aculeatus* L. J Fish Biol, 68: 1883–1890.

Brion, F., B.M. Nilsen, J.K. Eidem et al. 2002. Development and validation of an enzyme-linked immunosorbent assay to measure vitellogenin in the zebrafish (*Danio rerio*). Environ Toxicol Chem, 21: 1699–1708.

Brion, F., C.R. Tyler, X. Palazzi et al. 2004. Impacts of 17β-estradiol, including environmentally relevant concentrations, on reproduction after exposure during embryo-larval-, juvenile- and adult-life stages in zebrafish (*Danio rerio*). Aquat Toxicol, 68: 193–217.

Buhler, D.R. and J.L. Wang-Buhler. 1998. Rainbow trout cytochrome P450s: purification, molecular aspects, metabolic activity, induction and role in environmental monitoring. Comp Biochem Physiol, 121: 107–137.

Burdick, G.E., E.J. Harris, H.J. Dean et al. 1964. The bioaccumulation of DDT in lake trout and effect on reproduction. Trans Am Fish Soc, 93: 127–136.

Carlson, D.B., L.R. Curtis and D.E.Williams. 2000. Salmonid sexual development is not consistently altered by embryonic exposure to endocrine-active chemicals. Environ Health Perspect, 108: 249.

Chatain, B., E. Saillant and S. Peruzzi. 1999. Production of monosex male populations of European seabass, *Dicentrarchus labrax* L. by use of the synthetic androgen 17α-methyldehydrotestosterone. Aquaculture, 178: 225–234.

Cheek, A., T.H. Brouwer, S. Carroll et al. 2001. Developmental exposure to anthracene and estradiol alters reproductinve success in medaka (*Oryzias latipes*). Environ Sci, 8: 31–45.

Cheshenko, K., F. Pakdel, H. Segner et al. 2008. Interference of endocrine chemicals with aromatase *cyp*19 expression or activity and consequences for reproduction in teleost fish. Gen Comp Endocrinol, 155: 31–62.

Christiansen, L.B., A. Poulsen, S.N. Pedersen et al. 2000. A study of intersex in wild populations of roach and vitellogenin induction in caged rainbow trout (*Oncorhynchus mykiss*) in Danish Rivers. Proc Third SETAC World Cong, Brighton, UK, p. 136.

Clotfelter, E.D. and A.C. Rodriguez. 2006. Behavioral changes in fish exposed to phytoestrogens. Environ Pollut, 144: 833–839.

Coe, T.S., P.B. Hamilton, D. Hodgson et al. 2008. An environmental estrogen alters reproductive hierarchies, disrupting sexual selection in group-spawning fish. Environ Sci Technol, 42: 5020–5025.

Coe, T.S., M.K. Soffker, A.L. Filby et al. 2010. Impacts of early life exposure to estrogen on subsequent breeding behavior and reproductive success in zebrafish. Environ Sci Technol, 44: 6481–6487.

Cooper, K.R., J. Schell, T. Umbreit and M. Gallo. 1993. Fish-embryo toxicity associated with exposure to soils and sediments contaminated with varying concentrations of dioxins and furans. Mar Environ Res, 35: 177–180.

Couillard, C.M., P.J. Williams, S.C. Courtenay and G.P. Rawn. 1999. Histopathological evaluation of Atlantic tomcod (*Microgadus tomcod*) collected at estuarine sites receiving pulp and paper mill effluent. Aquat Toxicol, 44: 263–278.

Crain, D.A., M. Eriksen, T. Iguchi et al. 2007. An ecological assessment of bisphenol-A: evidence from comparative biology. Reprod Toxicol, 24: 225–239.

Cripe, G.M., B.L. Hemmer, S. Raimondo et al. 2010. Exposure of three generations of the estuarine sheep head minnow (*Cyprinodon variegatus*) to the androgen, 17β-trenbolone: Effects on survival, development, and reproduction. Environ Toxicol Chem, 29: 2079–2087.

Darrow, K.O. and W.A. Harris. 2004. Characterization and development of courtship in zebrafish, *Danio rerio*. Zebrafish, 1: 40–45.

DeLorenzo, M.E., L.A. Taylor, S.A. Lund et al. 2002. Toxicity and bioconcentration potential of the agricultural pesticide endosulfan in phytoplankton and zooplankton. Arch Environ Contam Toxicol, 42: 173–181.
Denslow, N. and M. Sepulveda. 2007. Ecotoxicological effects of endocrine disrupting compounds on fish reproduction. pp. 255–322. *In:* P.J. Babin, J. Ceard and E. Lubzens (eds.). *The Fish Oocyte: From Basic Studies to Biotechnological Applications*. Springer, Dordrecht, Netherlands.
Desbrow, C.E.J.R., E.J. Routledge, G.C. Brighty et al. 1998. Identification of estrogenic chemicals in STW effluent. 1. Chemical fractionation and *in vitro* biological screening. Environ Sci Technol, 32: 1549–1558.
Diekmann, M., V. Hultsch and R. Nagel. 2004. On the relevance of genotoxicity for fish populations: 1 Effects of a model genotoxicant on zebrafish (*Danio rerio*) in a complete life-cycle test. Aquat Toxicol, 68: 13–26.
Di Giulio, R.T. and D.E. Hinton. 2008. *The Toxicology of Fishes*. CRC Press, Boco Raton, FL.
Diniz, M.S., I. Peres and J.C. Pihan. 2005. Comparative study of the estrogenic responses of mirror carp (*Cyprinus carpio*) exposed to treated municipal sewage effluent (Lisbon) during two periods in different seasons. Sci Total Environ, 349: 129–139.
Donohoe, R.M. and L.R. Curtis. 1996. Estrogenic activity of chlordecone, o, p'-DDT and o, p'-DDE in juvenile rainbow trout: induction of vitellogenesis and interaction with hepatic estrogen binding sites. Aquat Toxicol, 36: 31–52.
Drastichova, J., Z. Svobodova and M. Groenland. 2005. Effect of exposure to bisphenol A and 17 beta-estradiol on the sex differentiation in zebrafish (*Danio rerio*). Acta Veterinaria Brno 74: 287–291.
Edmunds, J.S., R.A. McCarthy and J.S. Ramsdell. 2000. Permanent and functional male-to-female sex reversal in d-rR strain medaka (*Oryzias latipes*) following egg microinjection of o, p'-DDT. Environ Health Perspect, 108: 219–224.
Ensenbach, U. and R. Nagel. 1997. Toxicity of binary chemical mixtures: effects on reproduction of zebrafish (*Brachydanio rerio*). Arch Environ Contam Toxicol, 32: 204–210.
Fenske, M. and H. Segner. 2004. Aromatase modulation alters gonadal differentiation in developing zebrafish (*Danio rerio*). Aquat Toxicol, 67: 105–126.
Fenske, M., G. Maack, C. Schäfers and H. Segner. 2005. An environmentally relevant concentration of estrogen induces arrest of male gonad development in zebrafish, *Danio rerio*. Environ Toxicol Chem, 24: 1088–1098.
Ferreira, M., P. Antunes, O. Gil et al. 2004. Organochlorine contaminants in flounder (*Platichthys flesus*) and mullet (*Mugil cephalus*) from Douro estuary, and their use as sentinel species for environmental monitoring. Aquat Toxicol, 69: 347–357.
Filby, A.L., K.L. Thorpe and C.R. Tyler. 2006. Multiple molecular effect pathways of an environmental oestrogen in fish. J Mol Endocrinol, 37: 121–134.
Finn, R.N. 2007. The physiology and toxicology of salmonid eggs and larvae in relation to water quality criteria. Aquat Toxicol, 81: 337–354.
Fitzpatrick, M.S., C.B. Pereira and C.B. Schreck. 1993. *In vitro* steroid secretion during early development of mono-sex rainbow trout: sex differences, onset of pituitary control, and effects of dietary steroid treatment. Gen Comp Endocrinol, 91: 199–215.
Foran, C.M., B.N. Peterson and W.H. Benson. 2002. Transgenerational and developmental exposure of Japanese medaka (*Oryzias latipes*) to ethinylestradiol results in endocrine and reproductive differences in the response to ethinylestradiol as adults. Toxicol Sci, 68: 389–402.
Gale, W.L., R. Patiño and A.G. Maule. 2004. Interaction of xenobiotics with estrogen receptors α and β and a putative plasma sex hormone-binding globulin from channel catfish (*Ictalurus punctatus*). Gen Comp Endocrinol, 136: 338–345.
Gimeno, S., H. Komen, P.W. Venderbosch and T. Bowmer. 1997. Disruption of sexual differentiation in genetic male common carp (*Cyprinus carpio*) exposed to an alkylphenol during different life stages. Environ Sci Technol, 31: 2884–2890.

Gimeno, S., H. Komen, A.G. Gerritsen and T. Bowmer. 1998a. Feminisation of young males of the common carp, *Cyprinus carpio*, exposed to 4-*tert*-pentylphenol during sexual differentiation. Aquat Toxicol, 43: 77–92.

Gimeno, S., H. Komen, S. Jobling et al. 1998b. Demasculinisation of sexually mature male common carp, *Cyprinus carpio*, exposed to 4-*tert*-pentylphenol during spermatogenesis. Aquat Toxicol, 43: 93–109.

Gormley, K.L. and K.L. Teather. 2003. Developmental, behavioral and reproductive effects experienced by Japanese medaka (*Oryzias latipes*) in response to short term exposure to endosulfan. Ecotoxicol Environ Saf, 54: 330–338.

Graham, J.B. 1997. *Air-breathing Fishes*. Academics Press, San Diego, USA.

Gray, M.A., A.J. Niimi and C.D. Metcalfe. 1999a. Factors affecting the development of testis-ova in medaka, *Oryzias latipes*, exposed to octylphenol. Environ Toxicol Chem, 18: 1835–1842.

Gray, M.A., K.L. Teather and C.D. Metcalfe. 1999b. Reproductive success and behavior of Japanese medaka (*Oryzias latipes*) exposed to 4-tert-octylphenol. Environ Toxicol Chem, 18: 2587–2594.

Gray, L.E., V. Wilson, T. Stoker et al. 2005. Environmental androgens and anti-androgens: An expanding chemical universe. *In*: R.K. Naz (ed.). *Endocrine Disrupters: Effects on Male and Female Reproductive Systems*. CRC Press, Boca Raton, FL, p. 313.

Gronen, S., N. Denslow, S. Manning et al. 1999. Serum vitellogenin levels and reproductive impairment of male Japanese medaka (*Oryzias latipes*) exposed to 4-tert-octylphenol. Environ Health Perspect, 107: 385–396.

Guiguen, Y., J.F. Baroiller, M.J. Ricordel et al. 1999. Involvement of estrogens in the process of sex differentiation in two fish species: the rainbow trout (*Oncorhynchus mykiss*) and a tilapia (*Oreochromis niloticus*). Mol Reprod Dev, 54: 154–162.

Gunnarsson, L., E. Kristiansson, L. Forlin et al. 2007. Sensitive and robust gene expression changes in fish exposed to estrogen—a microarray approach. BMC Genomics, 8: 149. doi:10.1186/1471-2164-8-149.

Hagino, S., M. Kagoshima and S. Ashida. 2001. Effects of ethinylestradiol, diethylstilbestrol, 4-t-pentylphenol, 17β-estradiol, methyltestosterone and flutamide on sex reversal in S-rR strain medaka (*Oryzias latipes*). Environ Sci, 8: 75–87.

Hahlbeck, E., R. Griffiths and B.E. Bengtsson. 2004a. The juvenile three-spined stickleback (*Gasterosteus aculeatus* L.) as a model organism for endocrine disruption: I. Sexual differentiation. Aquat Toxicol, 70: 287–310.

Hahlbeck, E., I. Katsiadaki, I. Mayer et al. 2004b. The juvenile three-spined stickleback (*Gasterosteus aculeatus* L.) as a model organism for endocrine disruption: 2 Kidney hypertrophy, vitellogenin and spiggin induction. Aquat Toxicol, 70: 311–326.

Hamdoun, A. and D. Epel. 2007. Embryo stability and vulnerability in an always changing world. Proc Natl Acad Sci, USA, 104: 1745–1750.

Hano, T., Y. Oshima, T. Oe et al. 2005. Quantitative bio-imaging analysis for evaluation of sexual differentiation in germ cells of olvas-GFP/ST-II YI medaka (*Oryzias latipes*) nanoinjected in sex change in ova with ethynylestradiol. Environ Toxicol Chem, 24: 70–71.

Hano, T., Y. Oshima, S.G. Kim et al. 2007. Tributyltin causes abnormal development in embryos of medaka, *Oryzias latipes*. Chemosphere, 69: 927–933.

Hansen, P.D., H. Dizer, B. Hock et al. 1998. Vitellogenin—a biomarker for endocrine disruptors. TrAC Trends Analyt Chem, 17: 448–451.

Hansson, T., E. Lindesjoo, L. Forlin et al. 2006. Long-term monitoring of the health status of female perch (*Perca fluviatilis*) in the Baltic Sea shows decreased gonad weight and increased hepatic EROD activity. Aquat Toxicol, 79: 341–355.

Harries, J.E., T. Runnalls, E. Hill et al. 2000. Development of a reproductive performance test for endocrine disrupting chemicals using pair-breeding fathead minnows (*Pimephales promelas*). Environ Sci Technol, 34: 3003–3011.

Harshbarger, J.C., M.J. Coffey and M.Y. Young. 2000. Intersexes in Mississippi river shovelnose sturgeon sampled below Saint Louis, Missouri, USA. Mar Environ Res, 50: 247–250.

Hecker, M., C.R. Tyler, M. Hoffmann et al. 2002. Plasma biomarkers in fish provide evidence for endocrine modulation in the Elbe River, Germany. Environ Sci Technol, 36: 2311–2321.

Herman, R.L. and H.L. Kincaid. 1988. Pathological effects of orally administered estradiol to rainbow trout. Aquaculture, 72: 165–172.

Hill, R.L and D.M. Janz. 2003. Developmental estrogenic exposure in zebrafish (*Danio rerio*: I. Effects on sex ratio and breeding success. Aquat Toxicol, 63: 417–429.

Hill, A.J., H. Teraoka, W. Heideman and R.E. Peterson. 2005. Zebrafish as a model vertebrate for investigating chemical toxicity. Toxicol Sci, 86: 6–19.

Hirai, N., A. Nanba, M. Koshio et al. 2006. Feminization of Japanese medaka (*Oryzias latipes*) exposed to 17β-estradiol: Effect of exposure period on spawning performance in sex-transformed females. Aquat Toxicol, 79: 288–295.

Hiramatsu, N., T. Matsubara, T. Fujita et al. 2006. Multiple piscine vitellogenins: biomarkers of fish exposure to estrogenic endocrine disruptors in aquatic environments. Mar Biol, 149: 35–47.

Holbech, H., K. Kinnberg, G.I. Petersen et al. 2006. Detection of endocrine disrupters: evaluation of a Fish Sexual Development Test (FSDT). Comp Biochem Physiol Part C: Toxicol Pharmacol, 144: 57–66.

Hornung, M.W., K.M. Jensen, J.J. Korte et al. 2004. Mechanistic basis for estrogenic effects in fathead minnow (*Pimephales promelas)* following exposure to the androgen 17α-methyltestosterone: conversion of 17α-methyltestosterone to 17α-methylestradiol. Aquat Toxicol, 66: 15–23.

Howell, W.M., D.A. Black and S.A. Bortone. 1980. Abnormal expression of secondary sex characters in a population of mosquitofish, *Gambusia affinis holbrooki*: evidence for environmentally-induced masculinization. Copeia, 1980: 676–681.

Howell, W.M., R.N. Hunsinger and P.D. Blanchard. 1994. Paradoxical masculinization of female western mosquitofish during exposure to spironolactone. Progr Fish-Cult, 56: 51–55.

Hu, J., Z. Zhang, Q. Wei et al. 2009. Malformations of the endangered Chinese sturgeon, *Acipenser sinensis* and its causal agent. Proc Natl Acad Sci USA, 106: 9339–9344.

Hutchinson, T.H., G.T. Ankley, H. Segner and C.R. Tyler. 2006. Screening and testing for endocrine disruption in fish-biomarkers as "signposts", not "traffic lights", in risk assessment. Environ Health Perspect, 114: 106.

IDEA. 2003. Identification of endocrine disrupting effects in aquatic vertebrates. Ecotoxicol Environ Saf, 54: 302–314.

Ishibashi, H., M. Kobayashi, Y. Tomiyasu et al. 2004. Development of plasma vitellogenin assay for estrogenic effects of endocrine-disrupting chemicals using ovariectomized goldfish (*Carassius auratus*). J Health Sci, 50: 169–173.

Ishibashi, H., M. Hirano, N. Matsumura et al. 2006. Reproductive effects and bioconcentration of 4-nonylphenol in medaka fish (*Oryzias latipes*). Chemosphere, 65: 1019–1026.

Iwanowicz, L.R. and V.S. Blazer. 2011. An overview of estrogen-associated endocrine disruption in fishes: evidence of effects on reproductive and immune physiology. Proc Third Bilateral Conf, July, 2009, Shepherdstown, West Virginia, pp. 266–275.

Jacobsen, T., M.J. Fogarty, B.A. Megrey and E. Moksness. 2007. *Fish Reproduction Biology.* Wiley Blackwell.

Jezierska, B., K. Ługowska and M. Witeska. 2009. The effects of heavy metals on embryonic development of fish (a review). Fish Physiol Biochem, 35: 625–640.

Jobling, S. 1998. Natural and anthropogenic environmental oestrogens: the scientific basis for risk assessment: Review of suggested method for endocrine disrupting chemicals. Pure Appl Chem, 70: 1805–1827.

Jobling, S., J.P. Sumpter, D. Sheahan et al. 1996. Inhibition of testicular growth in rainbow trout (*Oncorhynchus mykiss*) exposed to estrogenic alkylphenolic chemicals. Environ Toxicol Chem, 15: 194–202.

Jobling, S., M. Nolan, C.R. Tyler et al. 1998. Widespread sexual disruption in wild fish. Environ Sci Technol, 32: 2498–2506.

Jobling, S., N. Beresford, M. Nolan et al. 2002a. Altered sexual maturation and gamete production in wild roach (*Rutilus rutilus*) living in rivers that receive treated sewage effluents. Biol Reprod, 66: 272–281.

Jobling, S., S. Coey, J.G. Whitmore et al. 2002b. Wild intersex roach (*Rutilus rutilus*) have reduced fertility. Biol Reprod, 67: 515–524.

Jobling, S., R. Williams, A. Johnson et al. 2005. Predicted exposure to steroid estrogens in UK rivers correlate with wide spread sexual disruption. Environ Health Perspect, 114: 32–39.

Jobling, S., R. Williams, A. Johnson et al. 2006. Predicted exposures to steroid estrogens in UK rivers correlate with widespread sexual disruption in wild fish populations. Environ Health Perspect, 114: 32–39.

Kaiser, J. 1996. Scientists angle for answers. Science, 274: 1837–1838.

Kang, I.K., H. Yokota, Y. Oshima et al. 2002. Effects of bisphenol A on the reproduction of Japanese medaka after continuous exposure to octylphenol. Ecotoxicol Environ Saf, 51: 187–196.

Kannan, K., S. Tanabe, H. Iwata and R. Tatsukawa. 1995. Butyltins in muscle and liver of fish collected from certain Asian and Oceanian countries. Environ Pollut, 90: 279–290.

Kavumpurath, S. and T.J. Pandian. 1992. Production of YY male guppy (*Poecilia reticulata*) by endocrine sex reversal and selective breeding. Asian Fish Sci, 57: 265–276.

Kavumpurath, S. and T.J. Pandian. 1993. Masculinisation of *Poecilia reticulata* by dietary administration of synthetic or natural androgens to gravid females. Aquaculture, 116: 83–89.

Kazeto, Y., A.R. Place and J.M. Trant. 2004. Effects of endocrine disrupting chemicals on the expression of CYP19 genes in zebrafish (*Danio rerio*) juveniles. Aquat Toxicol, 69: 25–34.

Khan, I.A. and P. Thomas. 1992. Stimulatory effects of serotonin on maturational gonadotropin release in the Atlantic croaker, *Micropogonias undulatus*. Gen Comp Endocrinol, 88: 388–396.

Khan, I.A., M.B. Hawkins and P. Thomas. 1999. Gonadal stage-dependent effects of gonadal steroids on gonadotropin II secretion in the Atlantic croaker (*Micropogonias undulatus*). Biol Reprod, 61: 834–841.

Kime, D.E. 1998. *Endocrine Disruption in Fish.* Kluwer, Boston.

Kime, D.E. 1999. Environmentally induced endocrine abnormalities in fish. *In*: R.E. Hester and R.M. Harrison (eds.). *Endocrine Disrupting Chemicals. Issues in Environmental Science and Technology*, The Royal Society of Chemistry, Cambridge, UK, 12: 27–48.

Kime, D.E. and J.P. Nash. 1999. Gamete viability as an indicator of reproductive endocrine disruption in fish. Sci Total Environ, 233: 123–129.

Kinnberg, K. and G. Toft. 2003. Effects of estrogenic and anti-androgenic compounds on the testis structure of the adult guppy (*Poecilia reticulata*). Ecotoxicol Environ Saf, 54: 16–24.

Kinnberg, K., B. Korsgaard and P. Bjerregaard. 2003. Effects of octylphenol and 17β-estradiol on the gonads of guppies (*Poecilia reticulata*) exposed as adults via the water or as embryos via the mother. Toxicol Pharmacol, 134: 45–55.

Kirankumar, S. and T.J. Pandian. 2002. Effect on growth and reproduction of hormone immersed and masculinised fighting fish *Betta splendens*. J Exp Zool, 293: 606–616.

Kitano, T., N. Yoshinaga, E. Shiraishi et al. 2007. Tamoxifen induces masculinisation of genetic females and regulates p450 aromatase and Mullerian inhibiting substances mRNA expression in Japanese flounder (*Paralichthys olivaceus*). Mol Reprod Dev, 74: 1171–1177.

Knapp, R., E. Marsh-Mathew, L. Vo and S. Rosencrans. 2012. Stress hormone masculinizes female morphology and behaviour. Biol Lett. doi:10.1098/rsbl.0514.

Kobayashi, H. and T. Iwamatsu. 2005. Sex reversal in the medaka *Oryzias latipes* by brief exposure of early embryos to estradiol-17β. Zool Sci, 22: 1163–1167.

Kobayashi, M. and N. Stacey. 1993. Prostaglandins-induced female spawning behavior in goldfish (*Carassius auratus*) appears independent of ovarian infl uence. Horm Behav, 27: 38–55.

Kobayashi, M., K. Aida and N.E. Stacey. 1991. Induction of testis development by implantation of 11-Ketotestosterone in female goldfish. Zool Sci, 8: 389–393.

Koger, C.S., S.J. Teh and D.E Hinton. 2000. Determining the sensitive developmental stages of intersex induction in medaka (*Oryzias latipes*) exposed to 17β-estradiol or testosterone. Mar Environ Res, 50: 201–206.
Kolok, A.S. and M.K. Sellin. 2008. The environmental impact of growth-promoting compounds employed by the United States beef cattle industry: history, current knowledge, and future directions. pp. 1–30. *In: Reviews of Environmental Contamination and Toxicology*. Springer Verlag, New York.
Kolok, A.S., D. Snow, D. Kohno et al. 2007. Occurrence and biological effect of exogenous steroids in the Elkhorn River, Nebraska, USA. Sci Total Environ, 388: 104–115.
Koplin, D.W., E.T. Furlong, M.T. Meyer et al. 2002. Pharmaceuticals, hormones, and other organic wastewater contaminants in US streams, 1999-2000: A national reconnaissance. Environ Sci Technol, 36: 1202–1211.
Kramer, V.J., S. Miles-Richardson, S.L. Pierens and J.P. Giesy. 1998. Reproductive impairment and induction of alkaline-labile phosphate, a biomarker of estrogen exposure in fathead minnows (*Pimephales promelas*) exposed to waterborne 17 β-estradiol. Aquat Toxicol, 51: 431–441.
Kuhl, A.J. and M. Brouwer. 2006. Anti-estrogens inhibit xenoestrogen-induced brain aromatase activity but do not prevent xenoestrogen-induced feminization in Japanese medaka (*Oryzias latipes*). Environ Health Perspect, 114: 500–506.
Lacerda, S.M.S.N., S.R. Batlouni, S.B.G. Silva et al. 2006. Germ cells transplantation in fish: the Nile-tilapia model. Anim Reprod, 3: 146–159.
Lahnsteiner, F., B. Berger, M. Kletzl and T. Weismann. 2005. Effect of bisphenol A on maturation and quality of semen and eggs in the brown trout, *Salmo trutta f. fario*. Aquat Toxicol, 75: 213–224.
Lance, V.A. and M.H. Bogart. 1991. Tamoxifen "sex reverses" alligator embryos at male producing temperatures, but is an antiestrogen in female hatchlings. Cell Mol Life Sci, 47: 263–266.
Länge, R., T.H. Hutchinson, C.P. Croudace et al. 2001. Effects of the synthetic estrogen 17α-ethinylestradiol on the life-cycle of the fathead minnow (*Pimephales promelas*). Environl Toxicol Chem, 20: 1216–1227.
Lange, I.G., A. Daxenberger, B. Schiffer et al. 2002. Sex hormones originating from different livestock production systems: fate and potential disrupting activity in the environment. Analyt Chim Acta, 473: 27–37.
Larkin, P., I. Knoebl and N.D. Denslow. 2003. Differential gene expression analysis in fish exposed to endocrine disrupting compounds. Comp Biochem Physiol Part B: Biochem Mol Biol, 136: 149–161.
Larsen, M.G. and E. Baatrup. 2010. Functional behavior and reproduction in androgenic sex reversed zebrafish (*Danio rerio*). Env Toxicol Chem, 29: 1828–1833.
Larsen, M.G., K. Bilbery and E. Baatrup. 2009. Reversibility of estrogenic sex changes in zebrafish (*Danio rerio*). Environ Toxicol Chem, 28: 1783–1785.
Larsen, M.G., K.B. Hansen, P.G. Henriksen and E. Baatrup. 2008. Male zebrafish (*Danio rerio*) courtship behaviour resists the feminising effects of 17α-ethinyloestradiol—morphological sexual characteristics do not. Aquat Toxicol, 87: 234–244.
Larsson, D.G.J., M. Adolfsson-Erici, J. Parkkonen et al. 1999. Ethinyloestradiol—an undesired fish contraceptive? Aquat Toxicol, 45: 91–97.
Leatherland, J.F. and P.T.K. Woo. 2010. *Fish Diseases and Disorders. 2. Non-infectious Disorders*, CAB International, Cambridge, MA, USA.
Leet, J.K., H.E. Gall and M.S. Sepúlveda. 2011. A review of studies on androgen and estrogen exposure in fish early life stages: effects on gene and hormonal control of sexual differentiation. J Appl Toxicol, 31: 379–398.
Leino, R.L., K.M. Jensen and G.T. Ankley. 2005. Gonadal histology and characteristic histopathology associated with endocrine disruption in the adult fathead minnow (*Pimephales promelas*). Environ Toxicol Pharmacol, 19: 85–98.

Leusch, F.D. and D.L. MacLatchy. 2003. *In vivo* implants of β-sitosterol cause reductions of reactive cholesterol pools in mitochondria isolated from gonads of male goldfish (*Carassius auratus*). Gen Comp Endocrinol, 134: 255–263.

Liao, T., Q.L. Guo, S.W. Jin et al. 2009. Comparative responses in rare minnow exposed to 17β estradiol during different life stages. Fish Physiol Biochem, 35: 341–349.

Lister, A.L. and G.J. Van der Kraak. 2001. Endocrine disruption: Why is it so complicated? Water Quality Res J Canada, 36: 175–190.

Liu, C., J. Deng, L. Yu et al. 2010. Endocrine disruption and reproductive impairment in zebrafish by exposure to 8: 2 fluorotelomer alcohol. Aquat Toxicol, 96: 70–76.

Loomis, A.K. and P. Thomas. 1999. Binding characteristics of estrogen receptor (ER) in Atlantic croaker (*Micropogonias undulatus*) testis: different affinity for estrogens and xenobiotics from that of hepatic ER. Biol Reprod, 61: 51–60.

Lye, C.M, C.L.J. Frid, M.E. Gill et al. 1999. Estrogenic alkylphenols in fish tissues, sediments, and waters from the UK Tyne and Tees estuaries. Environ Sci Technol, 33: 1009–14.

Maack, G. and H. Segner. 2004. Life-stage-dependent sensitivity of zebrafish (*Danio rerio*) to estrogen exposure. Comp Biochem Physiol Part C: Toxicol Pharmacol, 139: 47–55.

MacGregor, J.I. and V.C. Jordan. 1998. Basic guide to the mechanisms of antiestrogen action. Pharmocol Rev, 50: 151–196.

Makynan, E.A., M.D. Kahl, K.M. Jensen et al. 2000. Effects of the mammalian anti-androgen vinclozolin on development and reproduction of the fathead minnow (*Pimephales promelas*). Aquat Toxicol, 48: 461–475.

Malison, J.A., T.B. Kayes, C.D. Best et al. 1986. Sexual differentiation and use of hormones to control sex in yellow perch (*Perca flavescens*). Can J Fish Aquat Sci, 43: 26–35.

Marchand, M.J., G.M. Pieterse and I.E. Barnhoorn. 2008. Preliminary results on sperm motility and testicular histology of two feral fish species, *Oreochromis mossambicus* and *Clarias gariepinus*, from a currently DDT-sprayed area, South Africa. J Appl Ichthyol, 24: 423–429.

Matsubara, T. and Y. Koya. 1997. Course of proteolytic cleavage in three classes of yolk proteins during oocyte maturation in barfin flounder *Verasper moseri*, a marine teleost spawning pelagic eggs. J Exp Zool, 278: 189–200.

Matthiessen, P. 2003. Endocrine disruption in marine fish. Pure Appl Chem, 75: 2249–2261.

Matthiessen, P. and J.P. Sumpter. 1998. Effects of estrogenic substances in the aquatic environment. pp. 319–335. *In*: *Fish Ecotoxicology*. Birkhäuser Verlag, Basel.

Matthiessen, P., Y. Allen, S. Bamber et al. 2002. The impact of oestrogenic and androgenic contamination on marine organisms in the United Kingdom-summary of the EDMAR programme. Mar Environ Res, 54: 645–649.

Mazzoldi, C., M. Scagginate, E. Ambrosin and M.B. Rasola. 2000. Mating system and alternative male mating tactics in the grass goby *Zosterissor ophiocephalus* (Teleostei: Gobiidae). Mar Biol, 137: 1041–1048.

McAllister, B.G. and D.E. Kime. 2003. Early life exposure to environmental levels of the aromatase inhibitor tributyltin causes masculinisation and irreversible sperm damage in zebrafish (*Danio rerio*). Aquat Toxicol, 65: 309–316.

Metcalfe, C.D., K.A. Kidd and J.P. Sumpter. 2010. Chemically induced alterations to gonadal differentiation in fish. *In*: J.F. Leatherland and P.T.K. Woo (eds.). *Fish Diseases and Disorders*. CAB International, Wallinford, UK, 2: 144–165.

Metcalfe, T.L., C.D. Metcalfe, Y. Kiparissis et al. 2000. Gonadal development and endocrine responses in Japanese medaka (*Oryzias latipes*) exposed to o, p'-DDT in water or through maternal transfer. Environ Toxicol Chem, 19: 1893–1900.

Metcalfe, C.D., T.L. Metcalfe, Y. Kiparissis et al. 2001. Estrogenic potency of chemicals detected in sewage treatment plant effluents as determined by *in vivo* assays with Japanese medaka (*Oryzias latipes*). Environ Toxicol Chem, 20: 297–308.

Miles-Richardson, S.R., V.J. Kramer, S.D. Fitzgerald et al. 1999a. Effects of waterborne exposure of 17 β-estradiol on secondary sex characteristics and gonads of fathead minnows (*Pimephales promelas*). Aquat Toxicol, 47: 129–145.

Miles-Richardson, S.R., S.L. Pierens, K.M. Nichols et al. 1999b. Effects of waterborne exposure to 4-nonylphenol and nonylphenol ethoxylate on secondary sex characteristics and gonads of fathead minnows (*Pimephales promelus*). Environ Res Section A, 80: S122–137.
Miller, M.A. 1993. Maternal transfer of organochlorine compounds in salmonines to their eggs. Can J Fisher Aquat Sci, 50: 1405–1413.
Mills, L.J. and C. Chichester. 2005. Review of evidence: are endocrine-disrupting chemicals in the aquatic environment impacting fish populations? Sci Total Environ, 343: 1–34.
Mills, L.J., R.E. Gutjahr-Gobella, G.E. Zaroogiana et al. 2014. Modulation of aromatase activity as a mode of action for endocrine disrupting chemicals in a marine fish. Aquat Toxicol, 147: 140–150.
Minier, C., G. Caltot, F. Leboulanger and E.M. Hill. 2000. An investigation of the incidence of intersex fish in Seine-Maritime and Sussex regions. Analusis, 28: 801–806.
Ministry of the Environment, Japan (MoE). 2004. Fish testing results of endocrine disruptive chemicals. 2004. Environment Health and Safety Division, Endocrine Disrupters. http/ www.env.go/jp/en/chem./ed/approach/annex_7pdf.
Mochida, Y., K. Takeda, M. Saitoh et al. 2000. ASK1 inhibits interleukin-1-induced NF-κB activity through disruption of TRAF6-TAK1 interaction. J Biol Chem, 275: 32747–32752.
Mochida, K., K. Ito, K. Kono et al. 2010. Effect of tributyltin oxide exposure on the F_0 and F_1 generations of a marine teleost, the mummichog *Fundulus heteroclitus*. Fisher Sci, 76: 333–341.
Moncaut, N., F.L. Nostro and M.C. Maggese. 2003. Vitellogenin detection in surface mucus of the South American cichlid fish *Cichlasoma dimerus* (Heckel, 1840) induced by estradiol-17β. Effects on liver and gonads. Aquat Toxicol, 63: 127–137.
Nagler, J.J., J. Bouma, G.H. Thorgaard and D.D. Dauble. 2001. High incidence of a male-specific genetic marker in phenotypic female chinook salmon from the Columbia River. Environ Health Perspect, 109: 67.
Nakayama, K., Y. Oshima, T. Yamaguchi et al. 2004. Fertilization success and sexual behavior in male medaka, (*Oryzias latipes*) exposed to tributyltin. Chemosphere, 55: 1331–1337.
Nash, J.P., D.E. Kime, L.T. Van der Ven et al. 2004. Long-term exposure to environmental concentrations of the pharmaceutical ethynylestradiol causes reproductive failure in fish. Environmen Health Perspect, 112: 1725.
Navarro-Martin, L., M. Blazquez and F. Piferrer. 2009. Masculinization of the European sea bass (*Dicentrarchus labrax*) by treatment with an androgen or aromatase inhibitor involves different gene expression and has distinct lasting effects on maturation. Gen Comp Endocrinol, 160: 3–11.
Naylor, C.G. 1992. Environmental fate of alkylphenol ethoxylates. Soap Cosmet Chem Special, 68: 27–31.
Naylor, C.G. 1995. Environmental fate and safety of nonylphenol ethoxylates. Textile Chemist Colorist, 27: 29–33.
Naz, R.K. 2004. *Endocrine disrupters*: CRC Press, Boca Raton.
Njiwa, J.R.K. and P. Müller. 2002. Alteration in sperm release from zebrafish (*Brachydanio rerio*) exposed to DDT. J Health Sci, 48: 404–411.
Njiwa, J.R., P. Müller and R. Klein. 2004a. Binary mixture of DDT and Arochlor1254: effects on sperm release by *Danio rerio*. Ecotoxicol Environ Saf, 58: 211–9.
Njiwa, J.R.K., P. Miller and R. Klein. 2004b. Variations of sperm release in three batches of zebrafish. J Fish Biol, 64: 475–482.
Nimrod, A.C. and W.H. Benson. 1998. Reproduction and development of Japanese medaka following an early life stage exposure to xenoestrogens. Aquat Toxicol, 44: 141–156.
Nirmala, K., Y. Oshima, R. Lee et al. 1999. Transgenerational toxicity of tributyltin and its combined effects with polychlorinated biphenyls on reproductive processes in Japanese medaka (*Oryzias latipes*). Environ Toxicol Chem, 18: 717–721.
Ogawa, S., M. Akiyoshi, M. Higuchi et al. 2008. 'Post-sex differntiational' sex reversal in the female common carp (*Cyprinus carpio*). Cybium, 32: 102–103.

Orlando, E.F., Y. Katsu, S. Miyagawa et al. 2006. Cloning and differential expression of estrogen receptor and aromatase genes in the self-fertilizing hermaphrodite and male mangrove rivulus *Kryptolebias marmoratus*. J Mol Endocrinol, 37: 353–365.

Orn, S., A. Gessbo, A. Steinholz and L. Norrgren. 2000. Zebrafish (*Danio rerio*). A candidate to evaluate endocrine disrupting chemicals. Tema Nord, 555: 47–62.

Orn, S., H. Holbech, T.H. Madsen et al. 2003. Gonad development and vitellogenin production in zebrafish (*Danio rerio*) exposed to ethinylestradiol and methyltestosterone. Aquat Toxicol, 65: 397–411.

Oshima, Y., I.J. Kang, M. Kobayashi et al. 2003. Suppression of sexual behavior in male Japanese medaka (*Oryzias latipes*) exposed to 17β-estradiol. Chemosphere, 50: 429–436.

Pait, A.S. and J.O. Nelson. 2002. Endocrine disruption in fish: an assessment of recent research and results. NOAA Tech Memo, Silver Spring, p. 55.

Palace, V.P., K.G. Wautier, R.E. Evans et al. 2006. Biochemical and histopathological effects in pearl dace (*Margariscus margarita*) chronically exposed to a synthetic estrogen in a whole lake experiment. Environ Toxicol Chem, 25: 1114–1125.

Pandey, S. 1970. Effects of methallibure on the testes and secondary sex characters of the adult and juvenile guppy *Poecilia reticulata* Peters. Biol Reprod, 2: 239–244.

Pandey, S., N.S. Nagpure, R. Kumar et al. 2006. Genotoxicity evaluation of acute doses of endosulfan to freshwater teleost *Channa punctatus* (Bloch) by alkaline single-cell gel electrophoresis. Ecotoxicol Environ Saf, 65: 56–61.

Pandian, T.J. 2010. *Sexuality in Fishes*. Science Publishers, Enfield/CRC Press, Boca Raton, USA, p. 208.

Pandian, T.J. 2011. *Sex Determination in Fish*. Science Publishers, Enfield/CRC Press, Boca Raton, USA, p. 282.

Pandian, T.J. 2012. *Genetic Sex Differentiation in Fish*. CRC Press, Boca Raton, USA, p. 214.

Pandian, T.J. 2013. *Endocrine Sex Differentiation in Fish*. CRC Press, Boca Raton, USA, p. 303.

Pandian, T.J. and R. Bhaskaran. 1983. Uptake, accumulation and elimination of ^{14}C DDT in the fish *Channa striatus*. Indian J Exp Biol, 21: 88–91.

Pandian, T.J. and E. Vivekanandan. 1976. Effects of feeding and starvation on growth and swimming activity in an obligatory air-breathing fish. Hydrobiologia, 49: 33–39.

Panter, G.H., R.S. Thompson and J.P. Sumpter. 1998. Adverse reproductive effects in male fathead minnows (*Pimephales promelas*) exposed to environmentally relevant concentrations of the natural oestrogens, oestradiol and oestrone. Aquat Toxicol, 42: 243–253.

Panter, G.H., T.H. Hutchinson, K.S. Hurd et al. 2004. Successful detection of (anti-) androgenic and aromatase inhibitors in pre-spawning adult fathead minnows (*Pimephales promelas*) using easily measured endpoints of sexual development. Aquat Toxicol, 70: 11–21.

Panter, G.H., T.H. Hutchinson., K.S. Hurd et al. 2006. Development of chronic tests for endocrine active chemicals: Part 1. An extended fish early-life stage test for oestrogenic active chemicals in the fathead minnow (*Pimephales promelas*). Aquat Toxicol, 77: 279–290.

Papoulias, D.M., S.A. Villalobos, J. Meadows et al. 2003. In ovo exposure to o, p'DDE affects sexual development but not sexual differentiation in Japanese medaka (*Oryzias latipes*). Environ Health Perspect, 111: 29.

Parks, L.G., C.S. Lambright, E.F. Orlando et al. 2001. Masculinization of female mosquitofish in kraft mill effluent-contaminated Fenholloway River water is associated with androgen receptor agonist activity. Toxicol Sci, 62: 257–267.

Patyna, P.J., R.A. Davi, T.F. Parkerton et al. 1999. A proposed multigeneration protocol for Japanese medaka (*Oryzias latipes*) to evaluate effects of endocrine disruptors. Sci Total Environ, 233: 211–220.

Pawlowski, S., A. Sauer, J.A. Shears et al. 2004. Androgenic and estrogenic effects of the synthetic androgen 17α-methyltestosterone on sexual development and reproductive performance in the fathead minnow (*Pimephales promelas*) determined using the gonadal recrudescence assay. Aquat Toxicol, 68: 277–291.

Peters, R.E., S.C. Courtenay, S. Cagampan et al. 2007. Effect on reproductive potential and endocrine status after exposure to 17 α-ethylnylestradial in a short term reproductive bioassay. Aquat Toxicol, 85: 154–166.

Peters, R.E., S.C. Courtenay, L.M. Hewitt and D.L. MacLatchy. 2010. Effects of 17α-ethynylestradiol on early-life development, sex differentiation and vitellogenin induction in mummichog (*Fundulus heteroclitus*). Mar Environ Res, 69: 178–186.

Pickford, K.A., R.E. Thomas-Jones, B. Wheals et al. 2003. Route of exposure affects the oestrogenic response of fish to 4-tert-nonylphenol. Aquat Toxicol, 65: 267–279.

Piferrer, F. 2001. Endocrine sex control strategies for the feminization of teleost fish. Aquaculture, 197: 229–281.

Rasmussen, T.H., T.K. Andreassen, S.N. Pedersen et al. 2002. Effects of waterborne exposure of octylphenol and oestrogen on pregnant viviparous eelpout (*Zoarces viviparus*) and her embryos in ovario. J Exp Biol, 205: 3857–3876.

Reis-Henriques, M.A., M.M. Cruz and J.O. Pereira. 1997. The modulating effect of vitellogenin on the synthesis of 17beta-estradiol by rainbow trout *Oncorhynchus mykiss* ovary. Fish Physiol Biochem, 16: 181–186.

Robinson, C.D., E. Brown, J.A. Craft et al. 2003. Effects of sewage effluent and ethynyl oestradiol upon molecular markers of oestrogenic exposure, maturation and reproductive success in the sand goby (*Pomatoschistus minutus*, Pallas). Aquat Toxicol, 62: 119–34.

Rodgers-Gray, T.P., S. Jobling, S. Morris et al. 2000. Long-term temporal changes in the estrogenic composition of treated sewage effluent and its biological effects on fish. Environ Sci Technol, 34: 1521–1528.

Rodgers-Gray, T.P., S. Jobling, C. Kelly et al. 2001. Exposure of juvenile roach (*Rutilus rutilus*) to treated sewage effluent induces dose-dependent and persistent disruption in gonadal duct development. Environ Sci Technol, 35: 462–470.

Rougeot, C., B. Jacobs, P. Kestemont and C. Melard. 2002. Sex control and sex determinism study in Eurasian perch, *Perca fluviatilis*, by use of hormonally sex-reversed male breeders. Aquaculture, 211: 81–89.

Rurangwa, E., I. Roelants, G. Huyskens et al. 1998. The minimum effective spermatozoa: egg ratio for artificial insemination and the effects of mercury on sperm motility and fertilization ability in *Clarias gariepinus*. J Fish Biol, 53: 402–413.

Santos, E.M., G.C. Paull, K.J. Van Look et al. 2007. Gonadal transcriptome responses and physiological consequences of exposure to oestrogen in breeding zebrafish (*Danio rerio*). Aquat Toxicol, 83: 134–142.

Sarkar, A., D. Ray, A.N. Shrivastava and S. Sarker. 2006. Molecular biomarkers: their significance and application in marine pollution monitoring. Ecotoxicol, 15: 333–340.

Sarria, M.P., M.M. Santos, L.F. Castro et al. 2013. Estrogenic chemical effects are independent from the degree of sex role reversal in pipefish. J Hazard Mater. doi: 10.1016/j.jhazmat.2013.10.043.

Schiller, V., A. Wichmann, R. Kriehuber et al. 2013. Studying the effects of geistein on gene expression of fish embryos on an alternative approach for endocrine disruption. Comp Biochem Physiol Part C, 157: 41–53.

Scholz, S. and N. Kluver. 2009. Effects of endocrine disrupters on sexual, gonadal development in fish. Sex Dev, 3: 136–151.

Scholz, S. and H.O. Gutzeit. 2000. 17-α-ethinylestradiol affects reproduction, sexual differentiation and aromatase gene expression of the medaka (*Oryzias latipes*). Aquat Toxicol, 50: 363–373.

Schwaiger, J., U. Mallow, H. Ferling et al. 2002. How estrogenic is nonylphenol? A transgenerational study using rainbow trout (*Oncorhynchus mykiss*) as a test organism. Aquatic Toxicol, 59: 177–189.

Sebire, M. and I. Katsiadaki. 2008. The reproductive behaviour of the three spined stickleback as a novel bioassay for the detection of anti-androgen. Cybium, 32: 59–60.

Sebire, M., A.P. Scott, C.R. Tyler et al. 2009. The organophosphorous pesticide, fenitrothion, acts as an anti-androgen and alters reproductive behavior of the male three-spined stickleback, *Gasterosteus aculeatus*. Ecotoxicol, 18: 122–133.

Segner, H., K. Caroll, M. Fenske et al. 2003. Identification of endocrine-disrupting effects in aquatic vertebrates and invertebrates: report from the European IDEA project. Ecotoxicol Environ Saf, 54: 302–314.

Seki, M., H. Yokota, H. Matsubara et al. 2002. Effect of ethinylestradiol on the reproduction and induction of vitellogenin and testis-ova in medaka (*Oryzias latipes*). Environ Toxicol Chem, 21: 1692–1698.

Seki, M., H. Yokota, M. Maeda et al. 2003. Effects of 4-nonylphenol and 4-tert-octylphenol on sex differentiation and vitellogenin induction in medaka (*Oryzias latipes*). Environ Toxicol Chem, 22: 1507–1516.

Senthilkumar, K., C.A. Duda, D.L. Villeneuve et al. 1999. Butyltin compounds in sediment and fish from the Polish coast of the Baltic Sea. Environ Sci Poll Res, 6: 200–206.

Shane, B.S. 1994. *Introduction to Ecotoxicology*, CRC Press, Boca Raton, FL.

Shibata, N. and S. Hamaguchi. 1988. Evidence for the sexual bipotentiality of spermatogonia in the fish *Oryzias latipes*. J Exp Zool, 245: 71–77.

Shimasaki, Y., Y. Oshima, Y. Inoue et al. 2008. Distribution of tributyltin in tissues of mature Japanese whiting, *Sillago japonica* and their eggs. J Faculty Agricult-Kyushu Univ, 53.

Shioda, T. and M. Wakabayashi. 2000. Effect of certain chemicals on the reproduction of medaka (*Oryzias latipes*). Chemosphere, 40: 239–43.

Singh, P.B. and V. Singh. 2008. Bioaccumulation of hexachlorocyclohexane, dichlorodiphenyltrichloroethane, and estradiol-17β in catfish and carp during the pre-monsoon season in India. Fish Physiol Biochem, 34: 25–36.

Sloman, K.A. and R.W. Wilson. 2006. Anthropogenic impacts upon behaviour and physiology. pp. 413–160. *In*: K.A. Sloman, R.W. Wilson and S. Balshine (eds.). *Behavioural Physiology of Fish*. Elsevier, Amsterdam.

Snyder, S.A., T.L. Keith, D.A.V. Snyder et al. 1999. Analytical methods for detection of selected estrogenic compounds in aqueous mixtures. Environ Sci Technol, 33: 2814–20.

Soffker, M. and C.R. Tyler. 2012. Endocrine disrupting chemicals and sexual behaviors in fish—a critical review on effects and possible consequences. Crit Rev Toxicol, 42: 653–668.

Sohoni, P.C.R.T., C.R. Tyler, K. Hurd et al. 2001. Reproductive effects of long-term exposure to bisphenol A in the fathead minnow (*Pimephales promelas*). Environ Sci Technol, 35: 2917–2925.

Sone, K., M. Hinago, M. Itamoto et al. 2005. Effects of an androgenic growth promoter 17β-trenbolone on masculinization of mosquitofish (*Gambusia affinis affinis*). Gen Comp Endocrinol, 143: 151–160.

Sonnenschein, C. and A.M. Soto. 1998. An updated review of environmental estrogen and androgen mimics and antagonists. J Steroid Biochem Mol Biol, 65: 143–150.

Strmack, M. and T. Brauenbeck. 1999. Effect of triphenyltin acetate on survival, hatching success and liver intra structure of early life stages of zebrafish (*Danio rerio*). Ecotoxicol Environ Saf, 44: 25–29.

Suvaparp, R., V. Chawengsri, N. Tayaputch et al. 2001. Dissipation of endosulfan in the rice field. pp. 129–139. *In*: Proc Rice Res Inst, Dept of Agricult, Bangkok.

Suzuki, T., R. Matsuda and Y. Saito. 1992. Molecular species of tri-n-butyltin compounds in marine products. J Agricul Food Chem, 40: 1437–1443.

Swapna, I. and B. Senthilkumaran. 2009. Influence of ethynylestradiol and methyltestosterone on the hypothalmo-hypophyseal gonadal axis of adult air-breathing catfish. *Clarias gariepinus*. Aquatic Toxicol, 95: 222–229.

Swedenberg, E., J. Ruegg, S. Makela and I. Pongratz. 2009. Endocrine disruptive chemicals: mechanisms of action and involvement in metabolic disorders. J Mol Endocrinol, 43: 1–10.

Sweeting, R.A. 1981. Hermophrodite Roach in the River Lee, Thames Water, Lee Division, Swindon. UK.

Thibaut, R. and C. Porte. 2004. Effects of endocrine disrupters on sex steroid synthesis and metabolism pathways in fish. J Steroid Biochem Mol Biol, 92: 485–494.

Thibaut, R., L. Debrauwer, E. Perdu et al. 2002. Region-specific hydroxylation of nonylphenol and the involvement of CYP2K-and CYP2M-like iso-enzymes in Atlantic salmon (*Salmo salar*). Aquat Toxicol, 56: 177–190.

Thomas, P. 1989. Effects of Aroclor 1254 and cadmium on reproductive endocrine function and ovarian growth in Atlantic croaker. Mar Environ Res, 28: 499–503.

Thomas, P. 2008. The Endocrine System. pp. 457–488. *In*: R.T. Di Giulio and D.E. Hinton (eds.). *The Toxicology of Fishes*. CRC Press, Boca Raton.

Thorpe, K.L., R.I. Cummings, T.H. Hutchinson et al. 2003. Relative potencies and combination effects of steroidal estrogens in fish. Environ Sci Technol, 37: 1142–1149.

Todd, N.E. and M. Van Leeuwen. 2002. Effects of sevin (carbaryl insecticide) on early life stages of zebrafish (*Danio rerio*). Ecotoxicol Environ Saf, 53: 267–272.

Toft, G. and L.J. Guillette, Jr. 2005. Decreased sperm count and sexual behavior in mosquitofish exposed to water from a pesticide-contaminated lake. Ecotoxicol Environ Saf, 60: 15–20

Toft, G., T.M. Edwards, E. Baatrup and L.J. Guillette, Jr. 2003. Disturbed sexual characteristics in male mosquitofish (*Gambusia holbrooki*) from a lake contaminated with endocrine disruptors. Environ Health Perspect, 111: 695–701.

Tollefsen, K.E., J. Ovrevik and J. Stenersen. 2004. Binding of in xenoandrogens to the sex steroid-binding protein in plasma from Arctic charr (*Salvelinus alpinus*). Comp Biochem Physiol C Toxicol Pharmacol, 139: 127–133.

Turner, C.L. 1947. The rate of morphogenesis and regeneration of the gonopodium in normaland castrated males of *Gambusia affinis*. J Exp Zool, 106: 125–143.

Tyler, C.R., S. Jobling and J.P. Sumpter. 1998. Endocrine disruption on wild life: a critical review of the evidence. Environ Sci Technol, 37: 1142–1149.

Tyler, C.R., E.J. Routledge and R.v. Aerle. 2007. Estrogenic effect of treated sewage effluent on fish: steroids and surfactants in English Rivers. pp. 971–1002. *In*: T. Jakobsen, M.J. Fogarty, B.A. Megrey and E. Moksness (eds.). *Fish Reproductive Biology*. Wiley Blackwell.

Uchida, D., M. Yamashita, T. Kitano and T. Iguchi. 2002. Oocyte apoptosis during the transition from ovary-like tissue to testes during sex differentiation of juvenile zebrafish. J Exp Biol, 205: 711–718.

Ungerer, J.R. and P. Thomas. 1996. Role of very low density lipoproteins in the accumulation of o, p'-DDT in fish ovaries during gonadal recrudescence. Aquat Toxicol, 35: 183–195.

Urushitani, H., A. Shimizu, Y. Katsu and T. Iguchi. 2002. Early estrogen exposure induces abnormal development of *Fundulus heteroclitus*. J Exp Zool, 293: 693–702.

Urushitani, H., Y. Katsu, Y. Kato et al. 2007. Medaka (*Oryzias latipes*) for use in evaluating developmental effects of endocrine active chemicals with special reference to gonadal intersex (testis-ova). Environ Sci, 14: 211–233.

Vajda, A.M., L.B. Barber, J.L. Gray et al. 2011. Demasculinization of male fish by wastewater treatment plant effluent. Aquat Toxicol, 103: 213–221.

Valle, M., S. Orn, H. Holbech et al. 2001. Oral exposure in zebrafish: An alternative route of exposure to endocrine disrupting chemicals. pp. 42–51. *In*: G.I. Petersen, L. Norrgren, H. Holbech et al. (eds.). Suitability of Zebrafish as Test Organism for Detection of Endocrine Disrupting Chemicals. Teme Nord Serie.

van Aerle, R., M. Nolanusan, S. Jobling et al. 2001. Sexual disruption in a second species of wild cyprinid fish (the gudgeon, *Gobio gobio*) in United Kingdom freshwaters. Environ Toxicol Chem, 20: 2841–2847.

van Aerle, R., N. Pounds, T.H. Hutchinson et al. 2002. Window of sensitivity for the estrogenic effects of ethinylestradiol in early life-stages of fathead minnow, *Pimephales promelas*. Ecotoxicology, 11: 423–434.

Van den Belt, K., R. Verheyen and H. Witters. 2001. Reproductive effects of ethynylestradiol and 4t-octylphenol on the zebrafish (*Danio rerio*). Arch Environ Contam Toxicol, 41: 458–467.

Van den Belt, K., P.W. Wester, L. van der Ven et al. 2002. Effects of ethynylestradiol on the reproductive physiology in zebrafish (*Danio rerio*): time dependency and reversibility. Environ Toxicol Chem, 21: 767–775.

Van den Belt, K., P. Berckmans, C. Vangenechten et al. 2004. Comparative study on the *in vitro/in vivo* estrogenic potencies of 17β-estradiol, estrone, 17α-ethynylestradiol and nonylphenol. Aquat Toxicol, 66: 183–195.

van der Ven, L., P.W. Wester and J.G. Vos. 2003. Histopathology as a tool for the evaluation of endocrine disruption in zebrafish (*Danio rerio*). Environ Toxicol Chem, 22: 908–913.

Vethaak, A.D., J. Lahr, S.M. Schrap et al. 2005. An integrated assessment of estrogenic contamination and biological effects in the aquatic environment of The Netherlands. Chemosphere, 59: 511–524.

Viganò, L., A. Arillo, A. Bo Hero et al. 2001. First observation in intersex cyprinid in the Po River (Italy). Sci Total Environ, 269: 189–194.

Viganò, L., A. Mandich, E. Benfenati et al. 2006. Investigating the estrogenic risk along the river Po and its intermediate section. Arch Environ Contam Toxicol, 51: 641–651.

Vine, E., J. Shears, R. van Aerle et al. 2005. Endocrine (sexual) disruption is a prominent feature in the pike (*Esox lucius*) a top predator living in waters. Eniviron Toxicol Chem, 24: 1436–1443.

Vivekanandan, E. 1977. Ontogenetic development of surfacing behaviour in the obligatory air-breathing fish *Channa* (= *Ophiocephalus*) *striatus*. Physiol Behav, 18: 559–562.

Weber, L.P., R.L. Hill, Jr. and D.M. Janz. 2003. Developmental estrogenic exposure in zebrafish (*Danio rerio*): II. Histological evaluation of gametogenesis and organ toxicity. Aquat Toxicol, 63: 431–446.

Wibe, A.E., G. Rosenqvist and B.M. Jenssen. 2002. Disruption of male reproductive behavior three-spine stickleback *Gasterosteus aculeatus* exposed to 17β-estradiol. Environ Res, 90: 136–141.

Willey, J.B. and P.H. Krone. 2001. Effects of endosulfan and nonylphenol on the primordial germ cell population in pre-larval zebrafish embryos. Aquat Toxicol, 54: 113–123.

Yokota, H., M. Seki, M. Maeda et al. 2001. Life-cycle toxicity of 4-nonylphenol to medaka (*Oryzias latipes*). Environ Toxicol Chem, 20: 2552–2560.

Yambe, H., M. Shindo and F. Yamazaki. 1999. A releaser pheromone that attracts males in the urine of mature female masu salmon. J. Fish Biol, 55: 158–171.

Yamamoto, T. 1969. Sex differentiation. *In*: W.S. Hoar and D.J. Randall (eds.). *Fish Physiology*. Academic Press, New York, USA, 3: 117–175.

Zakes, K.D. and Z. Zakes. 1999. The effect of 11 beta hydroxyandrostenedione on pikefish *Stizostedion lucipercs* (L.). Aquacult Res, 30: 731–735.

Zerulla, M., R. Länge, T. Steger-Hartmann, G. Panter et al. 2002. Morphological sex reversal upon short-term exposure to endocrine modulators in juvenile fathead minnow (*Pimephales promelas*). Toxicol Lett, 131: 51–63.

Zhang, K., Y. Wan, J.P. Giesy et al. 2010. Tissue concentrations of polybrominated compounds in Chinese sturgeon (*Acipenser sinensis*): origin, hepatic sequestration, and maternal transfer. Environ Sci Technol, 44: 5781–5786.

Zhong, X., Y. Xu, Y. Liang et al. 2005. The Chinese rare minnow (*Gobiocypris rarus*) as an *in vivo* model for endocrine disruption in freshwater teleosts: a full life-cycle test with diethylstilbestrol. Aquat Toxicol, 71: 85–95.

5

Hypoxia and Anoxia

Introduction

The great atmospheric oxygenation event occurred some 2.3 billion years ago. Since then the atmospheric oxygen began to dissolve in water and the normoxic waters contain 6–8 mg O_2/l (see Diaz and Breitburg, 2009). Hypoxia is defined as Dissolved Oxygen (DO) of less than 2.8 mg O_2/l (≈2 ml O_2/l or 91.4 MM) (Diaz and Rosenberg, 1995) and anoxia means no oxygen. The amount of oxygen contained in a unit of water is only 1/30th of the amount present in the same of air. Further, the rate of diffusion of oxygen in water is 10,000 times slower than that in air. As temperature or concentration of salt increases, the amount of oxygen that can be dissolved in water is decreased. Understandably, the terrestrial mammals can tolerate only a narrow range of oxygen levels but fishes may survive wider and rapid fluctuations in DO (see Nikinmaa and Rees, 2005). For example, (i) air-breathing fishes may switch to acquire atmospheric oxygen (e.g., *Channa striatus*, Pandian and Vivekanandan, 1976), (ii) many gill-breathing fishes undertake aquatic surface respiration (Kramer and McClure, 1982), a behavior, in which fishes respire water in the upper well-oxygenated layer of the water column (e.g., *Pseudocrenilabrus multicolor victoriae*, Chi-Corrie et al., 2008; Reardon and Chapman, 2010; see also Pandian, 2010, pp. 27–28) or (iii) hold air bubbles in the buccal cavity to aerate water passing through the gills (Val, 1995) or hold the developing embryos in a bubble nest in land, as in *Hepsetus odoe* (see Kramer, 1978) or in surface water, as in many Anabantids (e.g., *Betta splendens*, Kirankumar and Pandian, 2002). Others like the hypoxic-tolerant cyprinids are periodic breathers in normoxic water and have low mean arterial Po_2 values (e.g., 3kPa in resting goldfish), in comparison to salmonids, which continuously ventilate their gills until their arterial Po_2 approaches that of surrounding water (15 kPa) (see Nikinmaa and Rees, 2005). Understandably, the cyprinids are amenable to crowding in aquaculture farms and others like the goldfish survive in indoor aquaria even when water is not changed for long periods. However, chronic exposure to hypoxia impairs reproductive performance of the cyprinids. Obligatory coral-inhabitant fishes also display remarkable tolerance to diurnal changes in DO levels. During nights, the DO levels outside and in between corals falls to 50 and 20% of air-saturation, respectively (Fig. 5.1). Interestingly, obligate coral fishes like *Gobiodon*, *Paragobiodon* and *Caracanthus* withstand the nocturnal hypoxia down to 8% saturation in their microhabitat (Nilsson and Ostlund-Nilsson, 2008).

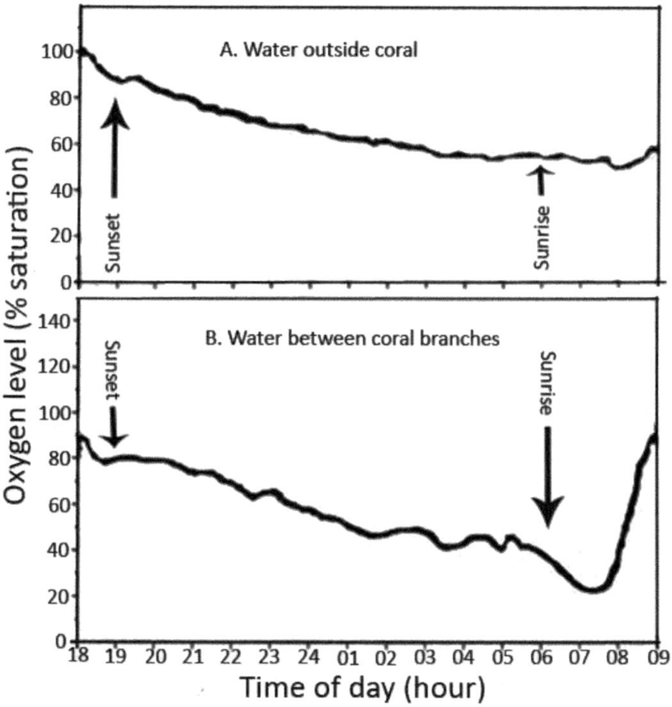

Fig. 5.1. Diurnal changes in oxygen levels outside the coral *Acropora nasuta* and in between the coral branches (source: Nilsson et al., 2004, modified).

5.1 Hypoxic habitats and fisheries

In aquatic habitats, wide diurnal fluctuations in DO is a regular phenomenon, due to photosynthetic release of O_2 by aquatic plants during day time (see Fig. 5.1). "In temperate areas, winter ice cover leads to seasonal hypoxia" (Nikinmaa and Rees, 2005). Upwelling areas can also develop extensive hypoxia, as the deep-water nutrients are added to the surface waters increasing production. Excessive input of nutrients, organics and others into water bodies with poor or limited circulation rapidly reduces DO in vast volumes of freshwater (e.g., Parvez et al., 2006) and seawater (e.g., Nixon, 1990; Wu, 1999) fast becoming hypoxic or anoxic. The global area, which is permanently hypoxic with 0.5 ml O_2/l in continental shelf and bathyal sea floor measures to over one million square kilometers (Helly and Levin, 2004). Woefully, the oxygen minimum and 'dead' zones are spreading (Diaz and Rosenberg, 2008) and expanding, especially in the tropical oceans (Stramma et al., 2008). The number of hypoxic aquatic systems has increased from < 50 in 1960 to about 400 at present (Diaz and Rosenberg, 2008). The seasonal hypoxia, which covers over 20,000 km^2 on the continental shelf of the Gulf of Mexico, did not occur annually prior to the 1950s (see Diaz and Breitburg, 2009).

Not surprisingly, not only the species richness but also the biomass of a fish species is significantly decreased in the hypoxic habitat; for example, the biomass and landing of the plaice and dab dramatically decrease from about 70 kg/hour catch in normoxic waters to about 15 and < 3 kg/hour catch in hypoxic waters containing 50 and 30% of the normoxic level, respectively (Fig. 5.2). Contrastingly, the complicated (e.g., Zhang et al., 2009) trends reported for the relative biomass of pelagic fishes indicate that there are two

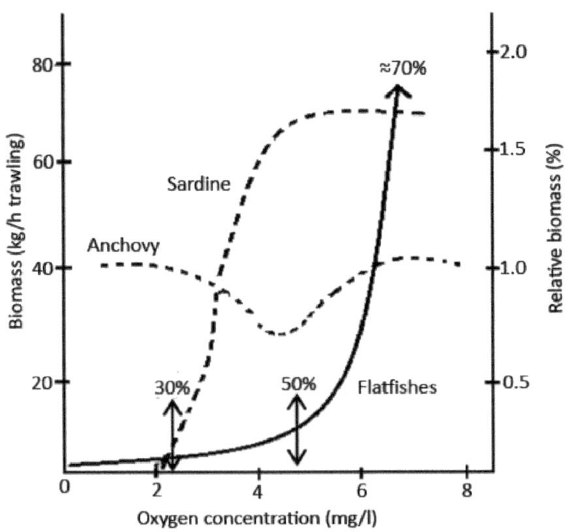

Fig. 5.2. The biomass of demersal flatfishes (kg/hour trawling) as a function of and oxygen concentration in the bottom waters of southeastern Kattegat during 1984–1990. The relative biomass of pelagic fishes as a function of oxygen concentration in southeastern tropical Pacific (compiled from Peterson and Pihl, 1995; Bertrand et al., 2011).

groups: 1. The sardine-like fishes, whose relative biomass remains equal in water containing oxygen saturation between about 40 and 70%. These fishes keep continuously moving to (i) irrigate their gills and (ii) filter-feed on plankton; they can adjust their swimming speed according to DO level. 2. The others like anchovies, in which the irrigation process of gills and feeding mode are separated, i.e., they may filter-feed or switch to particulate feed (see Pandian, 1975). With less than half of biomass in comparison to that of sardines (Fig. 5.2), they seem to possess an adaptive physiological mechanism, like the cyprinids and some coral fishes, to tolerate the hypoxic waters. The pelagic fishes seem to tolerate hypoxic waters containing 2.5–3.5 ml O_2/l (e.g., *Sardinops sagas*, *Katsuwonus pelamis*, *Engraulis capensis*, see Ekau et al., 2010). Clearly, DO levels have many implications to fishing and may serve as an important guide to trawlers for profitable capture of both demersal and pelagic fishes. However, more research is required on this aspect to promote the so called 'green fishing', i.e., capture more fishes at the reduced cost of fossil fuels.

5.2 Hypoxia and differentiation

The last two decades have witnessed fundamental and clinical research activity using zebrafish mutants as a model to study the hypoxia-induced gene expression (e.g., Nikinmaa and Rees, 2005), causing diseases like the Hippel-Lindau (VHL) tumor and others (van Rooijen et al., 2011). However, studies on hypoxia-induced disrupted sex differentiation and reproductive impairment in fishes are limited. The ensuing account is based on an excellent review by Wu (2009). Reproduction, development and sex differentiation involve a myriad of intricate process tightly controlled by hormones that are sensitive to environmental cues (Wu, 2009). In the long cascade of Hypothalamus-Pituitary-Gonad (HPG)-axis (Fig. 5.3), two biosynthetic processes vulnerable to hypoxia have thus far been identified: biosynthesis of (i) serotonin and (ii) p450 aromatase.

Lu et al. (2006, 2007) are perhaps the first to show sex dependent response to chronic exposure to hypoxia. They reared the marine medaka *Oryzias melastigma* exposing them to hypoxia (1.8 mg O_2/l, i.e., 28% saturation, see also Fig. 5.1) for a period of 3 months from 3–4 weeks old juvenile to sexual maturity. They observed a significant reduction in mRNA of tryptophan hydroxylase (5-HT), a rate limiting enzyme in synthesis of serotonin (Fig. 5.3) (see Pandian, 2013, p. 18) in the brain and FSH receptors in ovaries of females. However, no significant changes occurred in GnRH, GnRH receptors or FSH and LH of males. Artificial restoration of hypothalamic 5-HT levels recovered neuroendocrine function in the Atlantic croaker *Micropogonias undulatus* clearly indicating that the stimulatory serotogenic neuroendocrine pathway is the major site vulnerable to hypoxia-induced inhibition (Thomas

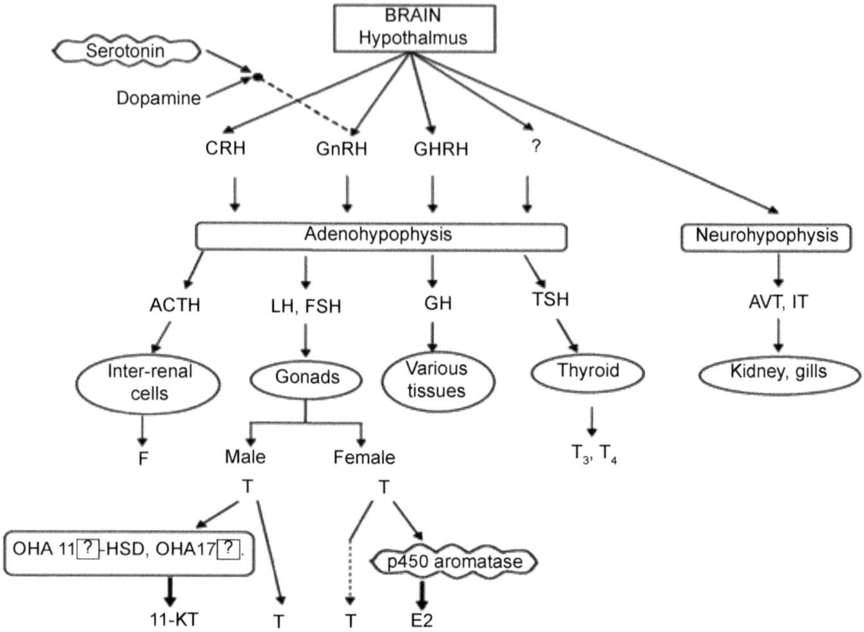

Fig. 5.3. Organization of the neuro endocrine cascade with major hypothalamic-hypophysial axis for the control of peripheral glands and tissues. Abbreviations: ? = unknown; CRH = corticotropin-releasing hormone; GHRH = growth hormone-releasing hormone; LH = luteinizing hormone; FSH = follicle stimulating hormone; TSH = thyroid stimulating hormone; ACTH = adrenocorticotropic hormone; GH = growth hormone; AVT = arginine-vasotocin; IT = isotocin; KT = 11-ketotestosterone; T = testosterone; P = progestogen; E_2 = estradiol; T_3 = triiodothyronine; T_4 = thyroxine. The vulnerable sites for hypoxia namely serotonin and p450 aromatase are shown by crinkled circles (© 2013 from Pandian, T.J. Endocrine Sex Differentiation in Fish, CRC Press, Boca Raton, reproduced by permission of Tayler and Francis Group, LLC, a division of Informa plc).

et al., 2007). From well designed study on tryptohan hydroxylase (Tph) of the brain of Nile tilapia, Sudhakumari et al. (2010) implicated a role for Tph in sex determination and differentiation (see also Pandian, 2013, pp. 18–19).

Steroidogeneis is another level vulnerable to hypoxia (Sanderson, 2006). Key enzymes in regulation of biosynthesis are (i) the first step in biosynthesis of steroidogenesis: cyp11a ($cyp450_{scc}$) converts cholesterol into progeneolone (see Fig. 5.4), and (ii) 3β-HSD (hydroxy dehydrogenases) are key enzymes regulating synthesis of sex hormones. Cytochrome p450 (cyp19) converts testosterone into estradiol. Its inhibition by androgen or aromatase inhibitors disrupt the sex differentiation process from genotypic females into phenotypic males in many fishes (e.g., *Oncorhynchus tshawytscha*, Piferrer et al., 1993, *Paralichthys olivaceus*, Kitano et al., 2000, *Coryphopterus nicholsii*, Kroon and Liley, 2000, *Danio rerio*, Fenske and Segner, 2004). The

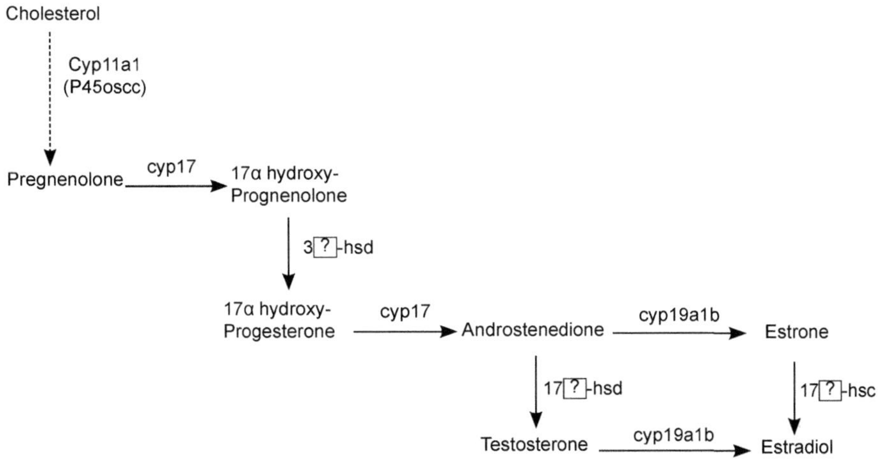

Fig. 5.4. Typical biosynthetic pathways of steroidogenesis in a teleostean fish (source: Diotel et al., 2011, modified).

cytochrome p450 enzymes demand oxygen (Nishimura et al., 2006). By disrupting normal steroidogenesis and by altering testosterone/estradiol (T/E_2) ratio, hypoxia interferes with sex differentiation (Wu, 2009).

In zebrafish, a secondary gonochore, all the presumptive males initially pass through a non-functional intersex stage, subsequently, an ovarian stage from 10–12 dpf onwards and finally the testicular development is initiated between 15 and 25 dpf in the presumptive males (see Pandian, 2013, pp. 52–55). Shang et al. (2006) exposed blastulae (30% epiboly) of zebrafish to normoxic (5.8 mg O_2/l) and hypoxic (0.8 mg O_2/l) waters at 28.5°C for 120 days. Hypoxia disrupted steroid sex hormones as early as 48 hpf, long before the commencement of sex differentiation (Shang and Wu, 2004). However, there was no mortality following 7 days exposure to hypoxia. Hence any alteration in sex ratio may not be due to sex-dependent mortality. In the 10-dpf zebrafish, hypoxia reduced specifically the biosynthesis of 3β-HSD, cyp11a and cyp19b and significantly affected oxygen-demanding key enzymes, which regulate steroidogenesis. Consequently, the levels of T, E_2 and 11-KT were significantly decreased. In *P. multicolor victoria*, T level is higher in hypoxic population than that in normoxic population, although E_2 level remain equal in both populations (Friesen et al., 2012). With the down-regulation of *cyp19a* and the consequent T/E ratio, hypoxia disrupted sex differentiation resulting in 74% males in the hypoxic group, in comparison to 62% males in the normoxic group. Similarly, 77% of genotypic XX females of medaka, following exposure to hypoxia, developed into phenotypic males (Cheung and Wu, 2006). Hence hypoxia is regarded as an endocrine disrupter.

5.3 Hypoxia and impairment

The synthesis of sex steroid hormones requires molecular oxygen (Raff et al., 2006). Understandably, hypoxia disrupts levels of sex hormones, vitellogenin (E_2) and triiodothyronine (T_3). For example, hypoxia (0.8 mg O_2/l) down-regulated significantly 3β-HSD, cyp11a and cyp19b in 10-dpf zebrafish (Shang et al., 2006). Similarly, chronic exposure to hypoxia (38% saturation) also significantly reduced E_2, T and 11-KT in the Atlantic croaker *Micropogonias undulatus* (Thomas et al., 2006, 2007). Thomas and Rahman (2011) undertook a field cum laboratory study on hypoxia-induced reproductive impairment in a marine fish *M. undulatus* (Table 5.1). In males and females of the croaker, hypoxia reduced GnRH mRNA level, a measure of neuroendocrine function. In turn, it reduced ovarian mRNA and plasma vitellogenin. Consequently, the development of 80–90% oocytes was inhibited at peri-nuclear stage, resulting in considerable reduction in GSI and fecundity. Besides, 21% females also underwent ovarian masculinization. Exposing carp *Cyrinus carpio* to hypoxia (1 mg O_2/l, ≈12% saturation) for 8 weeks, Wu et al. (2003) were perhaps the first to make a comprehensive study on the hypoxia-induced impairment at successive stages of reproduction (Table 5.2). There are also reports confirming the findings of Wu et al. (2003); for example, the daily egg production, following exposure to hypoxia (19% saturation) for 30 days was reduced to about 5% in *Fundulus grandis* (Cheung and Wu, 2006). Briefly, hypoxia reduced species richness and biomass of individual species by retarding gonadal development, reducing sperm and egg production as well as reducing fertilization and hatching success.

Table 5.1. Hypoxia (< 2.0–3.5 mg/l) induced reproductive impairment in Atlantic croaker *Micropogonias undulatus* from normoxic and hypoxic waters of Louisiana coast and Aranas Laboratory, USA. *approximate values (compiled from Thomas and Rahman, 2011).

Parameter	Normoxic water	Hypoxic water
Sex ratio	0.52♀ :0.48♂	0.39♀ : 0.61♂
	Female	
GSI	5.8	1.5
Fecundity (egg/female)	66150	1300
Ovarian masculinization (%)	0	21
Reductions in		
*brain aromatase mRNA	1.0	0.5
*ovarian aromatase mRNA	1.0	0.5
*plasma vitellogenin (ng/ml)	1.4	0.6
	Male	
GSI	3.5	0.6
Testicular tubule diameter (μm)	406	185
Relative sperm production (%)	75	10

Table 5.2. Hypoxia-induced reproductive impairment in adult carp *Cyprinus carpio* exposed to hypoxia (source: Wu et al., 2003). GSI = Gonado somatic index.

Parameter	**Impairment**
GSI	Reduced to 60% and 67% in males and females, respectively
Oogenesis	Yolk deposition impaired. 83% and 17% oocytes attained stage III and IV, respectively. Almost no oocyte attained stage V
Spermatogenesis	47% and 37% reduction in spermatocytes and spermatids, respectively
Spermiation	Reduced to 8%
Spawning*	Reduced to 53%
Fertilization success	Reduced to 56%
Hatching success	Reduced to 4%

*Information from Landry et al. (2007) for *Fundulus grandis* is also included.

5.4 Anoxia

Barring the air-breathing fishes belonging to 400–500 species (Graham, 1997), no gill-breathing fish is known to survive and flourish in anoxic waters. In zebrafish, not known to suffer from anoxia in its normal natural habitats, the responses to anoxic exposure vary; on exposure to anoxia from blastula to segmentation stage, 83–99% embryos survive for 24 hours. However, the survival begins to dramatically decrease to 64 and 4% during the early and middle straightening stages, respectively and finally to 0% in hatchling (Padilla and Roth, 2001). Early embryos of fishes are also known to withstand anoxic condition prevailing in ephemeral ponds at least for a few days. For example, the killifish *Austrofundulus limnaeus* has evolved a life history strategy similar to many aquatic invertebrate species, which includes the production of drought-tolerant diapausing embryos deposited into the muddy pond substratum (see Pandian, 2010 , p. 140). These embryos have developed tolerance to anoxia, and as they progress and enter into diapause II, during which metabolism is depressed by a coordinated down regulation of energy producing and energy consuming pathways, especially protein synthesis and ion pumping. Withstanding anoxia, they are able to completely recover after the 24 hours anoxia exposure for 30–40 days but succumb to it by the 60th day of exposure. The cause for their death is traced to the accumulation of toxic lactate and others like succinate and alanine (Podabsky et al., 2007), as in other invertebrates (see also Pandian, 1975). Understandably, eggs of many fluvial diadromous galaxiid species undergo development on land but hatch on being submerged by elevated stream level (McDowall and Charteris, 2006).

References

Bertrand, A., A. Chaigneau, S. Peratilla et al. 2011. Oxygen: A fundamental property regulating pelagic ecosystem structure in the coastal south eastern tropical Pacific. PLoS ONE, 6(12): e29558. Doi 101371/journal.pone.0029588.

Cheung, H.Y. and R.S.S. Wu. 2006. Effects of hypoxia on sex determination and differentiation. *In*: SETAC Asia/Pacific 2006. Growth with limit: The integration of Ecosystem Protection for Human Health Benefits. September 2006, Beijing, Abstract B4-2.

Chi-Corrie, L., L.J. Chapman and E.E. Reardon. 2008. Brood protection at a cost: mouth-brooding under hypoxia in an African cichlid. Env Biol Fish, 82: 41–49.

Diaz, R.J. and R. Rosenberg. 1995. Marine benthic hypoxia: a review of its ecological effects and the behavioural responses of benthic macro fauna. Oceanogr Mar Biol, 33: 245–303.

Diaz, R.J. and R. Rosenberg. 2008. Spreading dead zones and consequences for marine ecosystem. Science, 321: 926–929.

Diaz, R.J. and D.L. Breitburg. 2009. The hypoxic environment. *In*: Hypoxia, *Fish Physiology*, 27: 1–25.

Diotel, N., J.-L. Do Rogo, I. Anglade et al. 2011. Activity and expression of steroidogenic enzymes in the brain of adult zebrafish. Eur J Neuro Sci, 34: 45–56.

Ekau, W.H., H.O. Auel and D. Gilbert. 2010. Impacts of hypoxia on the structure and process in pelagic communities (zooplankton, macro-invertebrates and fish). Biogeosciences, 7: 1669–1699.

Fenske, M. and H. Segner. 2004. Aromatase modulation alters gonadal differentiation in developing zebrafish (*Danio rerio*). Aquat Toxicol, 67: 105–126.

Friesen, C.N., N. Aubin-Horth and L.J. Chapman. 2012. The effect of hypoxia on sex hormone in an African cichilid *Pseudocrenilabrus multicolor victoriae*. Comp Biochem Physiol, Part A, 162: 22–30.

Graham, J.B. 1997. *Air-Breathing Fishes*. Academic Press, San Diego, USA.

Helly, J.J. and L.A. Levin. 2004. Global distribution of naturally occurring marine hypoxia on continental margins. Deep Sea Res, Part 1 Oceanogr Res Pap, 51: 1159–1168.

Kirankumar, S. and T.J. Pandian. 2002. Effects on growth and reproduction of hormone immersed and masculinized fighting fish *Betta splendens*. J Exp Zool, 293: 606–616.

Kitano, T., K. Takemune, Y. Nagahamo et al. 2000. Aromatase inhibition and 17-α methyltestosterone cause sex reversal from genetic female to phenotypic males and suppression of p450 aromatase gene expression in Japanese flounder (*Paralichthys olivaceus*). Mol Reprod Dev, 56: 1–5.

Kramer, D.L. 1978. Terrestrial group spawning of *Brygon petrosus* (Pisces: Characidae). Copeia, 1978: 536–537.

Kramer, D.L. and M. McClure. 1982. Aquatic surface respiration, a widespread adaptation to hypoxia in tropical freshwater fishes. Env Biol Fish, 7: 47–55.

Kroon, F.J. and N.R. Liley. 2000. The role of steroid hormones in protogynesis sex change in black eye goby *Coryphopterus nicholsii* (Teleostei: Gobiidac). Gen Comp Endocrinol, 118: 273–283.

Landry, C.A., S.L. Steele, S. Manning and A.O. Cheek. 2007. Long term hypoxia suppresses reproductive capacity in the estuarine fish, *Fundulus grandis*. Comp Biochem Physiol, 148: 317–323.

Lu, S.X.Y., R.M.K. Yu, E.X.M. Chen et al. 2006. Molecular response to hypoxia along the HPG axis of the marine medaka *Oryzias melastigma*. *In*: SETAC Asia/Pacific 2006. Growth with a limit: The Integration of Ecosystem Protection for Human Health Benefits. September 2006, Beijing, Abstract B44.

Lu, S.X.Y., E.X.M. Chen, R.M.K. Yu et al. 2007. Molecular responses to hypoxia along the HPG axis of the zebrafish *Danio rerio*. *In*: Fifth Internatl Conf Mar Pollut Ecotoxicol June 2007, Hong Kong, Abstract 0-3.

McDowall, R.M. and S.C. Charteris. 2006. The possible adaptive advantages of terrestrial egg deposition in some fluvial diadromous galaxiid fishes (Teleostei: Galaxiidae). Fish Fisher, 7: 153–164.

Nikinmaa, M. and B.B. Rees. 2005. Oxygen-dependent gene expression in fishes. Am J Physiol Regul Integr Comp Physiol, 288: R1079–R1090.

Nilsson, G.E. and S. Ostlund-Nilsson. 2008. Hypoxic survival in two extremes: Under the ice and coral reefs. pp. 141–168. *In*: F. Sebert, D.W. Onyango and B.G. Kapoor (eds.). *Fish Life in Special Environments*. Science Publishers, Enfield.

Nilsson, G.E., J.-P. Hobbs, P.L. Munday and S. Ostlund-Nilsson. 2004. Coward or brave heart: external habitat fidelity through hypoxia tolerance in a coral dwelling goby. J Exp Biol, 207: 33–39.

Nishimura, R., Sakumoto, Y. Tatsukawa et al. 2006. Oxygen concentration is an important factor for modulating progesterone synthesis in bovine corpus luteum. Endocrinology, 147: 4273–4280.

Nixon, S. 1990. Marine eutrophication: A growing international problem. Ambo, 19: 101.

Padilla, P.A. and M.B. Roth. 2001. Oxygen deprivation causes suspended animation in the zebrafish embryo. Proc Natl Acad Sci USA, 98: 7331–7335.

Pandian, T.J. 1975. Mechanism of Heterotrophy. *In*: O. Kinne (ed.). *Marine Ecology*. John Wiley, 2(1): 61–250.

Pandian, T.J. 2010. *Sexuality in Fishes*. Science Publishers, Enfield, CRC Press, USA, p. 208.

Pandian, T.J. 2013. *Endocrine Sex Differentiation in Fish*. CRC Press, Boca Raton, USA, p. 303.

Pandian, T.J. and E. Vivekanandan. 1976. Effects of feeding and starvation growth and swimming activity in an obligatory air-breathing fish. Hydrobiologia, 49: 33–40.

Parvez, P., S. Pandey, M. Ali and S. Raisuddin. 2006. Biomarkers of oxidative stress in *Wallago attu* (Bl. And Sch.) during and after a fish kill episode at Panipet, India. Sci Tech Environ, 368: 327–363.

Peterson, J.K. and L. Pihl. 1995. Responses to hypoxia of plaice *Pleuronectes platesa* and dab *Limanda limanda* in the south east Kattegat. Distribution and growth. Env Biol Fish, 43: 311–321.

Piferrer, F., I.J. Baker and E.M. Donaldson. 1993. Effects of natural, synthetic aromatizable and non aromatizable androgens in inducing male sex differentiation in genotype female chinook salmon (*Oncorhynchus tshawystscha*). Gen Comp Endocrinol, 91: 59–65.

Podrabsky, J.E., J.P. Lopez, T.W.M. Fan et al. 2007. Extreme anoxia tolerance in embryos of the annual killifish *Austrofundulus limnaeus*: Insights from a metabolomics analysis. J Exp Biol 210: 2253–2266.

Raff, H., E.D. Bruder and St. Luke's Medical Center Adrenal Tumor Study Group. 2006. Steroidogenesis in human aldosterone-secreting adenomas and adrenal hyperplasia: Effects of hypoxia *in vitro*. Am J Physiol Endocrinol Metab, 290: E199–E203.

Reardon, E.E. and L.J. Chapman. 2010. Hypoxia and energetics of mouth brooding: Is parental care a costly affair? Comp Biochem Physiol Part A, 156: 400–406.

Sanderson, J.T. 2006. The steroid hormone biosynthesis pathway as a target for endocrine-disrupting chemicals. Toxicol Sci, 94: 3–21.

Shang, E.H.H. and R.S.S. Wu. 2004. Aquatic hypoxia is a teratogen and affects fish embryonic development. Env Sci Technol, 38: 4763–4767.

Shang, E.H.H., R.M.K. Yu and R.S.S. Wu. 2006. Hypoxia affects sex differentiation and development leading to a male dominated population in zebrafish (*Danio rerio*). Env Sci Technol, 40: 3118–3122.

Stramma, I., G.C. Johnson, J. Sprintall and V. Mohrholz. 2008. Expanding oxygen minimum zones in the tropical oceans. Science, 320: 655–658.

Sudhakumari, C.C., B. Senthilkumaran, K. Raghuveer et al. 2010. Dimorphic expression of tryptophan hydroxylase in the brain of XX and XY Nile tilapia during early development. Gen Comp Endocrinol, 166: 320–329.

Thomas, P. and M.S. Rahman. 2011. Extensive reproductive disruption, ovarian masculinization and aramatase suppression in Atlantic croaker in the northern Gulf of Mexico hypoxic zone. Proc R Soc B. doi: 10.1098/rspb.2011.0529.

Thomas, P., M.S. Rahman, J.A. Kummer and S. Lawson. 2006. Reproductive endocrine dysfunction in Atlantic croaker exposed to hypoxia. Mar Environ Res, 62: S249–S252.

Thomas, P., M.S. Rahman, I.A. Khan and J.A. Kummer. 2007. Widespread endocrine distruption and reproductive impairment in an estuarine fish population exposed to seasonal hypoxia. Proc Biol Sci, 274: 2693–2701.
Val, A.L. 1995. Oxygen transfer in fish: Morphological and molecular adjustments. Braz J Med Biol Res, 28: 1119–1127.
van Rooijen, E., K. Shanthakumar, I. Logister et al. 2011. A zebrafish model for VAL and hypoxia signaling. Methods Cell Biol, 105: 163–190.
Wu, R.S.S. 1999. Eutrophication, trace organics and water-borne pathogens. Pressing problems and challenge. Mar Poll Bull, 39: 11–22.
Wu, R.S.S. 2002. Hypoxia. From molecular response to ecosystem response. Mar Pollut Bull, 45: 35–45.
Wu, R.S.S. 2009. Effects of hypoxia on fish reproduction and development. Fish Physiol, 27: 79–141.
Wu, R.S.S., B.S. Zhou, D.J. Randall et al. 2003. Aquatic hypoxia is an endocrine disruptor and impairs fish reproduction. Environ Sci Technol, 37: 1137–1141.
Zhang, H., S.A. Ludsin, D.M. Mason et al. 2009. Hypoxia-driven changes in the behavior and spatial distribution of pelagic fish and macro zooplankton in the northern Gulf of Mexico. J Exp Mar Biol Ecol, 381: 580–591.

6

pH and Precipitation

Introduction

A water molecule may lose one of its hydrogen atoms to a neighboring one and the frequency of the loss of hydrogen atom in pure water at equilibrium is about one in 10 million (10^7). The resulting pair of charged particle, a hydrogen ion (H^+) and hydroxyl ion (OH^-) determines acidity and alkalinity, respectively. Even at a low concentration, they are of immense importance in biological systems. Most internal biological systems are highly sensitive to changes in H^+ ion concentration and function at best, when its concentration is in the range of 10^{-6}–10^{-8} molar or pH6–pH8. Biochemical processes occurring within intra- and extra-cellular compartments are protected against significant changes in pH by the presence of buffers or substances that minimize changes in H^+ and OH^- in solutions. By its change to carbonic acid or disassociation of carbonic acid, the bicarbonate, most commonly found in living organisms and in aqueous environment, accepts excess H^+ ions to prevent acidity or removes excess OH^- ions to prevent alkalinity (Eddy and Handy, 2012).

Carbon dioxide (CO_2) emission arising from the anthropogenic use of fossil fuels enters the surface ocean water from the atmosphere. Since pre-industrial times, the global aquatic surface pH has already decreased by more than 0.1 unit, at the rate of 0.02 units per decade during the last 30 years. The oceans have now absorbed about 30% of the CO_2 produced by anthropogenic activity. Thus they are progressively acidified; it is predicted that the decrease in global mean surface pH should touch pH 7.25 by 2050 primarily by the amount of atmospheric CO_2 absorbed by the ocean and not by the effects of climatic change. These apparently small decreases in pH result in a relatively large increase in the H^+ concentration of sea water causing substantial decreases in carbonate concentration (Denman et al., 2011).

6.1 pH and sex ratio

Fishes also thrive in waters of pH ranging from 4.5 to 9.5–9.8. For example, more than 10,000 t of endemic carp *Chalcaburnus tarichi* is captured from the Lake Van (3,738 km^2 area, 451 m depth at 1,650 m altitude in eastern Turkey), a lava-blocked lake with the pH as high as 9.5–9.8 and rich in sodium (Unal et al., 2007). On the other hand, Dedereu et al. (1986) reported the abundance and reproduction of the mudminnow *Umbra pygmaea* in the acidic waters with pH as low as 3.5–4.0 in the Netherlands. Rubin (1985) was the first to test the effect of pH on sex ratio in cichlid and poeciliid fishes. At low pH of 5–6, he found that the sex ratio was skewed in favor of males (up to 98%) but the ratio shifted to females (98%) at pH 7.8 in *Xiphophorus helleri*. Plasma cortisol level is reported to rapidly increase and peak in response to acidic stress in juvenile carp *Cyprinus carpio* exposed to pH 4.5 (Ikuta et al., 1997). Increase in cortisol level is known to increase male ratio in *Odontesthes bonariensis* (see Fig. 7.7, Hattori et al., 2009).

Reporting no reason for the selected pH range, Romer and Beisenherz (1996) reared the fishes at pH 4.5, 5.5 and 6.5—an unusually low pH to be tested. They reared eggs and larvae of 33 neotropical cichlid species belonging to the genus *Apistogramma*, one mouth-brooding African cichlid *Pseudocranilabrus multicolor victoriae* and a Jamaican poeciliid *Poecilia melanogaster* at 23°C. The experimental rearing period approximately corresponded with the normal sex differentiation period of these fishes. Among the fishes tested, the sex ratio of *A. borelli*, *A. eunotus*, *A. gaphra*, *A. longslot*, *A. meinkent*, *A. nijsseni* and *A. caetet* alone were affected by the tested pH. In the first six species *Apistogramma*, the male ratio decreased from about 50–60% at pH 4.5 to 40–50% at pH 6.5: this is quite opposite to the observation by Rubin (1985). Romer and Beisenherz did not explain how the tested pH influenced the sex differentiation process in these fishes. *A. caetet* may prove the ideal cichlid to study the role of pH in sex differentiation. Incidentally, decrease in pH from 8.2 to 7.1 in high density culture of *Dicentrarchus labrax* did not evoke any change in sex ratio (Saillant et al., 2003). To osmo-regulate their body fluids, marine fishes drink seawater. Drinking acidic/alkaline sea water may have a bearing on stomach acidity and pepsin digestion in marine fishes (see Pandian, 1975). Off from this, reduced pH ($\approx$ 6.0) also reduces sperm motility in *Merluccius australis* (Effer et al., 2013).

In an African cichlid *Pelvicachromis pulcher*, Reddon and Hurd (2013) investigated the effect of pH not only on male ratio but also male morph ratio. In this cichlid, males exist in two different morphs, which are readily distinguishable from their opercular color. The color of the male morph is genetically determined and fixed for life time (see Pandian, 2013, pp. 124–125). Yellow Opercular (YO) males are monogamous. But the Red Opercular (RO) males may breed monogamously or hold a harem of two or more females. Exposing the early developmental stage up to 30 dph either at pH 5.5 or 6.5, Redden and Hurd found that the male ratio increased from 0.4 at pH 6.5 to 0.6 at 5.5; pH 5.5 also induced a larger proportion of males to develop into RO morph from 18% at pH 6.5 to 67% at pH 5.5. Hence, the low pH induced not only sex reversal but also morph reversal in genetic females.

Baron et al. (2002) reared the female heterogametic molly *P. sphenops* at pH 6.23, 7.02 and 8.00. Figure 6.1 shows the male ratio as a function of pH at temperatures of 22.6, 26.0 and 29.7°C. At pH 7.00, male ratio was the lowest at the tested temperatures but it increased to 0.15–0.22 at pH 6.23 and to 0.24–0.41 at pH 8.00. Notably, the male ratio was lower (0.15–0.22) at the highest tested temperature (29.7°C) in all the tested pH range than those (0.21–0.49) at 26.0°C. This pattern of response of decreasing male ratio with increasing temperature is not in compliance with the generalization made by Ospina-Alvarez and Piferrer (2008). Fishes are considered to display a single sex ratio response pattern of increasing male ratio with increasing temperature. This has been discussed elsewhere (Chapter 7).

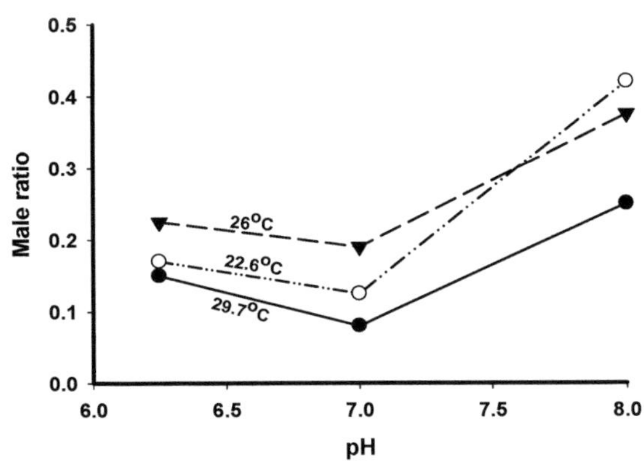

Fig. 6.1. Effect of pH at different temperature levels on male ratio of *Poecilia sphenops* (source: Baron et al., 2002).

6.2 Precipitation

Precipitation is not uniformly distributed either in time or space. However, monsoon, unique to tropical Asia, is temporally specific in its distribution pattern. The upwelling of nutrients, for example along Southwest coast of

India and productivity (e.g., freshwater productivity, Haniffa and Pandian, 1978) are bound by monsoon. Acidic rain water from montane terrains is neutralized during the course of its riverine passage (e.g., Murugavel and Pandian, 2000). In the Andaman Islands, agriculturists and aquaculturists neutralize acidic water by immersing injured banana stems; the oozing fluid from the injured banana stem neutralizes the acidic water.

In fishes, breeding is linked with the arrival of monsoon. For example, the response to hypophysation by female rohu *Labeo rohita* to spawning frequency, production of quality eggs and hatching success is significantly greater during monsoon than during the pre-monsoon season (Dasgupta et al., 2009). In a large country like India, the snakehead *Channa striatus* breed during different calendar months and facilitate the availability of seedling almost throughout the year (see Pandian, 2010, Fig. 36).

References

Baron, B.S., F.R. Buckle and S. Espina. 2002. Environmental factors and sexual differentiation in *Poecilia sphenops* Valenciennes (Pisces: Poeciliidae). Aquacult Res, 33: 615–619.

Dasgupta, S., S.K. Sarkar, N. Sarangi and S. Bhattacharya. 2009. Variation in spawning responses, eggs and larvae production from induced rohu (*Labeo rohita*) during pre-monsoon and monsoon seasons. Relationship with hormonal changes and oocyte responsiveness during final maturation. Aquaculture, 290: 320–326.

Denman, K., J.R. Chrishan, S.V. Steiner et al. 2011. Potential impacts of future ocean acidification on marine ecosystem and fisheries: Current knowledge and recommendations for future research. ICS J Mar Sci, 68: 1019–1029.

Dedereu, L.H.T., R.S.E.W. Leuven, S.E.W. Bonga and F.G.F. Oyen. 1986. Biology of the acid-tolerant fish species *Umbra pygmaea* (DeKay, 1842). J Fish Biol, 28: 307–327.

Eddy, F.B. and R.D. Handy. 2012. *Ecological and Environmentl Physiology of Fishes*. Oxford University Press.

Effer, B., F. Figueroa, A. Augsburger and I. Valdebenito. 2013. Sperm biology of *Merluccius australis*: Sperm structure, semen characteristics and effects of pH, temperature and osmolarity of sperm motility. Aquaculture, 408-409: 147–151.

Haniffa, M.A. and T.J. Pandian. 1978. Morphomentry, primary productivity and energy flow in a tropical pond. Hydrobiologia, 59: 23–48.

Hattori, R.S., J.I. Fernandino, A. Kishi et al. 2009. Cortisol-induced masculianization: Does thermal stress affect gonadal fate in pejerrey, a teleost fish with temperature-dependent sex determination? PLoS ONE, 4(8). doi:10.1371/journal:pone.0006548.

Ikuta, K., S. Kitamura, T. Yada et al. 1997. Development and assessment of system for the effects of acidification on inland water fish ecosystems in East Asia. Prog Rep Global Environ Res Acid-rain, 4: 62–66.

Murgavel, P. and T.J. Pandian. 2000. Effect of altitude on hydrology, productivity and species richness in Kodayar—a tropical peninsular Indian aquatic system. Hydrobiologia, 430: 33–57.

Ospina-Alvarz and F. Piferrer. 2008. Temperature-dependent sex determination in fish revisited: Prevalence, a single sex ratio response pattern and possible effects of climate change. PLoS ONE, 3(7): e2837. Doi: 10.1371/journal.pone.0002837.

Pandian, T.J. 1975. Mechanism of heterotrophy. *In*: O. Kinne (ed.). *Marine Ecology*. John Wiley, London, 2(1): 61–249.

Pandian, T.J. 2010. *Sexuality in Fishes*. Science Publishers, Enfield, CRC Press, p. 208.

Pandian, T.J. 2013. *Endocrine Sex Differentiation in Fish*. CRC Press, Boca Raton, USA, p. 303.

Reddon, A.R. and P.L. Hurd. 2013. Water pH during early development influences sex ratio and male morph in a West African cichlid fish *Pelvicachromis pulcher*. Zoology, 116: 139–143.
Romer, U. and W. Beisenherz. 1996. Environmental determination of sex in *Apistogramma* (Cichlidae) and two other freshwater fishes (Teleostei). J Fish Biol, 48: 714–725.
Rubin, D.A. 1985. Effect of pH on sex ratio in cichlids and a poeciliid (Teleostei). Copeia, 1985: 233–235.
Saillant, E., A. Fostier, P. Haffray et al. 2003. Effects of rearing density, size grading and parental factors on sex ratios of the sea bass (*Dicentrarchus labrax* L.) in intensive aquaculture. Aquaculture, 221: 183–206.
Unal, G., V. Turkoghu, A.R. Oguz and B. Kaptaner. 2007. Gonadal history and some biochemical characteristics of *Chalcaburnus tarichi* (Pallas, 1811) having abnormal gonads. Fish Physiol Biochem, 332: 153–165.

7

Temperature and Sex Ratio

Introduction

In vertebrates two general mechanisms of sex determination are recognized. They are (i) Genetic Sex Determination (*GSD*) and (ii) Environmental Sex Determination (*ESD*) (Crews and Bull, 2008). In *GSD* species, as in homeiothermic birds and mammals, sex is determined at fertilization and differentiation is driven by expression of a cascade of genes, whose expression is neither sensitive nor responsive to changes in the environment. In poikilithermic fishes and reptiles, sex is also determined by gene (e.g., *Dmy, Dmrt 1*) or combination of genes during embryonic and/or post-embryonic stage(s); in them the process of sex differentiation is, however, regulated by a host of genes, whose expression is sensitive and/or responsive to changes in environmental factors like temperature, pH and oxygen levels. As a consequence, environmental factors may more decisively regulate the sex differentiation process. The effects of changes in pH and oxygen levels are more important to aquatic fishes but may be less relevant to reptiles, whose eggs are incubated in land. The role of pH and hypoxia in disruption of genetic sex differentiation has already been discussed in the earlier chapters. Within Environmental Sex Differentiation (ESD), TSD is designated to convey Temperature-dependent and Thermo-labile Sex Differentiation. Unlike the usage by herpetalogists (e.g., Wibbels et al., 1994), the TSD is used here to convey temperature-dependent sex differentiation in fishes. To denote Genetic, Environmental and Temperature-dependent Sex Differentiation, the abbreviations, GSD, ESD and TSD are used to distinguish Genetic (*GSD*), Environmental (*ESD*) and Temperature-dependent (*TSD*) Sex Differentiation.

7.1 Identification of *GSD* and *TSD*

Ever since field and laboratory evidence for the occurrences of *TSD* in the Atlantic silverside *Menidia menidia* (Atherinidae) was reported by Conover and Kynard (1981), the *TSD* has been claimed in more than 59 species belonging to 13 families in four orders of fishes. Incidentally, the list of Ospina-Alvarez and Piferrer (2008) includes 33 species belonging to the cichlid genus *Apistogramma* alone. It excludes eels but includes sea bass, in both of which size is an important consideration in sex differentiation (see Chapters 1 and 3). Three patterns of shift in temperature-dependent sex ratio in fishes are recognized by Ospina-Alvarez and Piferrer (2008). For a change in temperature, male ratio increases with increasing temperature in *pattern* 1 (Fig. 7.1). But it decreases with increasing temperature in *pattern* 2. In *pattern* 3, it increases at extreme low and high temperatures. Patterns 1 and 2 are equivalent to patterns 1a and 1b of reptiles.

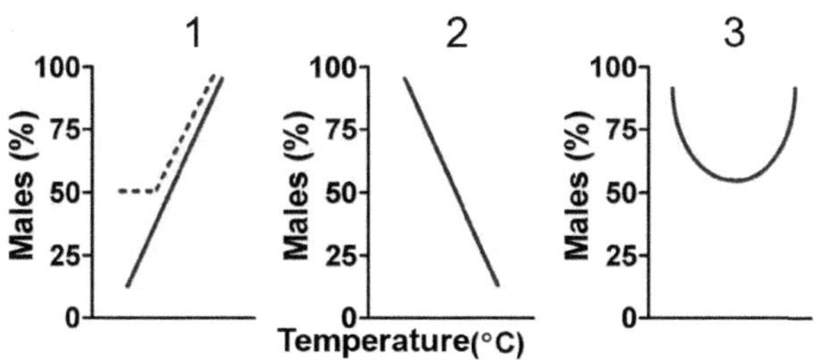

Fig. 7.1. ***Patterns*** of sex ratio response to change in temperature by fishes. ***Pattern 1***: With increasing temperature, percentage of males increases. ***Pattern 2***: With increasing temperature, percentage of males decreases. ***Pattern 3***: Percentage of males increases at the extreme low and high temperatures (permission by Dr. F. Piferrer from Ospina-Alvarez, N. and F. Piferrer, 2008. Temperature-dependent sex determination in fish revisited. Prevalence, a single sex ratio response pattern and possible effects of climate change. PLoS One, 3(7): e2837. doi: 10.1371/journal.pone 0002837).

To identify fishes truly having *TSD*, Ospina-Alvarez and Piferrer (2008) used the following three criteria and assigned the tested fishes to one of the three *patterns* after subjecting the reported data to selected statistical analyses. 1. Fishes with sex chromosome systems, constituting very strong evidence for the existence of *GSD* are excluded. It is extremely unlikely that 10 male heterogametic and three female heterogametic fishes have *TSD* (Table 7.1). 2. All those species, in which sex ratio shifts within the Range of Temperature-Sensitive Labile Period located within the larval Development stage (RTD) are considered to have *TSD*. RTD is located within the Range of Natural Temperature (RNT) in seasonally reproducing species of temperate latitudes. Incidentally, the temperature-sensitive labile period of many fishes (e.g., *Mugil cephalus*, *Siganus guttatus* and *Anguilla anguilla*) is located beyond

Table 7.1. ***Patterns*** of thermo-labile sex differentiation in gonochoric fishes (permission by Dr. F. Piferrer, from Ospina-Alvarez, N. and F. Piferrer. 2008. Temperature-dependent sex determination in fish revisited. Prevalence, a single sex ratio response pattern and possible effects of climate change. PLoS One, 3(7): e2837. doi: 10.1371/journal.pone 0002837.

Species	Sex chromosomes	Sex ratio shift within RTD	Diagnosis
	Male heterogametic		
Carassius auratus	Yes	No	***GSD + TE***
C. carassius	Yes	No	***GSD + TE***
Danio rerio?	Yes	No	***GSD + TE***
Pseudopleuronectes yokohamae	Yes	No	***GSD + TE***
Oncorthynchus nerka	Yes	No	***GSD + TE***
Gnathopogon caerulescens	Yes	Yes	***GSD + TE***
Misgurnus anguillicaudatus	Yes	Yes	***GSD + TE***
Oreochromis niloticus	Yes	Yes	***GSD + TE***
O. mossambicus	Yes	Yes	***GSD + TE***
Paralichthys olivaceus	Yes	Yes	***GSD + TE***
	Female heterogametic		
Ictalurus punctatus	Yes	No	***GSD + TE***
Oreochromis aureus	Yes	Yes	***GSD + TE***
Poecilia sphenops	Yes	Yes	***GSD + TE***
	Sex chromosomes not identified		
Dicentrarchus labrax	No	No	***GSD + TE***
Paralichthys lethostigma	No	No	***GSD + TE***
Sebastes schlegeli	No	No	***GSD + TE***
Verasper moseri	No	No	***GSD + TE***
Odontesthes hatcheri	No	No	***GSD + TE***
Menidia menidia	No	Yes	***TSD***
M. peninsulae	No	Yes	***TSD***
O. argentinensis	No	Yes	***TSD***
O. bonariensis	No	Yes	***TSD***
Limia melanogaster	No	Yes	***TSD***
Poeciliopsis lucida	No	Yes	***TSD***
Apistogramma spp. (33 species)	No	Yes	***TSD***
Hoplosternum littorale	No	Yes	***TSD***

the larval stage (see Pandian, 2013, Fig. 44). Hence the labile period rather than larval stage should have been considered. The fishes displaying shift in sex ratio within RNT but not within RTD (see Fig. 7.2B) are designated to have *GSD* + TE. 3. Those, in which sex chromosomes are absent but the shift in sex ratio occurs within RTD are designated as *TSD*.

Based on available information and considering the described three criteria, Ospina-Alvarez and Piferrer (2008) assigned 53 of 55, two of four and two of three species to *Patterns* 1, 2 and 3, respectively. According to them, *patterns* 2 and 3 may not exist in fishes. With reference to the initial exposure to low temperature inducing increased male ratio, i.e., the reported existence of *pattern* 2 in some fishes (e.g., *Ictalurus punctatus*,

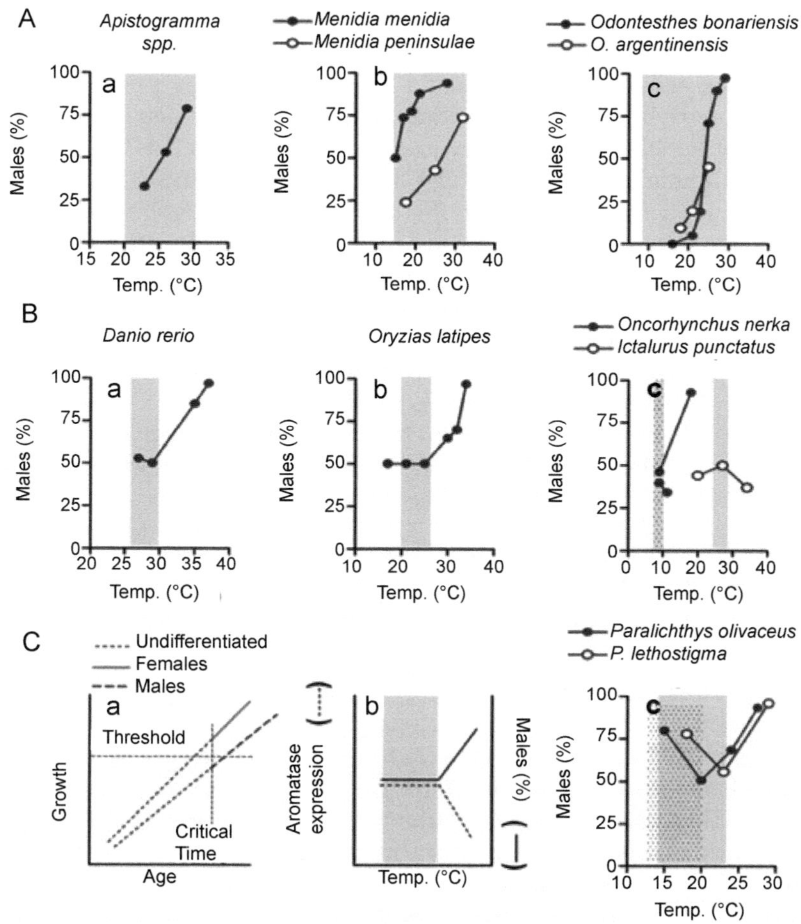

Fig. 7.2. *Patterns* of sex ratio response to temperature of fishes. A. Examples of *Pattern* 1 in which there are more males with increasing temperature. B. Examples for *TSD* in which sex ratio shifts only at extreme high temperatures and thus represent thermal effects on *GSD* (a, b). The fewer males at high temperatures, as in c is not supported by re-analysis of data. C. Formally proposed *Pattern* 3 in which more males occur at extreme low and high temperatures. However, these are proved to be wrong by subsequent publications (see Text) (permission by Dr. Piferrer, from Ospina-Alvarez, N. and F. Piferrer, 2008. Temperature-dependent sex determination in fish revisited. Prevalence, a single sex ratio response pattern and possible effects of climate change. PLoS One, 3(7): e2837. doi: 10.1371/journal.pone 0002837).

Color image of this figure appears in the color plate section at the end of the book.

Patino et al., 1996; *Oncorhynchus nerka*, Craig et al., 1996), Ospina-Alvarez and Piferrer (2008) provided the following explanation and evidence: (1) When the exposure was prolonged at low temperature, males became apparent in *I. punctatus* more due to slower growth (e.g., *Dicentrarchus labrax*,

Piferrer et al., 2005). (2) Additional experiments undertaken by Azuma et al. (2004) brought evidence for the existence of Pattern 1 in *O. nerka*. (3) The following analysis of Sullivan and Schultz (1986) may prove that the authors wrongly interpreted their data. Expectedly, the mean male ratio in strain 1 of *Poecilopsis lucida* increased from 0.41 at 24°C to 0.67 at 30°C. In strain 2, the mean ratio in Group 1 also increased from 0.37 at 24°C to 0.90 at 30°C. However, one male in this strain of Group 2A was allowed to pair with two females and the pairing produced progenies with the male ratios of 0.27 and 0.51 at 24°C. In Group 2B of this strain, the same male was paired with three other females yielding male ratios of 0.30, 0.43 and 0.68 at 30°. The respective means of Group 2A and Group 2B too suggest that the ratio mean of means increased from 0.37 at 24°C to 0.48 at 30°C. In fact, the results are a good example for the maternal effect on sex ratio rather than an insensitivity of a strain to temperature change.

Prior to conforming the generalization by Ospina-Alvarez and Piferrer (2008), caution may be exercised for following reasons: (1) For 29, 431 species of teleosts (see Pandian, 2011, p. 9), available information on changes in temperature altering sex differentiation in 59 species limited to 13 families in four orders is too small a sample, (2) *M. menidia* and *M. peninsulae* (Middaugh and Hammer, 1987) are the only two species, in which the existence of TSD has been demonstrated in the field. In all other species, data are reported from laboratory experiments designed to study the effects of temperatures, which the tested fishes may or may not experience in their respective natural habitats. (3) Evidence for the existence of TSD originates from a single laboratory study, e.g., 33 species of *Apistogramma* (Romer and Beisenherz, 1996), *Hoplosternum littorale* (Hostache et al., 1995). The male ratio of offspring of 32 *Apistogramma* species increased from 23°C to 29°C; however, *A. cacti* displayed a reversed response to temperature. (4) Male ratio of female heterogametic *Poecilia sphenops* decreases from about 0.2–0.4 at 26°C to 0.1–0.25 at 30°C in the tested pH range (Fig. 7.3) (Baron et al., 2002). Data reported for the male ratio of *P. melanogaster* by Khallaf et al. (2004) also confirm the observations of Baron et al. In *P. melanogaster*, the male ratio decreases from 1.0 at 19°C to 0.5 and 0.0 at 25.4 and 32°C, respectively, conforming *Pattern* 2. (5) In the Mexican poeciliid *P. lucida*, the evidence is not only from a single study but also from one strain that responded positively shifting sex ratio (Sullivan and Schultz, 1986). Incidentally, expressions of sex-linked allele like melanism may be constitutive (temperature-insensitive) in some populations but inducible (temperature-sensitive) in others (e.g., *Gambusia holbrooki*, Horth, 2006). Interestingly, populations of a single species with GSD and TSD are also known to co-exist (Conover, 2004). Remarkably, the TSD is reported to exist in the Carolina population but not in Nova Scotia population of *M. menidia* (Table 7.7) as well as in temperate population but not in sub-tropical population of *M. peninsulae*

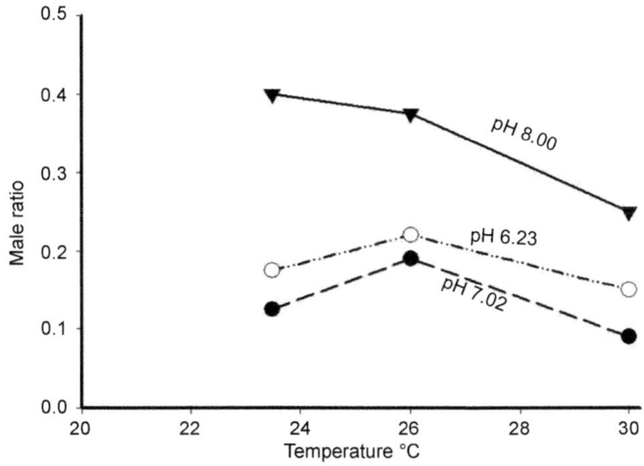

Fig.7.3. Effect of temperature at different pH levels on male ratio of *Poecilia sphenops* (source: Baron et al., 2002).

(p. 191). Further, as temperature-insensitive gene can selectively be inherited, the TSD may be lost within a few generations (Conover and Fleisher, 1986). Hence, TSD is not a fixed trait. At high temperature of about > 18°C, the male ratio may be increased in *Sander luciperca*, but post-pubertal maturation and gametogenesis do require lower temperature (Hermelink et al., 2013).

This presentation has clearly recognized that in all teleostean fishes, sex is determined by genes (GSD, see pp. 4–5, 257). However, a host of sex differentiation genes of fishes is sensitive and responsive to environmental factors. Within Sex Differentiation process there can be GSD, GSD + A and ESD (Fig. 1.1). Secondly, temperature is an important but one among the many environmental factors that disrupts GSD during the sex differentiation process. Hence, any hypothesis on temperature effects on sex ratio of fishes can be proposed only with reference to the possible sex reversal during the labile period in primary gonochoric fishes. In view of the fact that environmental factors like endocrine disrupting chemicals (EDCs including exogenous hormones and other compounds, Chapter 4, Pandian 2013, Chapter 7), hypoxia (Chapter 5), low pH (Chapter 6), and temperature can not induce sex reversal and hence alter sex ratio after the completion of the labile period prior to adult stage, i.e., RTD in primary gonochoric fishes (see Pandian, 2012, pp. 100–102). Hence it may be incorrect to propose that *Patterns* 2 and 3 do not exist in fishes. From the ensuing presentation, it may become obvious that the very criterion of a 'species' as a unit (see p. 181) chosen to assign the tested fishes to one of the three *patterns* is questionable. For, 1. There are temperature-sensitive and temperature-insensitive populations within a fish species, for example, *M. peninsulae* (p. 203) and

G. holbrooki (Horth, 2006, see also Table 7.5). 2. Within a population too, at least a few individuals both sexes are produced even at the most extreme temperatures encountered in nature (GSD + A, Fig. 1.1, see also p 202) and 3. There are a large number of non-additive parental genes that override temperature effects on sex differentiation and hence sex ratio. Therefor, the sex ratio response pattern of primary gonochoric fishes to a change in temperature or any other environmental factor during the labile period may vary from individual to individual, family to family, strain to strain as well as species to species. Hence, all of them are considered as criteria to redefine sex ratio response pattern to temperature change. Accordingly, *patterns* shall denote those of Ospinia-Alwarez and Piferrer (2008) but patterns represent those conforming the new definition.

7.2 Patterns 2 and 3

Whereas a single sex ratio response pattern 1 is known for many temperature-sensitive fishes, flounders and molly are reported to display a U-pattern of sex ratio response to changes in temperature. In the channel catfish *Ictalurus punctatus*, the ovarian differentiation becomes apparent on the 19th dph but that of testis only by the 90th dph. This prolonged temporal separation between the ovarian and testicular differentiation suggests that these two events may be mediated by independent signals (Patino et al., 1996). Besides, this catfish is not amenable to hormonal masculinization but readily undergoes paradoxical feminization (Goudie et al., 1983; Davis et al., 1992). Hence Patino et al. (1996) have investigated the temperature effects on sex differentiation in this catfish. With reference to the labile ovarian differentiation period, in a not well-designed experiment (cf. Kuhl and Brouwer, 2006), the catfish fry were initially reared at 27°C until the 10th dph, then at 20, 27 or 34°C from the 10th to 24th dph and finally all of them at 27°C until the 102nd dph. Thus the exposure of the fry to high temperature (34°C) covered the initial duration of the labile period of ovarian differentiation. But it did not last upto the 90th dph, when the commitment to the testicular differentiation is commenced. They found the sex ratio of the catfish exposed to 34°C for a short duration was skewed toward females, due to faster growth (?).

The southern flounder *Paralichthys lethostigma* is recorded to occur in temperature range from 5 to 35°C (Reagan and Wingo, 1985). In this temperature-sensitive flounder, Luckenback et al. (2003) exposed the juvenile flounders to 18, 23 or 28°C for a period of 245 days covering the entire stretch of the labile period. They reported the male ratios of 78% at 18°C and 96% at 28°C but 50% at the mid-temperature of 23°C. This U-pattern of sex ratio response to temperature changes, especially within the thermal range of natural habitat of the flounder is unique to fishes. Yamamoto et al. (1995) also reported the U pattern of response in its congener *P. olivaceus* with male ratios of 80% at 15°C, 51% at 20°C, 68% at 31.5°C and 93% at 27.5°C; Kitano et al. (1999) even obtained 100% male flounders at 27°C. The observations by Yamamoto et al. (1995, 1999) also confirm that in *P. lethostigma.* Patino et al. (1996) also brought evidence for the existence of pattern 2 in channel catfish *I. punctatus* with balanced sex ratio at low and intermediate temperatures and female-biased ratio at higher temperatures. However, these were considered by Ospinia-Alvarez and Piferrer (2008), who have brought explanation (see Fig. 7.2). Also the findings by Sun et al. (2013) support the hypothesis of Ospinia-Alvarez and Piferrer (2008). Sun et al. found the existence of pattern 1 in *P. olivaceus,* in which male sex ratio increased from 48% at 18–20°C to 61, 77 and 88% at 21, 24 and 28°C, respectively. However, it is not clear whether the threshold low temperature for increased male production in the southern and Japanese flounders lay between 15 and 18°C. If so, it is not clear why the sex ratio of Barfin flounder *Verasper moseri* is balanced at 14°C, while the acclimation of the flounder to 18°C within 62 days resulted in a all male population (Goto et al., 1999). Moreover, Baron et al. (2002) reported the highest male ratio in the female-heterogametic molly *Poecilia sphenops* at 26°C; but the ratio decreased at both 22.6°C and 29.7°C at all the tested pH from 6.23 to 8.0 (Fig. 7.3). This is an almost inverted U pattern of sex ratio response to changes in temperature. In the temperature sensitive European sea bass *Dicentrarchus labrax,* i.e., patterns 1 and 2 are shown to exist in the same species. Briefly, an increase in the male ratio is reported in a publication but decreased male ratio in another (Fig. 7.11).

Incidentally, Nomura et al. (1998) made an interesting study on temperature induced male ratio in gynogenic diploid loach *Misgurnus anguillicaudatus.* In normal diploids exposed to selected temperatures from the llth dph to 220–500th dph, the male ratio increased from 0.0% at 17°C to 100% at 30°C. When the treatment at 28°C was limited to different durations, the male ratio of gynogenic fry increased from 0.4 to 1.0 for the fry exposed for durations of 30 and 90 days, respectively. In the marbled sole *Limanda yokahamae* too, male ratio increased from 0.47 at 15°C to 0.82, when the sole was acclimatized to 25°C during the course of 115 days. But it decreased to 0.57 at 25°C, when the acclimatization period was prolonged to 144 days (Goto et al., 2000). Obviously, not only temperature but also acclimatization duration, presumably within the prolonged labile period, differently alters the male ratio. Hence, further investigations are required to confirm whether the sex ratio response pattern of flounders, molly, loach and channel catfish differs from the reported values.

7.3 Parental genetic role

A fact that has not been considered by Ospinia-Alvarez and Piferrer (2008) is the overriding paternal role on male ratio over the temperature effect.

As many as 52 publications have shown very interesting experiments to identify the paternal and maternal role in sex ratio overriding the temperature effect (Burt et al., 2011). These studies have demonstrated the paternal and maternal genetic traits overriding temperature effect per se at levels of individual (e.g., *Oreochromis niloticus*) and strain (e.g., *Oryzias latipes*). Surprisingly, a study has shown that the paternal trait has a greater overriding effect on temperature than the maternal trait in *Menidia menidia,* the first fish species to be identified to possess TSD.

The thermal sensitivity trait of Nile tilapia is not species specific but individual specific. Most male and female brooders generated highly temperature-sensitive progenies that differentiated into males, while others, for example in Manzala and Bouake (Table 7.2 marked by **) were

Table 7.2. Male ratios (%) of temperature (36°C)-treated and control (27–28°C) progenies form repeated matings of *Oreochromis niloticus* from Manzala, Bouake and Koka. * = significant difference, ** = insignificant difference.

Strain/Reference	Sire	Dame	Control (27–28°C)	36°C
Manzala, Tessema et al. (2006)	1	35	51	90*
	1	35	51	86*
	1	40	51	72*
	1	40	49	69*
	7	10	50	94*
	7	10	50	93*
	3	1	49	56**
	3	5	49	45**
	4	1	51	61*
	4	2	49	94*
	10	1	48	91*
Bouake, Baroiller and D'Cotta (2001)	1	2	64	73*
	1	2	64	76*
	2	3	58	80*
	2	3	65	83*
	3	4	49	63**
	3	4	50	59**
	4	5	59	65**
	4	5	60	65**
	1	2	70	82*
	1	3	68	73**
	3	2	67	98*
	3	3	56	85*
Koka, Bezault et al. (2007)	5	12	55	58**
	5	4	55	84*
	39	4	72	98*
	39	6	07	100*

insensitive to high temperature elevation and produced progenies with balanced sex ratio or with insignificant increase in male ratio. Clearly, the results summarized in Table 7.2 show the presence of overriding parental effects on temperature; in fact, the results reported by Baroiller et al. (1996), Baroiller and Clota (1998), Abucay et al. (1999), Baroiller and D'Cotta (2001), Tessema et al. (2006) and Bezault et al. (2007) reveal that the thermal sensitivity is more an individual specific trait; hence, there is a need to consider parental effect in the production process of monosex. Incidentally, parental effect on male ratio has also been reported in many fishes (e.g., goldfish, Fujioka, 2002; Goto-Kazeto et al., 2006). High temperature effects on sex differentiation have been studied in six wild populations of Nile tilapia adapted to different environmental conditions (Fig. 7.4). The existence of both sensitive and non-sensitive progenies in all the tested populations has been confirmed (Baroiller et al., 2009a).

The findings of Fujioka (2001) in *Gnathopogon caerulescens* are relevant to both inducing smaller body sized male differentiation and parental effect on sex ratio overriding temperature effects. From Table 7.3, the following may be noted: 1a. At 20°C, male ratio varies from 0.46 to 0.67 among progenies (Pair A, B and C) sired by the same father fertilizing the eggs of three mothers. 1b. However, the ratio at 20°C varied significantly from 0.50 to 0.89 among the progenies (Pair D and E) of the same mother, whose eggs were fertilized by different males. 2. The ratio did not differ

Fig. 7.4. Geographical map of Africa to show the locations of different lakes, in which the Nile tilapia is reported to flourish (source: Bezault et al., 2007, modified).

Table 7.3. Effects of parents and temperature on the male ratio of *Gnathopogon caerulescens*. Progenies of pairs A, B and C (in bold letters) arise from eggs of three mothers fertilized by one father. Progenies of pairs D, E and F arise from eggs of one mother fertilized by three fathers. *In these pairs alone, the observed increases in male ratio with increasing temperature are significant (source: Fujioka, 2001, modified).

Pair	**Temperature (°C)**			
	20	**25**	**30**	**34**
A	0.46	0.53	0.58	0.59
B*	0.67	0.74	0.90	0.91
C*	0.63	0.84	0.66	0.90
D*	0.50	0.72	0.74	0.83
E*	0.89	0.99	-	1.00
F*	-	0.83	0.91	0.86
Mean ♂ ratio	0.67	0.83	0.80	0.90
Mean size (cm)	3.2	3.3	3.2	2.5

significantly between the progenies of Pair A exposed to 20, 25, 30 and 34°C. This indicated the progenies of the Pair A inherited autosomal gene(s) responsible for insensitivity to temperature changes from one of the three females (GSD + A, see Fig. 1.1). Mean male ratio of the remaining pairs (B, C, D, E and F) significantly increased from 0.67 at 20°C to 0.90 at 34°C. 3. If smaller body size at the equal age is to induce male differentiation either at low or high temperature, it is not clear how the male ratio increased from 0.67 at 20°C to 0.80–0.83 at 25–30°C, despite the equal body size of 3.2 cm at 20 and 30°C. It is likely that sex differentiation induces size differences rather than size difference inducing sex differentiation (cf p. 213).

Continuing his findings in *G. caerulescens*, Fujioka (2006) undertook three series of experiments to demonstrate the overriding paternal genetic effect on temperature in sex differentiation. In series 1 (middle panel), the male ratios increased with increasing temperature, but remarkably at different levels, indicating the expression of different autosomal genes (GSD + A, Fig. 1.1) that modulate different sex ratios at the same temperature (Fig. 7.5). In series 2 (upper panel) too, the male ratio in the Lake Biwa, the natural habitat of the fish increased from about 0.5 during April-May to 0.68 during June; this observation parallels that of Conover and Heins (1987a) in that more females are produced during the beginning of the breeding season when the prevailing temperature is low (16.1–25.4°C) and more males during the late breeding season, when habitat temperatures are in the range of 20.5 to 28.2°C. More importantly, of the six pairs tested in series 3 (lower panel), progenies of the cross between DF1 ♀ x B5 ♂ showed balanced ratio at all the tested temperatures from 16 to 35°C. Apparently, the pair was insensitive to changes in temperature (GSD + A, Fig. 1.1). Interestingly, another pair did not produce males at any tested temperatures. Fujioka (2006) considered that the male (D4) was a sex reversed Δ neomale

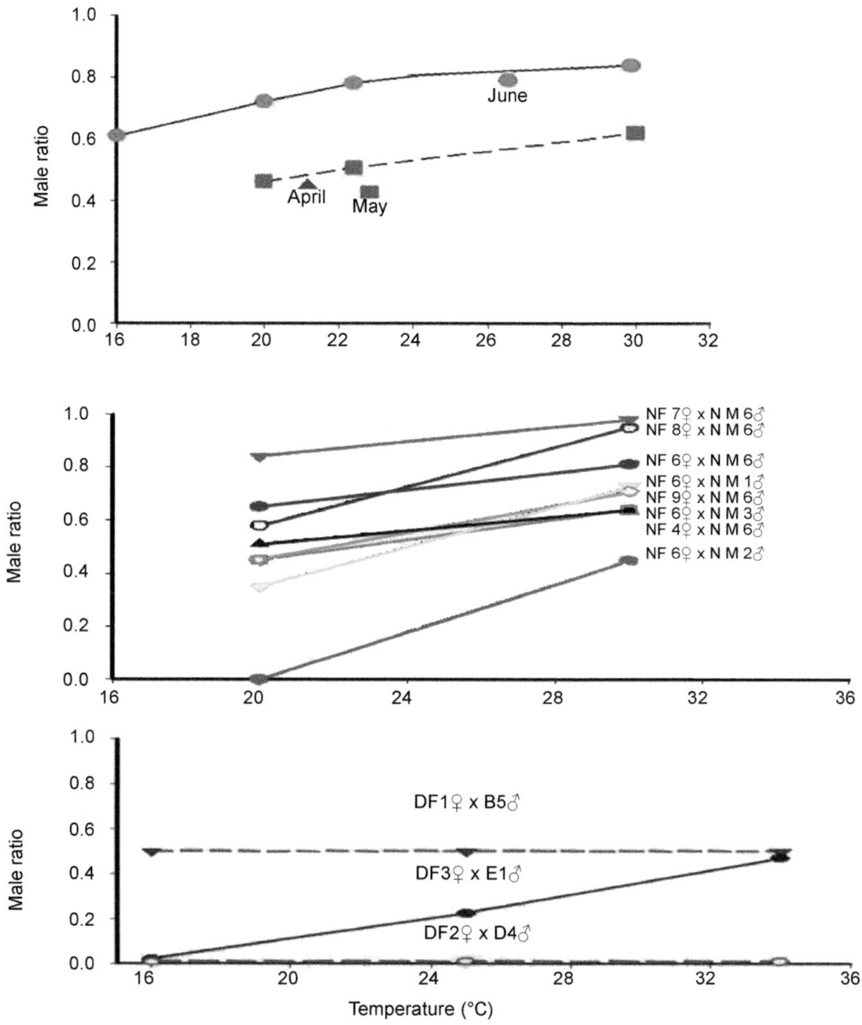

Fig. 7.5. Upper panel: male ratio of honmoroko *Gnathopogon caerulescens* in Lake Biwa during spring. Note the difference in male ratio response to increase in temperature by the progenies of early and late spring spawners. Middle panel: increasing trends in the male ratio of honmoroko with increase in temperature. Note the differences in the levels of responses by the progenies arising from different crosses. Lower panel: note that four of six crosses generated progenies that were insensitive to increase in temperature resulting in balanced sex ratio. One cross produced no male progenies, while the progenies of the sixth cross displayed increased male ratio with increasing temperature (compiled from Fujioka, 2006).

Color image of this figure appears in the color plate section at the end of the book.

carrying XX chromosomes. Abucay et al. (1999) also hinted that there were sex reversed XX Δ neomale and XY Δ neofemale Nile tilapias. This may have implications to the unexpected sex ratios observed in the catfish, flounders and molly, displaying pattern 2 and U-pattern 3. If Fujioka's (2006) suggestion is proved correct, which is likely (Winge, 1930; Aida, 1936), it may also clear apprehensions raised against the generalization of Ospina-Alvarez and Piferrer (2008) that fishes display only a single sex ratio response (Pattern 1) to temperature changes.

The Atlantic turbot *Scophthalmus maximus* do not have differentiated sex chromosomes (Bouza et al., 1994) and sex specific molecular markers (Husebye et al., 1994). Exposing the turbot fry (35 dpf) to 15, 17 and 24°C, Haffray et al. (2009) assessed the sex ratios in six families; unfortunately, the authors have not provided the actual data. However, it is possible to infer the following from their Table 2: 1. In three families, the sex ratio varied more widely but not significantly at 24°C than at 15 and 17°C. 2. In one family, the male ratio significantly increased from those at 15 and 17°C to those at 24°C. 3. In two families, the ratio significantly decreased from those at 15 and 17°C to those at 24°C. Presumably, sex ratio response to *Patterns* 1 and 2 seem to operate in different families of the turbot (GSD + A, Fig. 1.1).

Conover and Heins (1987a) undertook two series of diallel experiments in *M. menidia* collected from the Great South Bay, Long Island, New York: the first one involved two mothers (1, 2) and three fathers (A, B, C) and the second four mothers (3, 4, 5, 6) and three fathers (D, E, F). To keep uniformity, recalculated data on family male ratios are shown in Table 7.4. From the results, the authors concluded that the temperature-dependent sex differentiation has a genetic basis with (i) a highly significant paternal overriding effect on temperature effect and (ii) temperature effect on family sex ratio within a batch of eggs from each mother depends on the father. Further, (iii) fathers within mothers have a highly significant effect on the family ratio within each temperature. Sex ratio has a curvilinear response to temperature, suggesting the threshold response of sex ratio to temperature changes. The families show individualistic sex ratio response, displaying the increase in male ratio at different levels (GSD + A) with increasing temperature. Briefly, paternity has a stronger overriding influence on sex ratio response to temperature change. This observation may suggest that the sex chromosomes of *M. menidia* are perhaps on the verge of differentiation into male heterogametism. According to Conover and Heins (1987a), the results obtained by them amply demonstrated that the family sex ratios of *M. menidia* are clearly multimodal and suggest the expression of relatively a few major sex differentiation genes with non-additive (ef Kossowig, 1964) effect. "Thus the modes have a central tendency that corresponds with classic Mendelian ratios" (Conover and Heins, 1987a).

Table 7.4. Family male ratio (♂/♂+♀) of *Menidia menidia* among fathers within a single batch of eggs from different mothers. Progenies of each family were reared at 15 and 21°C. Data were subjected to multi-way contingency analysis to show the independent effect of father and temperature on male ratio within each mother. T = temperature, F = father, M = mother. (source: Conover and Heins, 1987a, modified).

Mothers	Fathers	15°C	21°C	Significance F vs. T	among Fs within T	among Ts within F
	A	0.44	-			
1	B	0.36	0.88	***	***	***
	C	0.42	0.64			
	A	0.48	0.74			
2	B	0.53	0.78	***	***	***
	C	0.28+	0.60			
	D	0.70+	0.75			
3	E	0.50	0.73	**	***	***
	F	0.43	0.78			
	D	0.50	0.54+			
4	E	0.46	0.58+	**	*	***
	F	0.40	0.69			
	D	0.61	0.72			
5	E	0.58	0.76	n.s.	n.s.	***
	F	0.50	0.79			
	D	0.61	0.70			
6	E	0.53	0.76	n.s.	n.s.	***
	F	0.50	0.73			

+ unexpected/unusual male ratios.

On the other hand, the female ratio of *Dicentrarchus labrax* depends strongly on the dame at low and high temperatures as well as at low and high rearing density. At high temperature, the sire-dame interaction was not significant, indicating that the female percentage is essentially additive (Saillant et al., 2002). Reared at low and high densities, Saillant et al. (2003a) also found the maternal effect, especially of one female on the female ratio of the progenies was very strong. A significant sire effect on female ratio was also found.

In medaka, *Oryzias latipes*, sex is determined by a single gene *Dmy* but its sex differentiation is derailed by temperature. Selim et al. (2009) reported the overriding effect of different strains on temperature. Females of Hd-rR strain are white in color (b/b, X^rX^r) and males are orange colored (b/b, X^rX^R). Hence sex can readily be identified. In *GfpTr* transgenics too, sex can easily be recognized by the presence of leucophores in XY male and green fluorescence in developing ovary in XX female. The HN1 is a wild strain. In two groups in each of these strains, hatchlings were continuously reared at 27°C or 32°C; in another two groups, one group was transferred from 27°C to 32°C and the other from 32°C to 27°C. They were transferred to selected temperature at the developmental stage 25 long before the 37th stage, when dimorphic proliferation of primordial germ cells (PGCs) in XX female and XY male became apparent.

At both 27°C and 32 to 27°C, no phenotypic sex reversal occurred in Hd-rR and *Gfp* Tg strains but it was 9 and 28% phenotypic males in the HN1 strain at 27°C and HN1 fry transferred from 32 to 32°C (Table 7.5). Both at 32°C and the HN1 fry transferred from 32 to 27°C, there were significantly more (> 55%) phenotypic males than in other strains. At 32°C, gonadal sex reversal occurred in 45% females in HN1 strain, against 15% in the Hd-rR strain. In the fry transferred from 27°C to 32°C, the observed difference per se was wider. With reference to gonadal sex reversal too, HN1 strain was more sensitive to temperature changes and transfer from 27 to 32°C. Briefly, the observations indicate that the wild HN1 strain is more sensitive to temperature changes than HdrR (Selim et al., 2009). Not only sensitivity to temperature shift varies from strain to strain but also sex within a strain. Sex determination and differentiation remained unaffected in XY medaka male embryos (stage 5) exposed to 17°C or 34°C. Conversely, the exposure of XX embryos at 5–6th stage from 27°C to 34°C induced sex reversal to phenotypic male upto 100% (Hattori et al., 2007).

Table 7.5. Effect of temperature treatment on sex reversal in genetic females of selected strains of medaka *Oryzias latipes* (source: Selim et al., 2009, modified).

Temperature	Phenotypic sex reversal (%)			Gonad sex reversal at 20 dph	
treatement	Hd-rR	GfpTg	HN1	HdrR	HN1
27°C	0	0	9	0	5
32°C	18.6	15.6	55.2	15	45
27°C to 32°C	3.8	3.7	58.2	5	75
32°C to 27°C	0	0	28.0	0	30

7.4 Insensitive fishes

Earlier an account was provided on the insensitive sex ratio response to temperature changes as an individual trait in *Oreochromis niloticus* and parental trait in *Gnathopogon caerulescens* and *Scophthalmus maximus*, maternal trait in *Dicentrarchus labrax* and paternal trait in *Menidia menidia*. Here temperature insensitivity at species level in fishes like *Hippoglossus hippoglossus* and *Thymallus thymallus* is described. In the context of climate change, these temperature-insensitive fishes are of great interest. Incidentally, the existence of temperature-insensitive fish species is another fact that was not considered by Ospina-Alvarez and Piferrer (2008).

Strussmann et al. (1996, 1997) were perhaps the first to discover that there can be a fish species, whose sex ratio is insensitive to changes in temperature. In natural populations of the South American pejerrey *Odontesthes hatcheri*, breeding occurs during a protracted spawning season in spring and again during a shorter period in autumn, thereby exposing its larvae to a wider range of temperature from 10 to 30°C. The pejerrey larvae survive between 13 and 27°C. Its sex is determined and differentiated earlier than its closely related thermo-sensitive atherinids like *O. bonariensis*. On treatment of *O. hatcheri* larvae at temperatures 20 and 28°C, the sex ratio remains around unity (see Fig. 7.9).

The Atlantic halibut *H. hippoglossus* attains sexual maturity after 4.5 years (55 cm TL, 1.7 kg) in males and 7 years (112 cm, 18 kg) in females (Jakupsstovu and Hang, 1988). Females grow much faster than males. To produce an all-female halibut, gynogenesis has been successfully induced and the induction study has also established female heterogamety (Tveldt et al., 2006). Exposing the halibut juveniles (3.8–5.0 cm) during the labile period (from 2.1 cm to 8.0 cm) to 7, 12 and 15°C, Hughes et al. (2008) found that the tested temperatures did not alter the sex ratio from unity. van Nes and Anderson (2006) too demonstrated a statistically non-significant trend toward male ratio, when the halibut juveniles were reared at 7, 10 and 13°C (see also van Nes et al., 2005). The halibut spawns in very deep cold waters, suggesting that larval and juvenile halibut may experience a more constant environment and may therefore rely solely on genetic sex differentiation. Clearly, the Atlantic halibut is one species, whose sex differentiation is insensitive to temperature changes.

To test the effect of temperature, Pompini et al. (2013) collected early as well as late spawners of grayling *T. thymallus* and brown trout *Salmo trutta* representing the genetic diversity within natural populations and exposed their progenies to warm and cold temperatures. Unfortunately, the authors have not reported the exact temperatures, to which the progenies of these fishes were exposed and the values obtained for the sex ratios. However, they reported that the "sex ratio of *T. thymallus* in all the 12 different thermal combinations did not significantly vary from unity". Also in *S. trutta*, the authors found "no evidence for temperature-induced sex reversal in any of the 14 ecologically relevant temperature experiments that the fishes were successfully exposed to over five consecutive years". The authors also found "no paternal or maternal effects on family sex ratio within samples of 26 genitors (6 females and 20 males)". Hence the sex ratio in these salmonids seems to be insensitive to cold and warm temperatures. Publications on insensitive response of sex ratio to temperature changes in fishes are beginning to appear and more of them are awaited.

7.5 Germ cells and aromatase

Germs cells: From their studies on the pejerries *Odontethes hatcheri* and *O. bonariensis* hatchlings exposed to 27 and 29°C, respectively for a shorter or longer duration, Strussmann et al. (1998, see also Strussman and Patino, 1995, 1999) were the first to show the high temperature induced deficiency in germ cells. In turn, the deficiency induced male differentiation or sterility. The process of differentiation of germ cells and germ cells supporting somatic cells (GCSSCs), though inter-related, is independent of each other. To begin with the somatic and germ cells components are known to possess bisexual potency (Nakamura et al., 2003). That the differentiation of somatic and germ cells are independent processes immediately suggests that they may (or may not) lose bisexual potency at different time points during ontogenetic development. This may support the conclusion of Pandian (2013, pp. 218–219) that gonadal differentiation means the loss of bisexual potency of GCSSCs in primary gonochoric fishes (Pandian, 2013, Fig. 5.3). However, bisexual potential of the germ cells is retained even after sexual maturity in them (Pandian, 2012, p. 100).

Continuing their work on dynamics of high temperature-induced deficiency of germ cells of *O. bonariensis* (Ito et al., 2002), Ito et al. (2003) reared hatchlings and juveniles, i.e., before and after the labile period from 5 to 7 wph (week post-hatching) at 17, 24 and 29°C; they also returned a group exposed to 29°C for 7 weeks to 24°C. Incidentally, histological sex differentiation has been observed during the 7th wph at 24°C but in the 4th wph at 29°C (Kurita, 1999) and functional sex differentiation has been shown to occur within 3–5 wph at 17°C but 1–4 wph at 27°C (Strussmann et al., 1997). Hence higher temperatures advance the events of both structural and functional sex differentiation processes.

The results reported for *O. bonariensis* by Ito et al. (2003, 2005) are simplified in Fig. 7.6. Their results may be summarized: 1. At 24°C, a 11-wph female had nearly six times more number of germ cells than a male. 2. At 24°C, gonadal area covered in a male was 1×10^4 µm^2, whereas that was as large as 5.7×10^5 µm^2 in a female (Fig. 7.6). 3. No obvious sexual dimorphism for thermal sensitivity was noted between putative male and female germ cells of the pejerry. Strikingly, sensitivity of putative female germ cells is quite a contrast to that in mammals. 4. Putative females reared at 17°C have no pyknotic or eosinophilic nucleated abnormal germ cells until the 7th wph. At 29°C, the abnormal germ cells per unit of gonadal area increased from about 0.5 in the 1st wph to about 1.25 during the 7th wph (see also Ito et al., 2008a,b); during the corresponding period, the number of normal germ cells, characterized by large, round and clear nuclei, decreased from about 0.40 to 0.25. The described histological features of abnormal germ cells are consistent with apoptosis (cf. Uchida et al., 2004) or

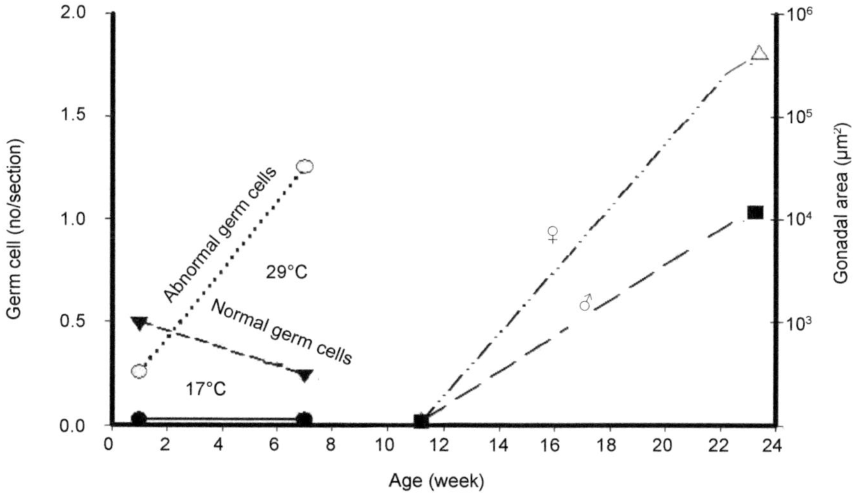

Fig. 7.6. Dynamics of germ cells in *Odontesthes bonariensis*. Left side: Note the decrease in number of normal germ cells and increase in the number of abnormal germ cells, when pejerrey is reared at 29°C. Note also no change in germ cell number in pejerry reared at 17°C. Right side: Gonadal area covered by female (Δ) and male (■) pejerries reared at 24°C (compiled from Ito et al., 2003).

programmed cell death, although these features alone may not fully explain the mechanism of degeneration of germ cells. For example, administration of exogenous estrogen induced feminization by completely preventing the male-producing temperature (MPT)-induced masculinization and reducing gonadal apoptosis (Yamamoto et al., 2013). 6. Surprisingly, the remaining 0.1 unit of germ cells in the gonad exposed to 29°C readily proliferated upon return to 24°C indicating that the germ cells were not completely damaged even after prolonged exposure at 29°C; however, they may do it over a longer period of exposure (cf p 187). 7. While pejerrey individuals left without remaining germ cells, as it happened in gonadectomized *Thalassoma bifasciatum* (see Pandian, 2013, p. 217), became permanently sterile. Not surprisingly, Cornejo (2002) recorded the occurrence of a large number of sterile *O. bonariensis*, wild-caught from Bueno-Aires lakes, in which wide oscillations in temperature is not uncommon. Incidentally, the dynamics and severity of temperature induced degeneration of germ cells in *O. bonariensis* were shown to depend on the magnitude of the thermal stress (Ito et al., 2008b). Interestingly, Karayucel et al. (2006) undertook progeny testing in temperature-treated guppy *Poecilia reticulata*; two of four males treated at 32–33°C did not produce any brood suggesting that they were sterile due to total loss of germ cells (see Strussmann et al., 1998).

Selim et al. (2009) estimated the number of germ cells in XX female and XY males of HdrR and HN1 strains of medaka on the 0 and 10th dph. From Table 7.6, the following may be summarized: 1. As expected the number of germ cells in males was less than that in females of both strains (see also Pandian, 2012, p. 47). The number of cells ranged from 75–300 in males but over a few hundreds to thousands in females. 2. On the 10th dph, the total number of germ cells in the HN1 strain was nearly 2–10-times more than those in Hd-rR strain. 3. On the same 10th dph, majority of the germ cells were also differentiated into one or more meiotic stages in females of either strain but in males, they remained undifferentiated or very few were differentiated.

Aromatase: The conversion of C_{19} androgens into C_{18} estrogens is facilitated by the enzyme complex comprising of mainly p450 aromatase. The

Table 7.6. Approximate number primordial germ cells PGCs on the 0 and 10th dph in Hd-rR and HN1 strains of *Oryzias latipes*. Numbers in bold letters indicate the germ cells undergoing meiosis (source: Selim et al., 2009).

Temperature treatment	Gonial cells (no)	
	0 dph	10dph
	Hd-rR strain	
XX at 27°C	140+**35**	700 + **900**
XX at 32°C	145	300
	125 + **5**	700 + **900**
XX at 27°C to 32°C	125	80
		700 + **1000**
XX at 32°C to 27°C	155	80
	145 + **10**	550 + **1000**
XY at 27°C	100	90
XY at 32°C	80	100
XY at 27–32°C	105	50
XY at 32–27°C	125	75
	HN1 strain	
XX at 27°C	140 + **30**	150
		1150 + **2000**
XX at 32°C	150	180
	120 + **5**	550 + **980**
XX at 27°C to 32°C	115 + **5**	150
		550 + **1000**
XX at 32°C to 27°C	160	100
	145+**5**	350 + **700**
XY at 27°C	155	300
XY at 32°C	105	200
XY at 27–32°C	155 + **10**	100
XY at 32–27°C	98	75

cytochromase p450 is the product of expression of the *cyp19A* gene (Simpson et al., 2002). Estrogens play an important role in the ovarian differentiation. Following the dimorphic proliferation of germ cells in *Oreochromis niloticus*, D'Cotta et al. (2001) observed strong expression of *cyp19A1* during the ovarian differentiation concomitant with high (465 fg/g) level of 17β-estradiol, in comparison to just 71 fg/g in the genetic male. High level (378 fg/g) of aromatase activity observed in the head/brain of a fry also established the dimorphic expression of *cyp19A1*. D'Cotta et al. (2001) also reported that the high temperature (35°C) repressed aromatase expression, for example to one third level in the brain aromatase activity. Karube et al. (2007) estimated the expression levels of *cyp19A1* in pejerrey *O. bonariensis* reared at the male-producing 29°C and mixed-sex producing temperatures of 24–25°C. RT-PCR analysis of these larvae revealed that *cyp19A1* expression at the mixed sex-producing temperature was not intermediate between male- and female-producing temperatures. But it was bimodal, with modal values seeming to match the values observed at 17 and 29°C. This observation suggests one of the two following possibilities: 1. The temperature influence on *cyp19A1* expression is not direct but mediated by the effects of other genes located upstream from *cyp19A1*. 2. The bimodal pattern of *cyp19A1* expression reflects the results of imperceptible differentiation earlier to the detectable histological differentiation. Research showing which of the possibility holds true may be rewarding. However, the observations by D'Cotta et al. and Karube et al. show that high temperature masculinizes genetic females by inhibition of estrogen synthesis.

Dietary administration of aromatase inhibitor like fadrozole results in repression of estrogen synthesis in the gonads and induces sex reversal in genetic females to phenotypic males in *O. niloticus* (Kwon et al., 1999, 2000) and *Danio rerio* (Uchida et al., 2004). For example, dietary administration of fadrozole to *O. niloticus* fry produced 99 and 100% males at 28 and 36°C, respectively (Kwon et al., 1999). Apparently, the response of genetic females to high temperature in these fishes is unidirectional, i.e., masculinization alone. Abucay et al. (1999) noted that the exposure of genetic Δ supermale YY Nile tilapia to high temperature unexpectedly induced paradoxical sex reversal in the opposite direction of feminization. Through a series of publications Kwon et al. (1999, 2000, 2002) confirmed that high temperature (36°C) treatment of YY Δ supermale fry induced significant differences in sex ratio, i.e., 100% males at 28°C but only 65% males at 36°C. Hence high temperature induced sex change in YY fry in the reverse direction. This observation may go against the hypothesis of Ospina-Alvarez and Piferrer (2008). Incidentally, survival and growth of these Δ supermales are lower than those of normal XY males (Mair et al., 1997), suggesting the loss of development stability in them. However, dietary administration of fadrozole to YY fry almost completely (♀ ratio = 0.01) suppressed feminization. This experimental evidence indicates that at the repression of aromatase activity and consequent excess amount of the substrate, androgens, triggered a negative feed-back system and rebalance the androgen-estrogen expression resulting in paradoxical feminization (see also Pandian, 2013, pp. 15–16).

Apoptosis: Uchida et al. (2004) were the first to show that aromatase inhibitor and high temperature blocked aromatase and eliminated oocytes by apoptosis. In the secondary gonochoric zebrafish, dietary administration of aromatase inhibitor fadrozole to genetic all-females at the dose of 100 or 1000 µg/g diet during the labile period from 15 to 40 dph produced 100% males. In these phenotypic males, apoptosis was found to have eliminated all the oocytes on the 40th dph. Similarly, gonadal masculinization of the genetic all-females increased from 0% at 28.5°C to 100% at 35°C. Apparently, high temperature, like fadrozole treatment, eliminated the oocytes by apoptosis. The process of apoptosis in some fishes like *Dicentrarchus labrax* is not complete resulting in the presence of scattered oogonia among testicular tissues (Saillant et al., 2003b). Incidentally, this observation may also explain the presence of undeveloped, previtellogenic or degenerated oocytes scattered within the testis of few cichlids (see Pandian, 2013, p. 137).

Among corticosteroids, cortisol is known to be a 'stress hormone' (see Pandian, 2013, Section 3.6). Stress interferes with the plasma level androgens, especially 11-Ketotestosterone (11-KT), as enzymes involved in their synthesis are also involved in glucocorticoid synthesis and inactivation (Berry and Grober, 2003). Cortisol-treatment of adult female *Gambusia affinis* during the post-labile period did induce the development of gonopodium-like structure but perhaps not testis (Knapp et al., 2012). However, the treatment of rainbow trout larvae, i.e., during the labile period with cortisol or cortisone inhibited the ovarian growth and caused an increase in male ratio (Van den Hurk and Van Oordt, 1985). Hattori et al. (2009) performed a series of experiments to assess the involvement of temperature stress as a modulator of gonadal sex differentiation and in particular as an agent of masculinization in the pejerrey *Odontesthes bonariensis*. In the first series, newly hatched larvae were reared at 17°C (females producing temperature, FPT), 24°C (mixed sex producing temperature MixPT) and 29°C (males producing temperature (MPT). In the second series, the larvae were reared at 24°C and fed with cortisol (0.4 mg/g diet) or Dexamthasone (0.8 mg/g diet). The final sex ratio in the first series was 100, 69 and 0% females at FPT, MixPT and MPT, respectively. In the first series, due to thermal stress, the whole body cortisol increased from about 9 ng/g at 17°C to about 24 ng/g at 29°C. Correspondingly, proportion of males also increased from 0% to 9 ng/g to 100% at ~24 ng/g at 29°C (Fig. 7.7). Analysis of mRNA transcript abundance showed that all the cortisol administered individuals displayed a male-like *amh* up-regulation and *cyp19a1a* down-regulation at

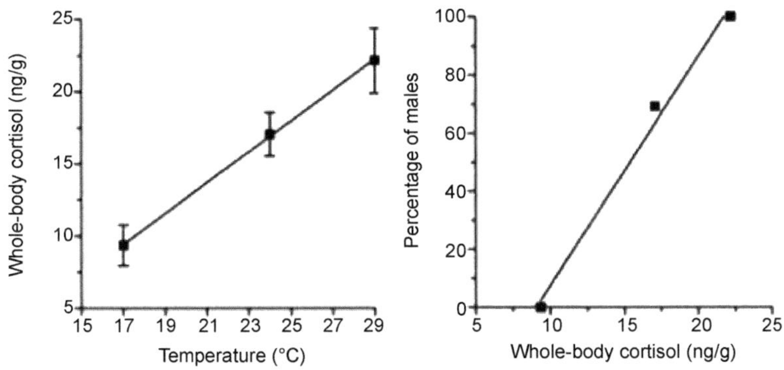

Fig. 7.7. Left panel: Effect of temperature on whole body cortisol in individual *Odontesthes bonariensis* larva during the labile period from 1st to 5th week after hatching. Right panel: Effect of whole body cortisol on percentage of males in *O. bonariensis* (from Hattori et al., 2009).

6 wph (week post hatching). Likewise, the TUNEL assay revealed a higher incidence of apoptosis in the gonads of cortisol treated fish, compared with controls at 6 wph. Clearly, high temperature-stress led to cortisol-induced masculinization of genotypic female pejerries into phenotypic males. Hence, cortisol plays an important role in high temperature-induced testicular differentiation in temperature-sensitive individuals and species of fishes, albeit during the labile period (Hattori et al., 2009; see also Yamamoto et al., 2013).

It is now possible to generalize the major pathway through which high temperature masculinizes the temperature-sensitive fishes. Figure 7.8 shows that in the temperature-sensitive genetic females, thermal stress-induced cortisol at temperatures is higher than the respective threshold temperature, which, in turn induces the development of abnormal germ cells and thereby reduces the number of normal germ cells, which, in turn, represses the *aromatase* expression, leading to the elimination of oocytes by apoptosis. The down-regulation of *cyp19A1* and up-regulation of *amh* facilitate testicular differentiation in phenotypic males. It is more likely that the processes involved in derailed sex differentiation of genetic females into phenotypic males may not act in a unidirectional pathway, as described in Fig. 7.8. They may act in multi-directional pathways more like a 'food web' rather than a 'food chain' (see Chapter 4).

7.6 Model fishes

Atherinidae is a primitive family. Following the seminal publication by Conover and Kynard (1981) reporting the existence of TSD in *Menidia menidia*, many contributions concerning the existence of TSD in silverside *M. peninsulae* (Middaugh and Hammer, 1987) on the North America Atlantic

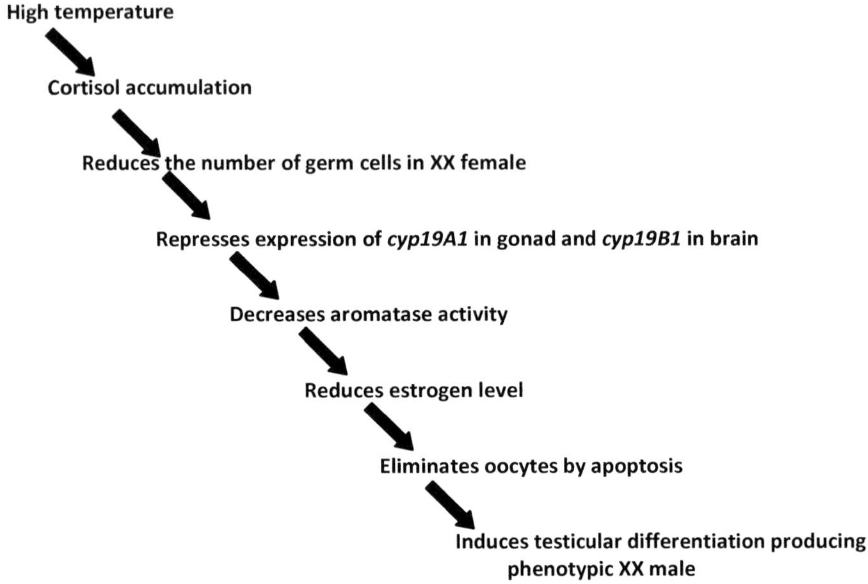

Fig. 7.8. Cortisol-germ cells-aromatase-apoptosis pathway, through which the elevated temperatures derail genetic sex differentiation of genetic females into phenotypic males in temperature-sensitive fishes.

Coast as well as the South American pejerries *Odontesthes argentienensis*, *O. bonariensis* and *O. hatcheri* have appeared, thanks to the groups led by D.O. Conover and C.A. Strussmann.

In natural habitats, most offspring produced by *M. menidia* during the early breeding season, when prevailing temperatures are low, differentiated into females. Conversely, most of those produced during late breeding season, when temperatures are higher, became males (cf. Fig. 7.5 upper panel). Hence females have a relatively longer duration to grow larger. Conover and Heins (1987b) found that the level of TSD is highly correlated with the length of the growing period (Table 7.7). Large females may have higher fecundity; however, the smaller size of a male may not affect its reproductive success. Hence the fitness differs with regard to size in males and females of *M. menidia*.

From an earlier publication (Conover and Heins, 1987a), it is noted: 1. At least a few individuals of both sexes are produced even at the most extreme temperatures encountered in nature (GSD + A, Fig. 1.1), suggesting that these individuals possess temperature-insensitive genes to realize genetic sex differentiation even at the temperature extremes. 2. There are large non-additive effects of parentage on family sex ratio and 3. Family sex ratios within temperatures tend to vary in a discrete Mendelian pattern. Hence

Table 7.7. Effects of temperature and growth period on male ratio of *Menidia menidia* at different locations in North America. T = temperature (compiled from Conover and Heins, 1987b).

Location	Latitude (N)	Male ratio at T (°C)		Growth period
		15	27	for ♀ (months)
Quebec	46° 10′	0.50	0.80	4
Nova Scotia	44° 10′	0.45	0.50	3
New York	40° 45′	0.50	0.75	5.5
Virginia	37° 33′	0.16	0.75	7
North Carolina	34° 43′	0.30	0.88	10
South Carolina	32° 35′	0.15	0.85	10

Conover et al. (1992) hypothesized the possibility of these temperature-insensitive genes of males and females being selectively inherited leading to the loss of TSD from the population after a few generations. To test this hypothesis, they reared four lines, two from New York and the others from South Carolina of *M. menidia* each at 17 and 28°C, considered to represent temperature extremes for *M. menidia*. As the offspring were hatched, grown and allowed to reproduce at constant temperatures of 17 and 28°C, the relative benefit of TSD affording longer growth period for females and shorter one for males was eliminated (see Conover and Heins, 1987b). All the four lines, that began with biased sex ratios, showed progressive increases in the 'minority sex' until a balanced sex ratio was established by the 6th generation at temperatures, at which each line was reared. Hence the effect of temperature on sex differentiation may not last beyond the life time of the the 6th generation of phenotypic individuals (Conover and Fleisher, 1986).

Whereas the distribution of *M. menidia* is limited to the temperate zone, that of *M. peninsulae* is extended from subtropics to temperate zone along the coast of Florida, USA. The transition from subtropics to temperate zone is followed by a shift from GSD to TSD, i.e., from temperature-insensitive populations to temperature-sensitive one (see also Horth, 2006). Along the coast of Florida TSD was detected in the Apalachee Bay population (29° 55′ N), wherein female ratio rapidly decreased with increase in temperature. The populations inhabiting southern coast of Florida between 24° 45′ N and 29° 05′ N were temperature insensitive and maintained balanced sex ratio at all temperatures ranging from 17 to 32°C. The shift from TSD to GSD with decreasing latitude occurs in the southern habitats, where the growing season is continuous and breeding occurs throughout much of the year (Yamashira and Conover, 2003).

The group led by C.A. Strussmann has elucidated the genetic-endocrine pathway of temperature effect on sex differentiation in the South American athernids, the pejerries. The major histological events of sex differentiation have been described for *O. argentienensis* (Strussmann et al., 1996) and *O. bonariensis* (Strussmann et al., 1997). In general, the ovarian differentiation precedes the testicular differentiation. With increasing temperature, the size and age at which sex differentiation occurs, are also advanced. In these pejerries, the male ratio increases with increasing temperature from about 0.05 at 17°C to about 0.55 at 25°C in *O. argentienensis*, this is also true of *O. hatcheri*. The ratio in *O. bonariensis* increases from almost 0.0 at 15°C to 100% at 29°C (Fig. 7.9). However, these values are unique

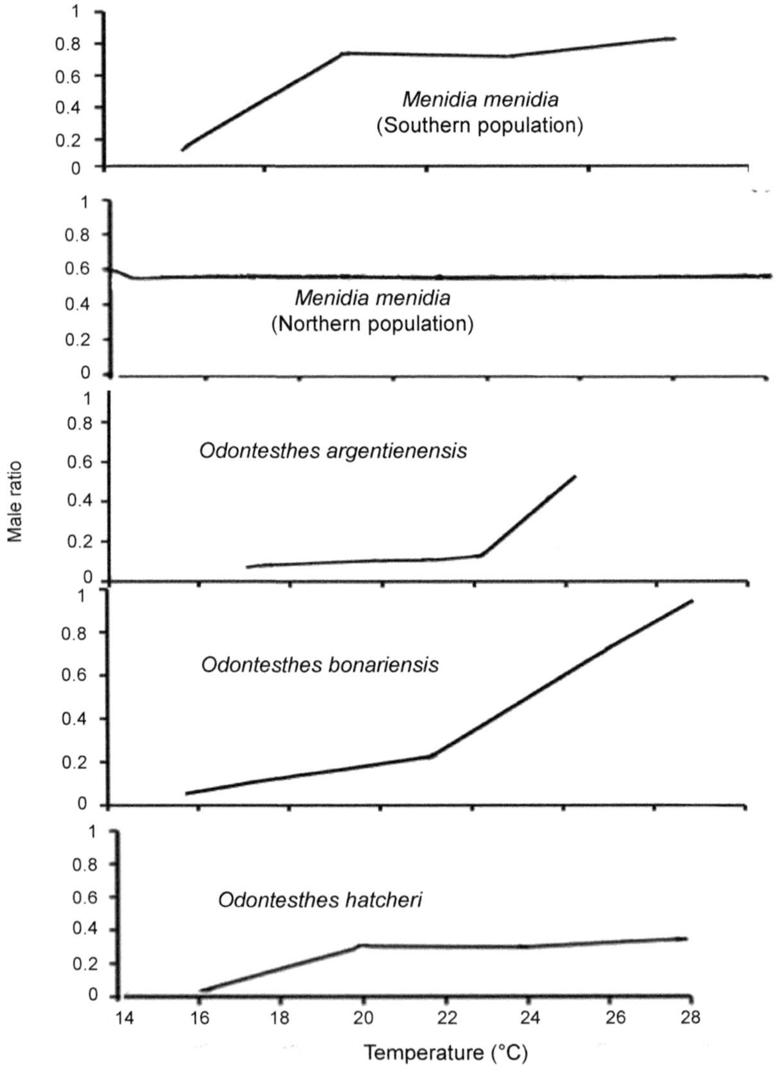

Fig. 7.9. Effect of temperature on male ratio of North and South American atherinids (source: Strussmann and Patino, 1995).

and difficult to comprehend; hence they call for more detailed studies. Not surprisingly, many publications are available on sex ratio of *O. bonariensis* (e.g., Strussmann et al., 1996; Fernandino et al., 2011).

A series of these publications have attempted to trace the temperature effects on the dimorphic expression of the proximate and peripheral hormones in the brain-pituitary-gonadal axis. Gonadotropins GtH1 and GtH2, known to be the earliest pituitary hormone to elicit gonadal differentiation, are shown to be associated with the initiation of gonadal differentiation in the pejerrey (Miranda et al., 2001, 2003). Shinoda et al. (2010) have reported a significant increase in expression of Follicle Stimulating Hormone Receptor (Fshr) gene at both 15–17 and 25°C in the female pejerrey. However, the Luteinising Hormone Receptor (Lhr) gene expression is low in the putative male. The authors suggest that temperature signals through differential expression of Fshr and Lhr affect the regulation of steroidogenesis during TSD process.

Blasco et al. (2010) have found a clear increase in *cyp11a1* in the gonads during the temperature-sensitive labile period but not in the trunk (brain) of the pejerry reared at the male-producing temperature (MPT). However, the dimorphic expression of *amh* seems to be more strongly associated with MPT than *cyp11a1*. The analysis of *amh* expression during gonadal differentiation of the pejerrey larvae reared at 17, 25 and 27°C pointed a clear temperature-dependent dimorphic expression pattern. Thus, the larva reared at MPT shows about two-times higher values than those at female-producing temperature (Fernandino et al., 2008). The higher expression of *pen-2* (gene known for regulation of apoptosis) and *ndrg* (gene known for regulation of spermatogenesis) in the gonads of the pejerrey larvae reared at the MPT has been confirmed by Fernandino et al. (2011). Most remarkably, the series of endocrine events thus far described occurs prior to histological sex differentiation supporting the Yamamoto's (1969) postulation that sex steroids are the inducers of natural sex differentiation.

Cichlidae, one of the most species-rich families of vertebrates, is composed of 3,000 species. About 1,000 cichlid species, representing 10% of the world's freshwater species, have arisen within the past million years in Lakes Malawi and Victoria. More than 1,500 cichlid species inhabiting the Great Lakes of eastern Africa are endemic. The cichlids are known for their impressive phenotypic diversity, explosive speciation rate and endemism (Allender et al., 2003). From these, tilapias, especially *Oreochromis niloticus* and *O. aureus*, are the second most farmed fishes with an annual production of 3.5 mt (Table 3.1). Remarkably, the Nile tilapia has successfully colonized a wide range of aquatic environments, varying from habitats with strong seasonal thermal oscillations to those with almost no oscillation. For example, the habitat (Fig. 7.4) ranges from West Africa alternating between 'hot' summer (28–34°C) and 'cold' winter (23–26°C) seasons

(Talling, 2001) to the Ethiopian high altitude (1,590 m) Lake Koka with constant low temperatures of 17–24°C (Admassu and Casselman, 2000) and to the hydrothermal springs (>40°C) (Trawavas, 1983) of Lake Metahara (955 m altitude) exclusively fed by resurgence of hydrothermal hot springs (43°C) (Bezault, 2005); the upper lethal limit of *Oreochromis* was fixed at 42°C by Denzer (1968). Not surprisingly, the largest number of publications concerning temperature effects on sex differentiation is available for the cichlids, especially *O. niloticus.*

The credit for identifying and establishing temperature as a major environmental factor that disrupts sex differentiation skewing it in favor of male ratio goes to Dr. J.F. Baroiller and his colleagues (Baroiller et al., 1995a,b, 1999, 2009a,b; Baroiller and Clota, 1998). Extending the findings of Baroiller et al. to another male heterogametic species *O. mossambicus*, Wang and Tsai (2000) have reported that male ratio increased to 75% at 28–32°C. That the male ratio increases with increasing temperature also holds true for the female heterogametic *O. aureus* (Desprez and Melard, 1998) and when subjected to cyclic thermal regime of 35°C during day and 27°C during night (Baras et al., 2000). The same is true of 33 species belonging to the cichlid genus *Apistogramma* (Romer and Beisenherz, 1996). Hence Baroiller's finding that increase in temperature increases male ratio may be a common trait of cichlids.

Almost all authors have found that an increase in temperatures, from ambient 25–28°C to 32–36°C increased male ratio to 89–98% in different strains of *O. niloticus*, e.g., Bouake, Baroiller et al. (1995a,b, 1996), Swansa, ICLARM and Ghana (Abucay et al., 1999), Koka, Metahara and Kpandu (Bezault et al., 2007) and Manzala (Baras et al., 2001) (Fig. 7.4). The threshold for the maximal thermal influence varies between species and strain; for example, it is ≈29°C for *A. trifasciata* (Romer and Beisenherz, 1996), ≈ 32°C for *O. mossambicus* (Wang and Tsai, 2000), 36.0°C for *O. niloticus* from Bouake (Baroiller et al., 1995b), 36.4°C for Swansa and Ghana (Abucay et al., 1999) and 37.0°C for Manzala (Baras et al., 2011). However, from their estimation of the sex ratio of *Poecilia reticulata* progenies treated at 19 to 35°C, Karayucel et al. (2006) noted that the male ratio gradually increased with increasing temperature rather than having a threshold response (see also Dzikowski et al., 2001). Incidentally, 6–39% of the treated Bouake phenotypic males possessed gonads essentially containing testicular tissue with scant oogonia (Baras et al., 2000).

Baroiller et al. (1995a) were the first to undertake progeny testing and show that the increase in temperature beyond the threshold level resulted in more and more genetic females differentiating into phenotypic males and thereby increasing male ratio. Male ratio of *O. niloticus* from Bouake increased from 53% at 27°C to 81% at 36°C. Four out of 10 randomly selected phenotypic males, when crossed with 'standard' females, sired progenies composed of an overwhelming majority of females. Hence higher temperatures disrupted sex differentiation in genetic females reducing the thermal-sensitive aromatase activity.

At high temperature, tilapia progenies are known to suffer from stunted growth. However exposure to cyclic thermal regimes between 27°C during night and 35°C during day may simulate more of the natural temperatures, to which the tilapias are exposed in their natural habitats. For example, the surface temperature of the Lake Koka varies between 21 and 26°C during the day and that of Lake Metahara between 32 and 39°C (Bezault et al., 2007). Exposing *O. aureus* to 35°C during day time and 27°C during night, Baras et al. (2000) found that male ratio increased to about 0.87; however, the cyclic thermal regime ensured higher survival and better growth of the blue tilapia.

Wessels and Horstgen-Schwark (2007) were the first to determine heritability of thermal sensitivity in Nile tilapia from Manzala. They provided evidence for thermal sensitivity as a quantitative trait. Covering two generations, around 14% of the tested families were selected to produce two distinct high and low lines. They realized heritability was 0.69 in the high line and 0.86 in the low line. In *Oncorhyncus mykiss* too, the thermal sensitivity was shown to be heritable and the heritability was 0.63 for the high line and 0.71 for the low line (Magerhans and Horstgen-Schwark, 2010).

The European sea bass *Dicentrarchus labrax* is a marine aquaculture species of great economic importance in the Mediterranean region. Female bass grows faster than male in farmed populations (Carrillo et al., 1995). The bass is not sexually dimorphic; hence its sex is not morphologically identifiable. As its gonads remain undifferentiated until 12 months of age (Blazquez et al., 1995; however, see Menu et al., 2005), its sex is not also distinguishable through gonadal histology. The karyotype of the bass consists of 44 small acentric chromosomes and sex chromosomes are not readily recognizable (Sola et al., 1993). No sex-specific molecular and protein markers are yet available to identify the genetic sex of the bass at an early stage of its life (Zanuy et al., 2001; Piferrer et al., 2005).

Indirect methods conventionally used to elucidate the sex determination mechanism in fishes have also failed to provide conclusive evidence for male or female heterogametic mechanism of sex determination in the bass. Meiogynogenics carry maternal chromosomes alone. Hence the expected female: male ratio is 1♀ : 0♂ in male (XX/XY) heterogametics but 0.5♀ : 0.5♂ in female (ZZ/ZW) heterogametics. However, the approximate sex ratios found by Felip et al. (2002) for the meiogynogenic bass were 0.33♀ : 0.67♂, 0.5♀ : 0.5♂ or 0.80♀ : 0.20♂. Induction of triploidy by cold shock produced mostly steriles and intersexuals. The bass is readily amenable to hormonal masculinization and yielded 100% males (Table 7.8) but feminization produced at the maximum 80 to 90% females only. The cause(s)

Table 7.8. Labile period chosen by authors to induce hormonal sex reversal in *Dicentrarchus labrax* (source: Piferrer et al., 2005).

Compound	Treatment duration (dpf)	Male ratio	Reference
MT	65–140	1.0	Gorshkov et al. (1999)
MT	86–106	1.0	Blazquez et al. (2001)
MDHT	84–116	1.0	Chatain et al. (1999)
AI	131–191	1.0	see Piferrer et al. (2005)
E_2	65–140	0.1	Gorshkov et al. (1999)
E_2	88–148	0.1	Gorshkov et al. (1999)
E_2	90–150	0.1	Saillant et al. (2001)
EE_2	60–260	0.2	Blazquez et al. (1998)

for the 10–20% non-responsive individuals to feminization are not yet known. Blazquez et al. (1999) generated 13 family lines crossing 13 sex reversed Δ neomales with two normal females. The reported sex ratios are 0.5♀: 0.5♂ in one family line, 0.33♀ : 0.67♂ in two family lines and 0.25♀ : 0.75♂ in the remaining 10 family lines, clearly indicating that these ratios can not be related to any known mechanism of sex determination.

Cloning the gonadal aromatase of the bass, Dalla Velle et al. (2002) observed its highest mRNA expression in the ovary but it was also expressed in the brain and testis. Evidence for the presence of brain aromatase was reported by Gonzalez and Piferrer (2002). Cloning it, Blazquez and Piferrer (2004) indicated that the neural aromatase appeared ahead of the ovarian aromatase. Socorro et al. (2007) detected the expression of *cyp11B* and *cyp19A* from the 4th dph; the expression of these genes was not different at 15 and 20°C. During the period from 1st to 6th dph, high temperature promotes masculinization. Throughout the period of gonadogenesis (60–300 dph), a high dimorphic pattern of *cyp11B* expression remained constant indicating its consistent role in ovarian development. However, the masculinizing effect of higher temperature during early larval development (prior to 60 dph) may not be mediated by the regulation of the expression of *cyp11B* or *cyp19A* but by upstream genes, whose expression is commenced prior to sex differentiation (Piferrer et al., 2005). Piferrer et al. (2005) comprehensively summarized relevant information reported from several publications on sex differentiation in the bass considering the durations of proliferation of germ cells, expression of gonadal aromatase, temperature effect and labile period as functions of age and body size. Nevertheless, their figure did not indicate which of the event(s) triggerred sex differentiation in one way or the other. Figure 7.10 shows the possibility that the thermal stress-induced cortisol concentration derails the sex differentiation process in genetic females by inhibiting ovarian differentiation during the labile period and thereby inducing 'sex reversal' from genetic females to phenotypic males (Figs. 7.7, 7.8) (see also Van den Hurk and Van Oordt, 1985).

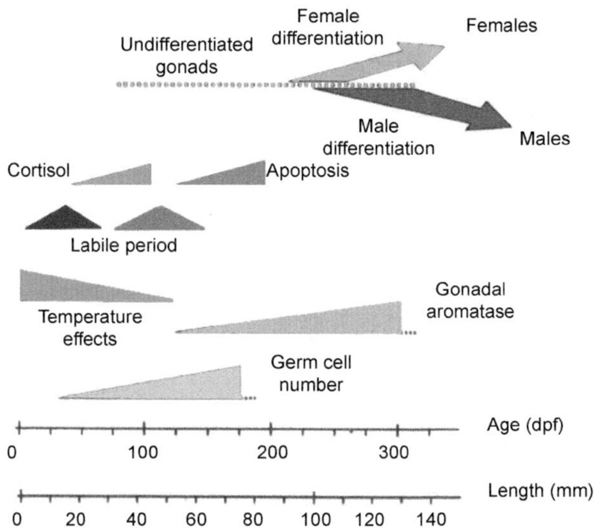

Fig. 7.10. The relationship between age and size on one hand and factors like germ cells gonadal aromatase and temperature on the other hand during the labile period of the European sea bass *Dicentrarchus labrax*. Note also the inclusion of locations of cortisol accumulation (Hattori et al., 2009, see also Fig. 7.7) and apoptosis (Uchida et al., 2004, see also Fig. 7.8) known for other fishes (source: Piferrer et al., 2005, modified).

Color image of this figure appears in the color plate section at the end of the book.

Blazquez et al. (2001) showed that the critical period of sex differentiation in the bass (86–126 dpf) is almost synchronized with that of rapid proliferation of primordial germ cells (PGCs) (Robin and Brusle, 1983). As shown earlier, the proliferation of PGCs is significantly affected by temperature. Hence it is to be expected that higher temperature advances the labile period. However, the values reported for the male ratio of the bass as functions of increasing temperature and exposure duration (Table 7.9) are highly complicated that it is difficult to make any generalization.

As sex of the bass is not identifiable morphologically, almost all authors have resorted to gonadal histology to recognize sex. The ovarian differentiation precedes the testicular differentiation; consequently, females can be recognized at the age of 9 months and size of 12.5 cm and males at 11.6 months and size of 12.8 cm (Blazquez et al., 1998). The temperature-treatment durations selected to manipulate sex ratio vary widely: it ranges from the longest duration from post-fertilization to > 149 dpf (Saillant et al., 2002) to the shortest duration from half epiboly to mid-metamorphosis (≈60 dpf) (Koumoundouros et al., 2002) (Fig. 7.11). The duration selected by Blazquez et al. (1995) corresponds with the labile period (80–137 dph), estimated for hormonal manipulation of sex in the bass. The high and

Table 7.9. Effect of temperature and exposure duration on the male ratio of *Dicentrarchus labrax*. T = temperature (source: Piferrer et al., 2005).

T (°C)	Exposure duration (dpf)	♂ ratio (%)	References
13	1–93	27	Pavlidis et al. (2000)
13	1–346	89	Saillant et al. (2002)
15	1–11	62	Koumoundouros et al. (2002)
15	1–38	53	Koumoundouros et al. (2002)
15	1–64	34	Koumoundouros et al. (2002)
15	1–120	65	see Piferrer et al. (2005)
20	1–56	82	Koumoundouros et al. (2002)
20	1–65	74	Pavlidis et al. (2000)
20	19–149	77	Saillant et al. (2002)
21	15–120	88	see Piferrer et al. (2005)
22	15–57	94	Blazquez et al. (1998)

low temperatures selected by the authors also fall within (13 and 25°C) tolerance range of the bass (Picket and Pawson, 1994). Remarkably, Greek scientists who treated the bass for a shorter duration from early embryonic stage to mid metamorphosis (≈5 cm SL), obtained the highest 66–73% females at 13–15°C, against 7–11% of females at the same temperatures by the French-Spanish groups, who selected longer durations of treatment. Hence, the existence of two temporally separated labile periods becomes apparent, the first one from half epiboly to mid-metamorphosis, i.e., from 1 to ≈60 dpf (Koumoundouros et al., 2002) and the second from 80 to 137 dph (Blazquez et al., 1995). The existence of temporally separated two labile periods is not uncommon in fishes; it exists in medaka (cf. p. 97) and eel (cf. p. 84). Incidentally, Diaz et al. (2013) reported the appearance of the first known aromatase expression in the bass of 4.6 cm SL (cf. p. 84). The first period may represent the ovarian differentiation and the second testicular differentiation. If so, how the differentiated ovary transdifferentiates into testis (cf. Kuhl and Brouwer, 2006) is not comprehensible. Nevertheless, this temporal difference in labile periods, the first during the embryonic to post-hatching stage and the second during the post-hatch to juvenile stage, calls for further study. It is not clear whether the sex ratio response to temperature changes by the Greek strain of the bass differs from that of French and Spanish strain. However, it may also be noted that against the treatment at low and high temperatures of the north-western and south-eastern strains of the bass in Israel, the sex ratio response was similar (Mylonas et al., 2005).

Understandably, due to thermal stress-induced accumulation of cortisol, more and more sensitive genetic females 'change sex' by derailing sex differentiation process in to phenotypic males. Consequently, the percentage of the female bass decreased with increasing temperatures in the Greek strain

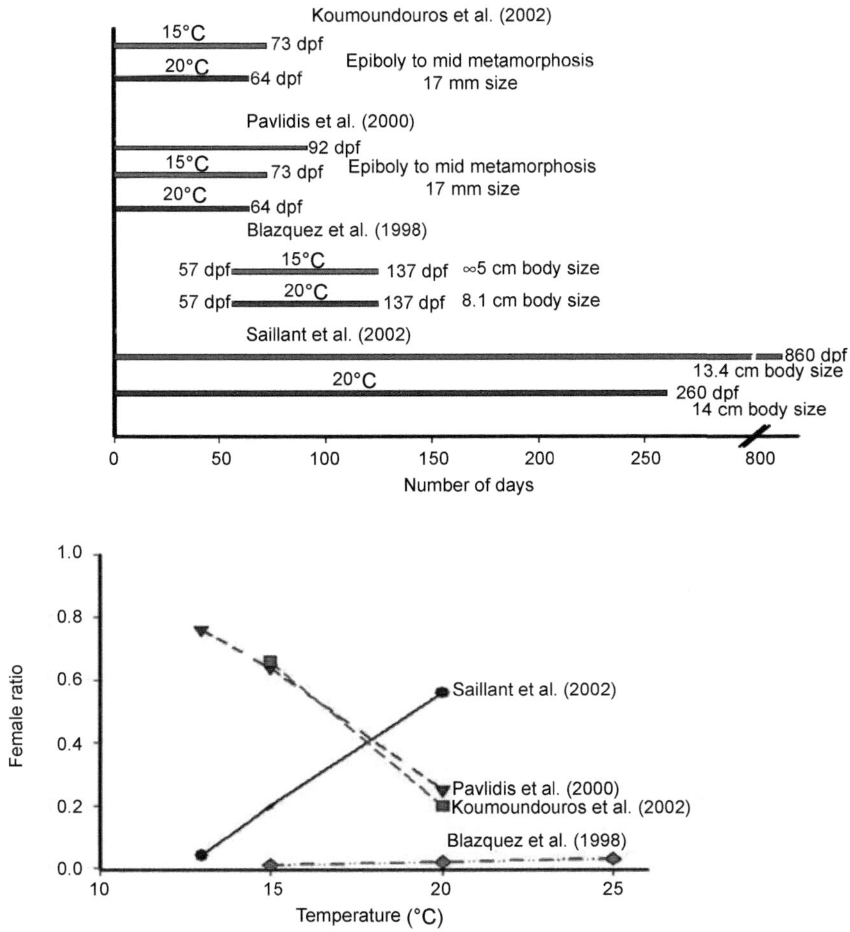

Fig. 7.11. Upper panel: labile period selected for temperature treatment of *Dicentrarchus labrax* by different European authors. Lower panel: Female ratio response to different temperatures. Note the completely different trends of female ratio response to temperature changes by the sea bass in relation to the difference in labile period shown above.

Color image of this figure appears in the color plate section at the end of the book.

(Pavlidis et al., 2000; Koumoundouros et al., 2002) (Fig. 7.11). Remarkably, the unexpected increase in females with increasing temperature, as reported by Saillant et al. (2002) is also difficult to comprehend. Notably, the treatment duration lasted from fertilization to 149 dpf (cf. p. 178).

Transfer treatments from low to high and high to low temperatures revealed that such shifts, when imparted after 149 dpf (8.1 cm size) failed to evoke the sex ratio response. Evidently, the sex differentiation process is completed before the 149th dpf (Saillant et al., 2002). However, the shift made at alevin (7 dph, 5.3 mm) from 15 to 20°C or at flexion stage (34 dph, 10 mm) from 15 to 20°C yielded 38 and 47% females, respectively, against 66 or 18% females at 15 or 20°C. Clearly, the shifts in treatment temperatures at alevin and flexion stages induce significant changes in percentage of females (Koumoundouros et al., 2002). Clearly, sex differentiation in the bass is commenced from epiboly. In a viviparous dwarf perch *Micrometrus minimus* too, environmental temperature plays a decisive role in sex differentiation even from the time of conception (Schultz, 2008).

With reference to temperature-size interaction, Blazquez et al. (1998) reported that the body length of a female was longer by 0.2 and 0.5 cm than a male at the ages of 11.6 and 21.4 months, respectively. Koumoundouros et al. (2002) also found that a female was larger by 1 cm in body length and 10 g in weight at the age of 11.4 months. During the first sampling on 308 dph, Pavlidis et al. (2000) noted significant difference in size between those reared at 15°C (17.5 cm, 176 g) and those (≈17 cm, ≈166 g) at 13 and 20°C, despite attaining recognizable female sex at the ages of 64, 73 and 92 dph at 20, 15 and 13°C, respectively. However the authors found that in all these groups, females were larger than males, a difference that was statistically significant on the final sampling at the ages between 10 and 19 months. At the age of 14 months, i.e., long after sexual differentiation is completed around 9–12 months, females reared at low and high densities were significantly heavier than the males (Saillant et al., 2003a). Vinagre et al. (2009) also reported that growth of bass after the 100th dph increases from 0.48 mm/day at latitude 33.90°N (≈18.5°C) to 0.56 mm/day at 37.5°N (≈ 20.0°C) with the consequent higher temperature at lower latitude. Evidently, the size difference between female and male is commenced at the onset of sex differentiation and is progressively increased with advancing age. It appears that it is the sex that induces differences in size rather than size inducing sex differentiation (see also Fujioka, 2001; Saillant et al., 2003a). From a series of experiments on the relationship between growth, i.e., size and sex differentiation in the sea bass, Diaz et al. (2013) also concluded that growth rate and size do not influence sex differentiation either during the embryonic stage or on the growing stage and identified the 3–4 cm SL as a size range at which molecular biologist may study the process of sex differentiation.

7.7 Prevalence and patterns

A revisit to Ospina-Alvarez and Piferrer (2008) has become necessary. Both fishes and reptiles are poikilothermic vertebrates. Irrespective of aquatic or terrestrial, reptilian eggs are incubated in land but the embryos, larvae and juveniles of almost all fishes (cf. Chapter 5.4) are aquatic. Diurnal and seasonal amplitudes in aquatic habitats are known to be far narrower than those in terrestrial environment. Despite the limitations listed on p. 184–186, the analyses of Ospina-Alvarez and Piferrer (2008) have made a contribution. 1. The existence of TSD is more prevalent among reptiles, including crocodilians, a majority of turtles, some lizards (e.g., Gekkonidae, Wibbels et al., 1994) and snakes but the TSD is less prevalent in fishes and is limited to 59 species belonging to 13 families in just four orders only. 2. The prevalence of TSD in many reptiles including the archiaic two tuatara *Sphenodon* (Mitchell and Mitchell, 2006) suggests that it is originated rather than derived (see also Pieau and Dorizzi, 2004). But in fishes the TSD is derived rather than of ancestral origin and 3. All the three sex ratio response *Patterns* 1a, 1b and 2 are known to exist in reptiles. *Pattern* 1a occurs mostly in species, in which adult females are larger than adult males and in pattern 1b the reverse holds good (Ewert and Nelson, 1991). For example, the turtle *Emys orbicularis* represents pattern 1b. Incubated at 25°C, its eggs yield 100% males but 100% females at 30°C (Raynaud and Pieau, 1985). In the American alligator *Alligator mississippiensis*, egg incubation temperatures 30°C and below produce 100% females and 100% males at 33°C but ≈90% females at 35°C and thereby the alligator represents a modified pattern 2 (Ferguson, 1985). In fishes the TSD was first reported in 1981 for the silverside *Menidia menidia* 16 years after the first discovery of TSD in the rainbow lizard *Agama agama* by Charnier in 1966. Understandably, a more voluminous body of literature and information on the mechanisms of shift in sex ratio are available for TSD in reptiles than for fishes.

This chapter has revealed the possible existence of temporally separated two labile periods in fishes. As sex can not be reversed after the completion of the labile period prior to adult stage in primary gonochoric fishes, the sex ratio response patterns are redefined. Within the labile period, the sex ratio response per se may follow different trends corresponding to the original *patterns* 1, 2 and 3. As the potency and dose of sex reversing EDCs including exogenous hormones as well as exposure duration singly or jointly decide the sex ratio, not only by the dose, namely temperature level but also the exposure duration play a role in deciding sex reversal and hence sex ratio. The parental (GSD + A) genetic factors override and alter the temperature effect on sex ratio in some individuals, families, strains and species. Hence, there are four different trends/patterns of sex ratio response of fishes to change in temperature.

References

Abucay, J.S., G.C. Mair, D.O.F. Skibinski and J.A. Beardmore. 1999. Environmental sex determination: the effect of temperature and salinity on sex ratio in *Oreochromis niloticus*. Aquaculture, 173: 219–234.

Admassu, D. and J.M. Casselman. 2000. Otolith age determination for adult tilapia *Oreochromis niloticus* L. from Lake Awassa (Ethiopian Rift Valley) by interpreting biannuli and differentiating biannual recruitment. Hydrobiologia, 418: 15–24.

Aida, T. 1936. Sex reversal in *Aplocheilus latipes* and a new explanation of sex differentiation. Genetics, 21: 136–153.

Allender, C.J., O. Seehausen, M.E. Knight et al. 2003. Divergent selection during speciation of Lake Malawi cichlid fishes inferred from parallel radiations in nuptial colouration. Proc Natl Acad Sci USA, 100: 14074–14079.

Azuma, T., K. Takeda, T. Doi et al. 2004. The influence of temperature on sex determination in sockeye salmon *Oncorhynchus nerka*. Aquaculture, 234: 461–473.

Baras, E., C. Prignon, G. Gohoungo and C. Melard. 2000. Phenotypic sex differentiation of blue tilapia under constant and fluctuating thermal regimes and its adaptive and evolutionary implication. J Fish Biol, 57: 210–223.

Baras, E., B. Jacobs and C. Melard. 2001. Effect of water temperature on survival, growth and phenotypic sex of mixed (XX-XY) progenies of Nile tilapia *Oreochromis niloticus*. Aquaculture, 192: 187–199.

Baroiller, J.F. and F. Clota. 1998. Interaction between temperature effects and genotypes on *Oreochromis niloticus* sex determination. J Exp Zool, 281: 507.

Baroiller, J.F. and H. D'Cotta. 2001. Environmental and sex determination in farmed fish. Comp Biochem Physiol, 130: 399–409.

Baroiller, J.F., Y. Guiguen and A. Fostier. 1999. Endocrine and environmental aspects of sex differentiation in fish. Cell Mol Life Sci, 55: 910–931.

Baroiller, J.F., D. Chourrout, A. Fostier and B. Jalabert. 1995a. Temperature and sex chromosomes govern sex ratios of the mouthbrooding cichlid fish *Oreochromis niloticus*. J Exp Zool, 273: 216–223.

Baroiller, J.F., F. Clota and E. Geraz. 1995b. Temperature and sex determination in two tilapias *Oreochromis niloticus* and the red tilapia (Red Florida strain): effect of high and low temperatures. Proc Fifth Internatl Symp Reprod Physiol Fish. Austin, Texas, pp. 158–160.

Baroiller, J.F., A. Fostier, C. Cauty et al. 1996. Effects of high rearing temperature on the sex ratio of progeny from sex reversal males of *Oreochromis niloticus*. *In*: R.S. Pullin, M. Lazard, M. Legendre et al. (eds.). Third Internatl Symp Tilapia Aquacult. ICLARM Conf Proc, 4: 246–256.

Baroiller, J.F., Y. Guiguen and A. Fostier. 1999. Endocrine and environmental aspects of sex differentiation in fish. Cell Mol Life Sci, 55: 910–931.

Baroiller, J.F., H. D'Cotta, E. Bezault et al. 2009a. Tilapia sex determination. Where temperature and genetics meet. Comp Biochem Physiol, 153A: 30–38.

Baroiller, J.F., H. D'Cotta and E. Saillant. 2009b. Environmental effects on fish sex determination and differentiation. Sex Dev, 3: 118–135.

Baron, B., R.F. Buckle and S. Espina. 2002. Environmental factors and sexual differentiation in *Poecilia sphenops* Valenciennes (Pisces: Poeciliidae). Aquacult Res, 33: 615–619.

Berry, A.N. and M.S. Grober. 2003. A model for social control of sex change: interactions of behaviour, neuropeptides, glucocorticoids and sex steroids. Horm Behav, 43: 31–36.

Bezault, E. 2005. Etude du system de determinisme du sexe an sein de populations naturelles de tilapia du Nil *Oreochromis niloticus*: Importance des composanties genetiques et environmentales. Ph.D. Thesis, Universite Paris.

Bezault, E., F. Clota, M. Devivaz et al. 2007. Sex determination and temperatue-induced sex differentiation in three natural populations of Nile tilapia (*Oreochromis niloticus*) adapted to extreme temperature condition. Aquaculture, 27s1: S3–S16.

Blasco, M., J.I. Fernandino, L.G. Guilgur et al. 2010. Molecular characterization of *cyp11a1* and *cyp11b1* and their gene expression profile in pejerrey (*Odontesthes bonariensis*) during early gonadal development. Comp Biochem Physiol, 156A: 110–118.

Blazquez, M. and F. Piferrer. 2004. Cloning, sequence analysis, tissue distribution and sex specific expression of neural form of p450 aromatase in juvenile sea bass (*Dicentrarchus labrax*). Mol Cell Endocrinol, 219: 83–94.

Blazquez, M. and G.M. Somoza. 2010. Fish with thermolabile sex determination (TSD) as models to study brain sex differentiation. Gen Comp Endocrinol, 166: 470–477.

Blazquez, M., F. Piferrer, S. Zanuy et al. 1995. Development of sex control techniques for European sea bass (*Dicentrarchus labrax* L.) aquaculture: effects of dietary 17α–methyltestosterone prior to sex differentiation. Aquaculture, 135: 329–342.

Blazquez, M., S. Zanuy, M. Carrillo and F. Piferrer. 1998. Effects of rearing temperature on sex differentiation in the European sea bass (*Dicentrarchus labrax* L.). J Exp Zool, 281: 207–217.

Blazquez, M., M. Carrillo, S. Zanuy and F. Piferrer. 1999. Sex ratios in offspring of sex reversal sea bass and the relationship between growth and phenotypic sex differentiation. J Fish Biol, 55: 916–930.

Blazquez, M., A. Felip, S. Zanuy et al. 2001. Critical period of androgen-inducible sex differentiation in a teleost fish, the European sea bass (*Dicentrarchus labrax*). J. Fish Biol, 58: 342–358.

Bouza, C., L. Sanchez and P. Martinez. 1994. Karyotypic characterization of turbot (*Scophthalmus maximus*) with conventional fluorochrome and endonuclease-banding techniques. Mar Biol, 120: 609–613.

Brusle, J. and C. Roblin. 1984. Sexuality of sea bass *Dicentrarchus labrax* under controlled fish farm condition. pp. 33–43. *In*: G. Barnabe and R. Billard (eds.). *L' Aquaculture Du Bar Des Sparids*. INRA, Paris.

Burt, J.M., S.G. Hinch and D.A. Patterson. 2011. The importance of parentage in assessing temperature effects on fish early life history: a review of the experimental literature. Rev Fish Biol Fisher, 21: 377–406.

Carrillo, M., S. Zanuy, M. Blazquez et al. 1995. Sex control and ploidy manipulation in sea bass. pp. 125–143. *In*: Environmental Impacts of Aquatic Biotechnology UECD document, Paris.

Charnier, M. 1966. Action de la temperature sur la sex ratio chex l'embryon d'*Agama agama* (Agamidae Lacertilien). CR Seanc Soc Biol, 160: 620–622.

Chatain, B., E. Saillant and S. Perruzi. 1999. Production of monosex male populations of European sea bass *Dicentrarchus labrax* L. by use of the synthetic androgen 17–methldehydrotestosterone. Aquaculture, 178: 225–234.

Conover, D.O. 2004. Temperature-dependent sex determination in fishes. pp. 11–20. *In*: N. Valenzuela and V. Lance (eds.). *Temperature Dependent Sex Determination in Vertebrates*. Smithsonian Books, Washington.

Conover, D.O. and B.E. Kynard. 1981. Environmental sex determination: Interaction of temperature and genotype in a fish. Science, 213: 577–579.

Conover, D.O. and M.H. Fleisher. 1986. Temperature-sensitive period of sex determination in the Atlantic silverside *Menidia menidia*. Can J Fish Aquat Sci, 43: 514–520.

Conover, D.O. and S.W. Heins. 1987a. Environmental and genetic components of sex ratio in *Menidia menidia* (Pisces: Atherinidae). Copeia, 1987: 732–745.

Conover, D.O. and S.W. Heins. 1987b. Adoptive variation in environmental and genetic sex determination in a fish. Nature, 326: 496–498.

Conover, D.O., D.A. Van Voorhess and A. Ehtisham. 1992. Sex ratio selection and the evolution of environmental sex determination in laboratory population of *Menidia menidia*. Evolution, 46: 1722–1730.

Cornejo, A.M. 2002. Gonadal abnormalities in the silverside *Odontesthes bonariensis* (Pisces: Atherinidae) in Buenos Aures lakes, Argentina. Biosciencias, 10: 43–56.

Craig, J., C. Foote and C. Wood. 1996. Evidence for temperature-dependent sex determination in sockeye salmon (*Orcorhynchus nerka*). Can J Fish Aquat Sci, 53: 141–147.

Crews, D. and J.J. Bull. 2008. Sex determination: Some like it hot (and some do not). Nature, 7178: 527–528.

Davis, K.B., C.A. Goudie, B.A. Simco et al. 1992. Influence of dihydrotestosterone on sex determination in channel catfish and blue catfish. Gen Comp Endocrinol, 86: 147–151.

Dalla Valle, L., L. Lunardi, L. Colombo and P. Belvedere. 2002. European sea bass (*Dicentrarchus labrax* L.) cytochrome p450 arom: DNA cloning, expression and genomic organization. J Steroid Biochem Mol Biol, 80: 25–34.

D'Cotta, H., A. Fostier, Y. Guiguen et al. 2001. Aromatase plays a key role during normal and temperature-induced sex differentiation of tilapia *Oreochromis niloticus.* Mol Reprod Dev, 59: 265–276.
Denzer, H.W. 1968. Studies on the physiology of young *Tilapia.* FAO Fish Report. 44: 356–366.
Desprez, D. and C. Melard. 1998. Effect of ambient water temperature on sex determination in the blue tilapia *Oreochromis aureus.* Aquaculture, 162: 79–84.
Diaz, N., L. Ribas and F. Piferrer. 2013. The relationship between growth and sex differentiation in the European sea bass (*Dicentrarchus labrax*). Aquaculture, 408-409: 191–202.
Dzikowski, R., G. Hulata, I. Karplus and S. Horpaz. 2001. Effect of temperature and dietary L. caratinine supplementation on reproduction performance of female guppy (*Poecilia reticulata*). Aquaculture, 199: 323–332.
Ewert, M.A. and C.E. Nelson. 1991. Sex determination in turtles: Diverse patterns and some possible adaptive values. Copeia, 1991: 50–69.
Felip, A., F. Piferrer, M. Carrillo and S. Zanuy. 2002. Growth, gonadal development and sex ratios of meiogynogenetic diploid sea bass. J Fish Biol, 61: 347–359.
Ferguson, M.W.J. 1985. The reproductive biology and embryology of crocodelians. *In*: C. Gans, G. Billet and P.F.A. Maderson (eds.). *Biology of the Reptiles*. John Wiley, New York, 14: 329–491.
Fernandino, J.I., R.S. Hattori, H. Kimura et al. 2008. Expression profile and estrogenic regulation of anti-Mullerian hormone during gonadal development in pejerrey *Odontesthes bonariensis,* a teleost fish with strong temperature-dependent sex determination. Dev Dyn, 237: 3192–3199.
Fernandino, J.I., J.T. Popesku, B. Paul-Prasanth et al. 2011. Analysis of sexually dimorphic expression of genes at early gonadogenesis of pejerrey *Odontesthes bonariensis* using a heterologous microarray. Sex Dev, 5: 89–101.
Fujioka, Y. 2001. Thermolabile sex determination in honmoroko. J Fish Biol, 59: 851–861.
Fujioka, Y. 2002. Effect of hormone treatments and temperature on sex reversal of Nigoroubuna *Carassius carassius grandoculis.* Fisher Sci, 68 : 889–893.
Fujioka, Y. 2006. Pattens of sex ratio response to temperature during sex determination in honmoroko *Gnathopogon caerulescens.* Fisher Sci, 72: 1034–1041.
Gonzalez, A. and F. Piferrer. 2002. Characterization of aromatase activity in the sea bass: Effects of temperature and different catalytic properties of brain and ovarian homogenates and microsomes. J Exp Zool, 293: 500–510.
Gorshkov, S., G. Gorshkova, W. Knibb and H. Gordin.1999. Sex ratios and growth performance of European sea bass (*Dicentrarchus labrax* L.) reared in mariculture in Eilat (Red Sea). Isr J Aquacult. Bamidgeh, 51: 91–105.
Gorshkov, S., G. Gorshkova, I. Meiri and H. Gordin. 2004. Culture performance of different strains and crosses of the European sea bass (*Dicentrarchus labrax*) reared under controlled conditions at Eilat, Isreal. J Appl Ichthyol, 20: 194–203.
Goto, R., T. Mori, K. Kawamata et al. 1999. Effects of temperature on gonadal sex determination in Barfin flounder *Verasper moseri*. Fisher Sci, 65: 884–887.
Goto, R., T. Kayaba, S. Adachi and K. Yamaguchi. 2000. Effect of temperature on sex determination in marbled sole *Limanda yokohamae*. Fisher Sci, 66: 400–402.
Goto-Kazeto, R., Y. Abe, K. Massai et al. 2006. Temperature-dependent sex differentiation in goldfish: establishing the temperature-sensitive period and effect of constant and fluctuating temperatures. Aquaculture, 254: 617–624.
Goudie, C.A., B.D. Redner, B.A. Simco and K.B. Davis. 1983. Feminization of channel catfish by oral administration of steroid sex hormones. Trans Am Fish Soc, 112: 670–672.
Haffray, P., E. Lebegue, S. Jeu et al. 2009. Genetic determination and temperature effects on turbot *Scophthalmus maximus* sex differentiation: An investigation using steroid sex-inverted males and females. Aquaculture, 294: 30–36.
Hattori, R.S., R.J. Gould, T. Fujioka et al. 2007. Temperature-dependent sex determination in Hd-rR medaka *Oryzias latipes*: gender sensitivity, thermal threshold, critical period, and DMRT1 expression profile. Sex Dev, 1: 138–146.

Hattori, R.S., J.I. Fernandino, A. Kishi et al. 2009. Cortisol-induced masculianization: Does thermal stress affect gonadal fate in pejerrey, a teleost fish with temperature-dependent sex determination? PLoS ONE, 4 (8). doi.10.1371/journal:pone.0006548.

Hermelink, B., S. Wuertz, B. Rennet et al. 2013. Temperature control of pikeperch (*Sander luciperca*) maturation in recirculating aquaculture system—induction of puberty and course of gametogenesis. Aquaculture, 400-401: 36–45.

Horth, L. 2006. A sex-linked allele autosomal modifiers and temperature-dependence appear to regulate melanism in male mosquitofish (*Gambusia holbrooki*). J Exp Biol, 209: 4938–4945.

Hostache, G.M. Pascal and C. Tessier. 1995. Influence de la temperature d'incubation sur le rapport male, female chez l'atipa, *Hoplosternum littorale* Hancock (1828). Can J Zool, 73: 1239–1246.

Hughes, V., T.J. Benfey and D.J. Martin-Robichaud. 2008. Effect of temperature on sex ratio in juvenile Atlantic halibut, *Hippoglossus hippoglossus*. Env Biol Fish, 81: 415–419.

Husebye, H., S. Lund, M. Moller et al. 1994. A Bkm-releated DNA sequence gives individual DNA fingerprints in turbot (*Hippoglossus maximus*) but neither Bkm-related human *SRY* or human *ZFY* probes detect genetic sex differences. Comp Biochem Physiol, 107B: 69–73.

Ito, I.S., M. Yamashita and C.A. Strussmann. 2002. Dynamics of heat induced germ cell loss in pejerry *Odontesthes bonariensis*. Fisher Sci, 68: 1313–1314.

Ito, L.S., M. Yamashita and C.A. Strussmann. 2003. Histological process and dynamics of germ cell degeneration in pejerrey *Odontesthes bonariensis*. Larvae and juveniles during exposure to warm water. J Exp Zool, 297A: 169–179.

Ito, L.S., M. Yamashita, F. Takahashima and C.A. Strussmann. 2005. Dynamics and histological characteristics of gonadal sex differentiation in pejerry (*Odontesthes bonariensis*) at feminizing and masculinizing temperatures. J Exp Zool A Comp Exp Biol, 303: 504–514.

Ito, L.S., C. Takahashi, M. Yamashita and C.A. Strüssmann. 2008a. Warm water induces apoptosis, gonadal degeneration, and germ cell loss in subadult pejerrey *Odontesthes bonariensis* (Pisces, Atheriniformes). Physiol Biochem Zool, 81: 762–774.

Ito, L.S., A.M. Cornejo, M. Yamashita and C.A. Strüssmann. 2008b. Thermal threshold and histological process of heat-induced sterility in adult pejerrey (*Odontesthes bonariensis*): a comparative analysis of laboratory and wild specimens. Physiol Biochem Zool, 81: 775–784.

Jakupsstovu, S.H.J. and T. Hang. 1988. Growth, sexual maturation and spawning season of Atlantic halibut *Hippoglossus hippoglossus* in Faroese waters. Fisher Sci, 6: 201–215.

Karayucel, I., O. Ak and S. Karayucel. 2006. Effect of temperature on sex ratio in guppy *Poecilia reticulata* (Peters, 1860). Aquacult Res, 37: 139–150.

Karube, M., J.I. Fernandino, P. Strobl-Mazzulla et al. 2007. Characterization and expression profile of the ovarian cytochrome P-450 aromatase (*cyt19A1*) gene during thermolabile sex determination in pejerrey *Odontesthes bonariensis*. J Exp Zool, 307A: 625–636.

Khallaf, E.A., D. Passia, A.M. Abdeen and F.A.El. Mesedy. 2004. The effect of temperature variation on the sex composition and growth of *Poecilia melanogaster*. Egypt J Aquat Biol Fisher, 5: 29–43.

Kitano, T., K. Takamune, T. Kobayashi et al. 1999. Suppression of p450 aromatase gene expression in sex reversed males produced by rearing genetically female larvae at a higher temperature during a period of sex differentiation in the Japanese flounder (*Paralichthys olivaceus*). J Mol Endocrinol, 23: 167–176.

Knapp, H., E. Marsh-Mathew, L. Vo and S. Rosevcrans. 2012. Stress hormone masculianizes female morphology and behavior. Biol Lett, 7: 150–152.

Kosswig, C. 1964. Polygenic sex determination. Experientia, 20: 190–199.

Koumoundouros, G., Pavlidis, L. Anezaki et al. 2002. Temperature sex determination in the European sea bass *Dicentrarchus labrax* (L. 1758) (Teleostei, Perciformes, Moronidae): critical sensitive ontogenetic phase. J Exp Zool, 292: 573–579.

Kuhl, A.J. and M. Brouwer. 2006. Antiestrogens inhibit xenoestrogen-induced brain aromatase activity but do not prevent xenoestrogen-induced feminization in Japanese medaka (*Oryzias latipes*). Environ Health Perspect, 114: 500–506.

Kurita, M. 1999. Effect of water temperature on the process of gonadal sex differentiation in pejerry. M.Sc. Thesis, Tokyo University of Fisheries.

Kwon, J.Y., B.J. McAndrew and D.J. Penman. 1999. Inhibition of aromatase activity suppresses high-temperature feminization of genetic male Nile tilapia *Oreochromis niloticus*. *In*: Proc Sixth Internatl Symp Reprod Physiol Fish. Berger, p. 268.

Kwon, J.Y., V. Haghpanah, L.M. Kogson-Hurtado et al. 2000. Masculinization of genetic female Nile tilapia (*Oreochromis niloticus*) by dietary administration of an aromatase inhibitor during sexual differentiation. J Exp Zool, 287: 46–53.

Kwon, J.Y., B.J. McAndrew and D.J. Penman. 2002. Treatment with an aromatase inhibitor suppresses high-temperature feminization of genetic male (YY) Nile tilapia. J Fish Biol, 60: 625–636.

Lozano, C.A., B. Gjerde, J. Odegard and H.B. Bensten. 2013. Heritability estimates for male proportion in the GIFT Nile tilapia (*Oreochromis niloticus* L.). Aquaculture, 372-375: 137–148.

Luckenbach, J.A., J.G. Harry, H.V. Daniels and R.J. Borski. 2003. Gonadal differentiation and effect of temperature on sex determination in southern flounder (*Paralichthys lethostigma*). Aquaculture, 216: 315–327.

Magerhans, A. and G. Horstgen-Schwark. 2010. Selection experiments to alter the sex ratio in rainbow trout (*Oncorhynchus mykiss*) by means of temperature treatment. Aquaculture, 306: 63–67.

Mair, G.C., E.J. Morales, J. Mongkonswad et al. 1997. Properties of novel YY male Nile tilapia and implications for technology dissemination in the Philippines. Aquaculture, 73: 12.

Menu, B., S. Perruzzi, A. Vergnet et al. 2005. A short method for sexing juvenile European sea bass *Dicentrarchus labrax* L. Aquacult Res, 36: 41–44.

Middaugh, D.P. and M.J. Hammer. 1987. Influence of environmental temperatures on sex ratios in the tidewater silverside *Menidia peninsulae* (Pisces: Atherinidae). Copeia, 1987: 958–964.

Miranda, L.A., C.A. Strussmann and G.M. Somoza. 2001. Immuno cytochemical identification of GtH1 and GtH2 cells during the temperature sensitive period for sex determination of pejerrey *Odontesthes bonariensis*. Gen Comp Endocrinol, 124: 45–52.

Miranda, L.A., P.H. Strobl-Mazzulla, C.A. Strussmann et al. 2003. Gonadotropin-releasing hormonal neuronal development during the sensitive period of temperature sex determination in the pejerrey fish *Odontesthes bonariensis*. Gen Comp Endocrinol, 132: 444–453.

Mitchell, N.J. and N.J. Mitchell. 2006. Support for a rare pattern of temperature-dependent sex determination in archaic reptiles: evidence from two species of tuatara (*Sphenodon*). Front Zool, 3: 9. Doi:10.1186/1742-9994-3-9.

Mylonas, C., C.L. Anezaki, P. Divanach et al. 2005. Influence of rearing temperature during the larval and nursery periods on growth and sex differentiation in two Mediterranean strains of *Dicentrarchus labrax*. J Fish Biol, 67: 652–668.

Nakamura, M., R.H. Bhandari and M. Higa. 2003. The role estogens play in sex differentiation and sex change of fish. Fish Physiol Biochem, 28: 113–117.

Nomura, T., K. Arai, T. Hayashi and R. Suzuki. 1998. Effect of temperature on sex ratio of normal and gynogenetic diploid loach. Fisher Sci, 64: 753–758.

Ospina-Alvarez, N. and F. Piferrer. 2008. Temperature-dependent sex determination in fish revisited. Prevalence, a single sex ratio response pattern and possible effects of climate change. PLoS One, 3(7): e2837. doi: 10.1371/journal.pone 0002837.

Pandian, T.J. 2011. *Sex Determiantion in Fish*. Science Publishers, Enfield/CRC Press, Boca Raton, USA, p. 282.

Pandian, T.J. 2012. *Genetic Sex Differntiation in Fish*. CRC Press, Boca Raton, USA, p. 214.

Pandian, T.J. 2013. *Endocrine Sex Differentiation in Fish*. CRC Press, Baca Raton, USA, p. 303.

Patino, R., K.B. Davis, J.E. Schoore et al. 1996. Sex differentiation in channel catfish gonads. Normal development and effects of temperature. J Exp Zool, 276: 209–218.

Pavlidis, M., G. Koumoundouros, A. Sterioti et al. 2000. Evidence of temperature-dependent sex determination in the European sea bass (*Dicentrarchus labrax* L.). J Exp Zool, 287: 225–232.

Picket, G.D. and M.G. Pawson. 1994. *Sea Bass: Biology, Exploitation and Conservation*. Chapman and Hall, London.

Pieau, C. and M. Dorrizzi. 2004. Oestrogens and temperature-dependent sex determination in reptiles: all is in the gonad. J Endocrinol, 181: 367–377.

Piferrer, F., M. Blazquez, L. Navarro and A. Gonzalez. 2005. Genetic, endocrine and environmental components of sex determination and differentiation in the European sea bass (*Dicentrarchus labrax* L.). Gen Comp Endocrinol, 142: 102–110.

Pompini, M., A.M. Buser, M.R. Thali et al. 2013. Temperature-induced sex reversal is not responsible for sex ratio distortions in grayling *Thymallus thymallus* or brown trout *Salmo trutta*. J Fish Biol, 83: 404–411.

Raynaud, A. and C. Pieau. 1985. Embryonic development of the genital system. *In*: C. Gans and G. Billet (eds.). *Biology of the Reptiles*. John Wiley, New York, 15: 149–300.

Reagan, R.E., Jr. and W.M. Wingo. 1985. Species profiles: life histories and environmental requirements of coastal fishes and invertebrates (Gulf of Mexico) southern flounder. US Fish Wild Ser Biol Rep, 82: 9.

Robin, C. and J. Brusle. 1983. Ontogenese gonadique et differentiation sexualle du loup *Dicentrarchus labrax* en conditions d' elevage. Reprod Nut Dev, 23: 115–127.

Romer, U. and W. Beinsenherz. 1996. Environmental determination of sex in *Apistogramma* (Cichlidae) and two freshwater fishes (Teleostei). J Fish Biol, 48: 714–725.

Saillant, E., A. Fostier, B. Menu et al. 2001. Sexual growth dimorphism in sea bass (*Dicentrarchus labrax*). Aquaculture, 202: 371–387.

Saillant, E., A. Fostier, P. Haffray et al. 2002. Temperature effects and genotype-temperature interactions on sex determination in the European sea bass (*Dicentrarchus labrax* L.). J Exp Zool, 292: 494–505.

Saillant, E., A. Fostier, P. Haffray et al. 2003a. Effects of rearing density, size grading and parental factors on sex ratios of the sea bass (*Dicentrarchus labrax* L.) in intensive aquaculture. Aquaculture, 221: 183–206.

Saillant, E., B. Chatain, B. Menu et al. 2003b. Sexual differentiation and juvenile intersexuality in the European sea bass (*Dicentrarchus labrax*). J Zool Lond, 260: 53–63.

Sato, T., T. Endo, K. Yamahira et al. 2005. Induction of female-to-male sex reversal by high temperature treatment in medaka *Oryzias latipes*. Zool Sci, 22: 985–988.

Schultz, E.T. 2008. A sex difference in seasonal timing of birth in a live bearing fish. Copeia, 2008: 673–679.

Selim, K.M., A. Shinomiya, H. Otake et al. 2009. Effects of high temperature on sex differentiation and germ cell population in medaka *Oryzias latipes*. Aquaculture, 289: 340–349.

Shinoda, T., L.A. Miranda, K. Okuma et al. 2010. Molecular cloning and expression analysis of *Fshri* and *Lhr* in relation to *Fshb* and *Lhb* subunits during the period of temperature-dependent sex determination in pejerrey *Odontesthes bonariensis*. Mol Reprod Dev, 77: 521–532.

Simpson, E.R., C. Colyne, G. Rubin et al. 2002. Aromatase—a brief overview. Annu Rev Physiol, 64: 93–127.

Socorro, S., R.S. Martins, L. Deloffre et al. 2007. A $_{C}$DNA for European sea bass (*Dicentrarchus labrax*) 11p-hydroxylase gene expression during thermo-sensitive period and gonadogenesis. Gen Comp Endocrinol, 150: 164–173.

Sola, I., S. Brassonello, A.R. Rossi et al. 1993. A karyotype analysis of the genus *Dicentrarchus* by different staining techniques. J Fish Biol, 43: 329–337.

Strussmann, C.A. and R. Patino. 1995. Temperature manipulation of sex differentiation in fish. pp. 153–157. *In*: F.W. Goetz and P. Thomas (eds.). Proc Fifth Internatl Symp Reprod Physiol Fish. Austin, Texas.

Strussmann, C.A. and R. Patino. 1996. Evidence of thermo-labile sex determination in pejerrey. J Fish Biol, 48: 643–651.

Strussmann, C.A. and R. Patino. 1999. Sex determination, Environmental. *In*: E. Knobil and J.D. Neil (eds.). *Encyclopedia of Reproduction*. Academic Press, San Diego, 4: 402–409.

Strussmann, C.A., C. Joue, C. Cota et al. 1996. Temperature effects on sex differentiation of two South American atherinids *Odontesthes argentienensis* and *Patagonina hatcheri*. Env Biol Fish, 47: 143–154.

Strussmann, C.A., T. Saito, M. Usui et al. 1997. Thermal thresholds and critical period of thermo-labile sex determination in two atherinid fishes (*Odontesthes bonariensis* and *Patagonina hacheri*. J Exp Zool, 278: 167–177.

Strussmann, C.A., T. Saito and F. Takashima. 1998. Heat-induced germ cell deficiency in the teleosts *Odontesthes bonariensis* and *Patagonina hatcheri*. Comp Biochem Physiol, 119A: 637–644.

Sullivan, J.A. and R.J. Schultz. 1986. Genetic and environmental basis of variable sex ratio in laboratory strains of *Poeciliopsis lucida*. Evolution, 40: 152–158.

Sun, P., F. You, D. Ma et al. 2013. Sex steroid changes during temperature-induced gonadal differentiation in *Paralichthys olivaceus* (Temminck and Schlegeli, 1846). J Appl Ichthyol, 29: 886–890.

Talling, J.F. 2001. Environmental controls on the functioning of shallow tropical lakes. Hydrobiologia, 458: 1–8.

Tessema, M., Muller-Belecke and G. Horstgen-Schwark. 2006. Effect of rearing temperatures on the sex ratio of *Oreochromis niloticus* populations. Aquaculture, 258: 270–277.

Trewavas, E. 1983. Tilapine fishes of the genera *Sarotherodon*, *Oreochromis* and *Danakilia*. British Museum (Natural History), London, p. 583.

Tveldt, H.B., T.J. Benfey, D.J. Martin-Robichaud et al. 2006. Gynogenesis and sex determination in Atlantic halibut (*Hippoglossus hippoglossus*). Aquaculture, 252: 573–583.

Uchida, D., M. Yamashita, T. Kitano and T. Iguchi. 2004. An aromatase inhibitor or high water temperature induces oocyte apoptosis and depletion of p450 aromatase activity in the gonads of genetic female zebrafish during sex reversal. Comp Biochem Physiol, 137A: 11–20.

Van den Hurk, R. and Van Oordt. 1985. Effects of natural androgens and corticosteroids on gonad differentiation in the rainbow trout, *Salmo gairdneri*. Gen Comp Endocrinol, 57: 216–222.

van Nes, S. and O. Anderson. 2006. Temperature effects on sex determination and ontogenetic gene expression of the aromatase *cyp19a* and *cyp19b* and the estrogen receptors *esr1* and *esr2* in Atlantic halibut *Hippoglossus hippoglossus*. Mol Reprod Dev, 73: 1481–1490.

van Nes, S., M. Moe and O. Anderson. 2005. Molecular characterization and expression of *cyp19* (p45 aromatase) genes in embryos, larval and adults of Atlantic halibut (*Hippoglossus hippoglossus*). Mol Reprod Dev, 72: 437–449.

Vinagre, C., T. Ferreira, L. Motos et al. 2009. Latitudinal gradients in growth and spawning of sea bass *Dicentrarchus labrax* and their relationship with temperature and photoperiod. Estua Coast Shelf Sci, 81: 375–380.

Wang, L.-H. and C.-L. Tsai. 2000. Effects of temperature on the deformity and sex differentiation of tilapia *Oreochromis mossambicus*. J Exp Zool, 286: 534–537.

Wessels, S. and G. Hortgen-Schwark. 2007. Selection experiments to increase the proportion of males in Nile tilapia (*Oreochromis niloticus*) by means of temperature treatment. Aquaculture, 272S1: S80–S87.

Wibbels, T., J.J. Bull and D. Crews. 1994. Temperature-dependent sex determination: a mechanistic approach. J Exp Zool, 270: 71–78.

Winge, O. 1930. On the occurrence of XX males in *Lebistes* with some remarks on Aida's so called " Non-disjunctional" males in *Aplocheilus*. J Genet, 23: 69–76.

Yamamoto, T. 1969. Sex differentiation. *In*: W.S. Hoar and D.J. Randall (eds.). *Fish Physiology*. Academic Press, New York, 3: 117–175.

Yamamoto, E. 1995. Studies on sex-manipulation and production of cloned populations in hirame flounder *Paralichthys olivaceus* (Temminck et Schlegeli). Bull Tottori Prof Fish Exp Stn, 34: 1–145.

Yamamoto, E. 1999. Studies on sex manipulation and production of cloned populations in *Paralichthys olivaceus* (Temminck et Schlegeli). Aquaculture, 173: 235–246.

Yamamoto, Y., R.S. Hattori, A. Kitahara et al. 2013. Thermal and endocrine regulation of gonadal apoptosis during sex differentiation in pejerrey *Odontesthes bonariensis*. Sex Dev, 7: 316–324.
Yamashira, K. and D.O. Conover. 2003. Interpopulation variability in temperature-dependent sex determination in the tidewater silverside *Menidia peninsulae* (Pisces: Atherinidae). Copeia, 2003: 155–159.
Zanuy, S., M. Carrillo and A. Felip et al. 2001. Genetic hormonal and environmental approaches for the control of reproduction in the European sea bass (*Dicentrarchus labrax* L.). Aquaculture, 202: 187–203.

8

Climate Change and Differentiation

Introduction

Since 1750, the level of atmospheric carbon dioxide has been rising from 280 ppm to 385 ppm in 2010 and is predicted to increase to 560 ppm by 2050 (Table 8.1). During the last 250 years, the levels of other greenhouse gases, methane and nitrous oxide have also increased from 715 ppb to 1774 ppb and from 270 ppb to 320 ppb, respectively (IPCC, 2013). As a consequence, the global climate is gradually warming the habitats including oceans, seas, lakes, rivers and ponds inhabited by fishes and other organisms. Global mean temperature has increased by about 0.2°C/decade over the last 30 years with most of the added energy absorbed by surface waters of oceans (up to 700 m depth), where temperature increased by about 0.6°C over the 100 years and is continuing to increase. Global spring season is advanced by 2.3 days per decade and species range is shifted by 6.1 km/decade toward the poles (Parmesan and Yohe, 2003). This undesired change in temperature of aquatic systems may have a dramatic effect on geographic distribution of fishes (e.g., cod being replaced by sea bass, Eddy and Handy, 2012; see also Perry et al., 2005), shift in spawning season (e.g., threadfin bream, Vivekanandan, 2013b) and change in sex ratio (e.g., Atlantic silverside, Conover and Kynard, 1981). Long term monitoring of the health and reproductive status of *Perca fluviatilis* females in the Baltic Sea, Hansson et al. (2006) reported decreases in age from 6.0 years in 1988 to 4.2 years in 2000 and gonado somatic index from 6.75 to 4.5. During these years, the

Table 8.1. Changes in climate features during the last 30–50 years and predicted changes by 2050 (source: Diaz and Rosenberg, 2008; Vivekanandan, 2011; Eddy and Handy, 2012; IPCC, 2013).

Climate features	Last 30–50 years	2050
Atmospheric CO_2 (ppm)	385	560
pH of oceanic waters	–0.1	–0.1 to –35
Sea surface temperature (°C)	+0.4	+1.5
Catastrophic coral bleaching (times)	+2	+15 to 25
Sea level rise (mm/year)	1	8*
Hypoxic aquatic system (no)	400	+280
Wind speed increase	3.5%/1°C increase	

*Predicted during 2080s.

authors noted not only temperature increased but also continuous supply of hydrocarbons to the Baltic Sea.

Another effect of global warming is the melting of polar ice resulting in expansion of ocean volume, elevating the sea levels (Table 8.1) and dilution of sea water. Global climate models predict 40% reduction in winter ice cover by 2050 (Pfeiffer and Haynie, 2012). Consequently, stratification of oceanic waters and re-routing of currents may reduce mixing in some areas, which may lead to reduced nutrient availability and primary production resulting in 'ocean deserts' (Eddy and Handy, 2012). The oceans have absorbed about 30% of the CO_2 produced by human activity (Eddy and Handy, 2012). Consequent to the use of seas as sink for CO_2, the acidification has reduced the pH of ocean waters (Table 8.1). Denman et al. (2011) have shown that the decrease in pH is more critical than that of temperature but this area of research has attracted far less investigations than that of temperature (Chapter 6, 7). An account on the effect of increasing temperature on reproductive impairment of fishes is described.

8.1 Model systems

While climatic change has inflicted negative effects on organisms and ecosystems, specific negative effects encountered by selected ecosystems differ from each other. Some of these ecosystem specific negative effects are described below:

Freshwater systems: With increase in temperature, freshwater ecosystems suffer from 'oxygen squeeze' and changes in algal assemblages. The impacts of increasing global temperature range from rapidly drying smaller ponds and puddles to sinking of epilimnion to deeper waters in large lakes (Ficke et al., 2007). More pronounced and long-lasting stratification may reduce the amount of oxygen exchange to the hypolimnion from oxygen-rich epilimnion. When oxygen demand in the hypolimnion exceeds supply, hypoxic or anoxic conditions prevail. Fishes inhabiting such hypolimnion encounter 'temperature-oxygen-squeeze'. The effects of hypoxia on sex ratio and impairment of reproduction have been described in Chapter 5. Tropical freshwater fishes live in systems, where diurnal thermal fluctuations usually approach their incipient upper lethal limits (Irion and Junk, 1997). Hence tropical freshwater fishes may not endure even a slight increase in regional temperatures (Roessig et al., 2004). However, elevated temperatures seem to equally affect temperate fishes like *Menidia menidia*, *Odontesthes bonariensis*, *Hoplosternum littorale* and tropical fish like *Poeciliopsis lucida*, in which male ratio is increased over 0.75 at elevated temperatures (Fig. 8.1, Table 8.2).

Thermal stratification is a major driving force in determining algal-assemblages (Ficke et al., 2007). Longer periods of stratification promote the blooming of blue-green algae, which have the ability to fix nitrogen in amictic, nutrients-limited conditions (George et al., 1990). Blue-green algae are inedible to most species of zooplankton that planktivorous fishes eat (Kangur et al., 2002). Besides some algae like *Anabaena flos-aquae* produce toxic alkaloids (de Souza et al., 1998).

Marine ecosystem: Every species has its temperature range and typically, energy allocated towards growth and reproduction declines at temperatures near the extremes (Sogard and Olla, 2002; Sogard and Spencer, 2004). The capture-release mortality of fishes also increases within species-specific

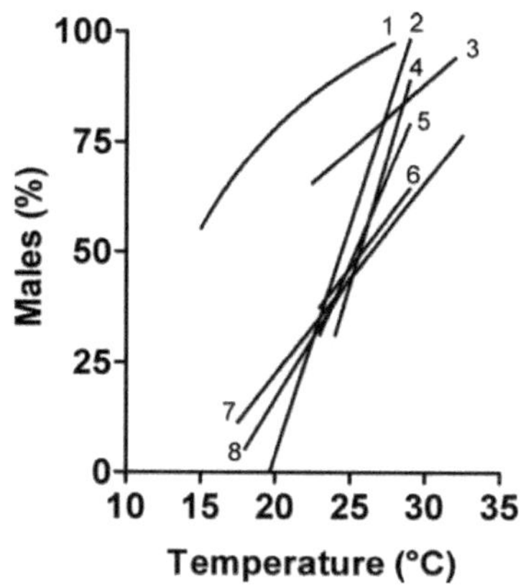

Fig. 8.1. Patterns of sex ratio response to fishes characterized by Temperature-dependent Sex Differentiation (TSD). In all these fishes, proportion of males increases with increasing temperature. Key: 1 *Menidia menidia*, 2. *Odontesthes banariensis*, 3. *Hoplosternum littorale*, 4. *Poeciliopsis lucida*, 5. Mean of 33 *Apistogramma* spp., 6. *Limia melanogaster*, 7. *M. peninsulae*, 8. *Odontesthes argentiensis* (permission by Dr. F. Piferrer, from Ospina-Alvarez, N. and F. Piferrer. 2008. Temperature-dependent sex determination in fish revisited. Prevalence, a single sex ratio response pattern and possible effects of climate change. PLoS One, 3(7): e2837. doi: 10.1371/ journal.pone 0002837).

thermal preferenda (Gale et al., 2013). Lower and upper incipient lethal temperatures for the Atlantic salmon *Salmo salar* range between –0.5 and +25°C (Wallace, 1993). Within its optimal range 13–17°C, the salmon is usually found in cooler habitat. Food-deprived *Gasterosteus aculeatus* prefers cooler seawater (15.9°C) than well-fed fish (20.1°C), when both are earlier acclimatized to the same temperatures at 10–12°C (Magee et al., 1999). *Gadus morhua* prefers cooler conditions, when fed lower ration (Despatie at al., 2001). Similarly, the intertidal fish *Girella larveifrons* selects cooler temperatures, when fed poor quality food (Pulgar et al., 2003). Hence fishes seem to prefer a cooler habitat, where they can reduce the energy cost of metabolism.

Phytoplankton records in the North Sea showed a regime break in 1988, causing expanded distribution of phytoplankton and associated zooplankton. The planktivorous horse mackerel (*Trachus trachura*) also expanded its migratory pattern following food availability (Reid et al., 2001). Hence, temperature-induced availability of quality and/or quantity of food may lead to shift in expanded distribution pattern of the fish to cooler waters (e.g., Perry et al., 2005). Incidentally, laboratory experiments on seven species of phytoplankton showed that the rate of multiplication was faster and cell density was greater at 29°C than those at 24°C. However, the decay set in earlier, for instance, the decay of microalgae was on day 12 at 24°C but on day 10 at 29°C. Further, the composition of this algal assemblage was also altered. Briefly, these laboratory experiments indicate the potential response in growth rate, species composition and longevity of phytoplankton to higher temperature (CMFRI, 2009).

Polar regions: Polar marine habitats are characterized by well oxygenated, ice-laden waters with narrow cold temperature ranges (Rose, 2000). For example, the upper lethal temperature of *Trematous hansoni* is 5°C (Somero and DeVries, 1967). The fish can live in super-cooled waters at –1°C below the ice sheet. To withstand the super-cooled waters, many polar fishes have antifreeze peptides or glycopeptides in their blood and tissues, which allow their body temperature sinks to –0.8°C but without freezing (DeVries and Cheng, 2005). Possessing these antifreeze molecules, these fishes are capable of tolerating –1.9°C (Portner, 2002). Fishes in the Antarctic region produce these compounds year-round, while those in the Arctic Ocean produce them only during winter, as regulated temperature and photoperiod (Koch, 1992). Antarctic fishes (e.g., Nototheriidae) lack heat shock proteins, that repair damaged proteins from cellular insults like heat (Hofmann et al., 2000). Cold temperatures decrease oxygen requirements of fishes. The hematology of these fishes in oxygen-rich habitat, for example, *Pleurogramma antarcticum* is very fascinating (Mesa and Eastman, 2012). Eggs and larvae of the Antarctic silverfish with an intact chorion can survive and remain viable at temperatures as low as –9°C (Cziko et al., 2006).

Food chain in the polar regions are short and depend on relatively a few species. For example, the Antarctic silverfish inhabiting both open waters and areas of pack ice feeds on three calanoid copepods and two species of euphausiids (Mesa and Eastman, 2012). Hence the polar regions represent an extremely fragile ecosystem at weak equilibrium. Specializations to colonize the polar regions with unusual environmental constraints have taken million of years of evolution. A sudden change in environmental conditions driven by the current rapid climate change and overfishing (e.g., *Dissostichus mansoni*, Ainley et al., 2013) could negatively affect this weak equilibrium, with a catastrophic cascade effects on life in the polar regions.

Coral reefs: Mean sea temperatures in the vicinity of coral reefs are predicted to increase over the course of this century. Corals are known for their extreme sensitivity to changes in temperature (Vivekanandan et al., 2009). Exposure to higher ocean temperature just a few degrees above the long-term average at any location can cause coral to become stressed, bleach and die (Hoegh-Guldberg, 1999). The number of decadal catastrophic bleaching events is predicted to increase from 0 during 2000–2009 to 8 during 2080–2089 (Vivekanandan et al., 2009). Coral reefs are also threatened by ocean acidification, which reduces the ability of scleractinian coral to form carbonate skeleton (Kleypas et al., 2006). More dangerously, the acidification negatively interferes with the formation of skeletal system in fish larvae and calcification of external skeleton of crustacean larvae. Figure 8.2 provides an

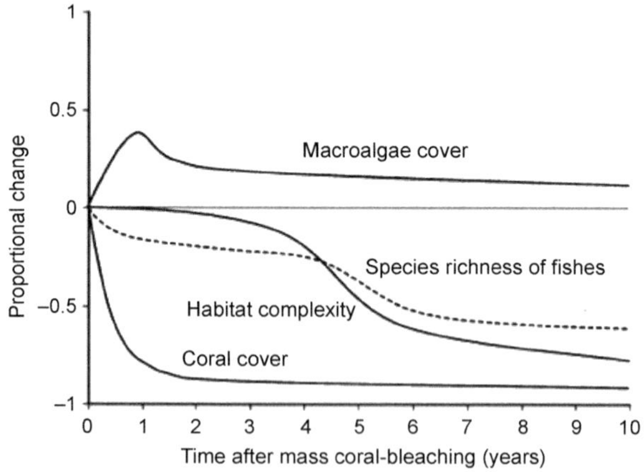

Fig. 8.2. Proposed declines in key attributes of coral reef habitats namely coral cover, habitat complexity and macro algae cover (solid lines) and fish communities, i.e., species richness (from Munday et al., 2008).

idea of changes in key attributes of coral reef habitats and fish communities following severe coral bleaching.

Over 4,000 species of teleost fishes are associated with coral reefs (Allen, 2008). Declines in live coral cover and changes in the species composition of the remaining corals and the loss of habitat complexity will decrease the diversity and abundance of reef fish assemblages and alter their species composition (Pratchett et al., 2008). Coral dependent species include obligate coral-dwellers, corallivores and others, who settle into live corals (Munday et al., 2008). Approximately, 10% coral reef fishes are obligatory coral dependent (Pratchett et al., 2008) and they are the ones most immediately affected by coral loss (Kokita and Nakazona, 2000; Munday, 2004; Pratchett et al., 2006). Especially, those that inhabit or consume just one or a few coral species often suffer the greatest declines, as they are unable to use alternative coral resources (Munday, 2004). Butterflyfishes (Chaetodontidae), cardinalfishes (Apogonidae) and gobies (Gobiidae) are some coral reef families that are more reliant on corals than others; they are directly at risk from loss of corals (Pratchett et al., 2008). Among others, 75% of the coral fish species declined in abundance, with 50% of species declining more than 50% following severe loss of live corals on coral reefs off Papua New Guinea (Jones et al., 2004).

Slower growth of herbivorous damselfishes during warm El Nino conditions in the Galapagos Islands has been traced to the replacement of the preferred green and red algae by the brown algae of lower nutrient value (Meekan et al., 1999). Most coral reef fishes have a life cycle that includes a pelagic larval stage lasting for a few weeks, more or less coinciding with the temperature-sensitive labile period for sex differentiation. With climate change, the embryos and larval stages associated with corals are more vulnerable. Unlike the corals, most reef fishes are not living very close to their lethal thermal limits (Mora and Ospina, 2001). Nevertheless, they are reported to suffer from slower growth and impaired reproduction, in response to the increasing temperature in the coral reefs.

8.2 Impairment and distribution

Temperature is the single most important environmental factor that directly influences molecular, biochemical and physiological processes, especially in poikilothermic fishes. The previous chapter described how elevated temperatures derail sex differentiation in genetic females to 'change sex' to phenotypic males and skewed sex ratios in temperature-sensitive individuals, families, stocks, strains and species of fishes. In response to elevated temperatures, the fishes may also shorten or expand or shift spatial distribution to a cooler environment. Temperature may also impair reproduction by disruption of endocrine process.

Impairment of differentiation: As in other vertebrates, the reproductive cycle of fishes is controlled by brain-pituitary-gonadal axis. Gonadotropic hormones, namely Follicular Stimulating Hormone (FSH) and Luteinizing Hormone (LH) are the key hormones that regulate gametogenesis, maturation and spawning/spermiation (Pandian, 2013, Chapters 2 and 3). Exposure of Tasmanian stock of *Salmo salar* to temperatures of 14, 18 and 22°C from midsummer to late autumn reduced egg size at 22°C with corresponding reductions in egg fertility and survival (King et al., 2003). Incidentally, this observation confirms an earlier report by Campbell et al. (1994) in *S. trutta* and in non-salmonids like *Anarhichus lupus* (Tveiten and Johnsen, 2001). Exposure of salmonids to high temperature during autumn delayed or inhibited ovulation. In *Salvelinus alpinus*, elevated temperatures > 8°C delayed ovulation and that at 11°C completely inhibited it (Gillet, 1991). Maintanance of the Tasmanian *S. salar* females at 16°C for longer duration (from spring to autumn) completely inhibited ovulation (King and Pankhurst, 2000). More experiments involving exposure at 22°C for a longer duration and initial exposure to 22°C and subsequent transfer to 14°C showed that (i) the follicles of *S. salar* held at 22°C have low capacity to secrete estradiol (E_2) in response to steroid precursors and gonadotropin (hCG), (ii) consequently, the impaired E_2 secretion reduces vitellogenesis and finally egg size. However, Baroiller et al. (1995) successfully undertook progeny testing by crossing temperature-induced sex reversed phenotypic (XX) Δ neomale with a normal XX female. Clearly, spermatogenesis in the phenotypic male *Oreochromis niloticus* was not affected by elevated temperatures of 36°C.

In many Arctic, temperate and tropical fishes, elevated temperatures reduce or inhibit (i) the release of GnRH, (ii) steroidogenesis, (iii) proportion of ovulating females, and (iv) impair and delay maturation and fertility of females, as well (Table 8.2). In some, they also reduce fecundity and survival of progenies. A climax is the European sea bass, which does not spawn at temperatures above 16°C. Corresponding information is awaited for tropical fishes but a single publication is available on tropical *Hippocampus kuda* (Lin et al., 2006). A comparative account on the effects of elevated temperatures in a temperate pollock and tropical seahorse shows that elevated temperatures affect more or less equally selected parameters of reproduction in both the temperature and tropical fishes (Table 8.3).

The mean air temperature recorded in Chascomus County, Argentina increased by 1.74°C between 1968 and 2007 (40 years). Hence it may be quite interesting to know the elevated temperature effect in an Argentinean *Odontesthes bonariensis*. In this pejerrey, reproductive activity begins in austral spring at 13–15°C and peaks at 18–20°C; natural spawning occurs at 19°C (Strussmann, 1989). Soria et al. (2008) exposed the adult pejerrey to 23 and 27°C to a short period of 8 days; that reduced gonado somatic

Table 8.2. Effect of elevated temperatures on steroidogenesis and reproductive performance of sub-arctic, temperate and tropical fishes.

Species/Authors	Reported observations
	Sub-arctic
Salvelinus fontinalis Gillet and Breton (2009)	Inhibition of LH and ovulation at 10–11°C
Anarhichus lupus Tveiten et al. (2001)	Reduced testosterone and estradiol levels. Ovulation is delayed at 8°C and 12°C; survival of incubated eggs at 6°C is significantly decreased
	Cold temperate
Coregonus larvetus Wahl and Loffler (2009)	Delayed ovulation and spawning at higher temperature
Salmo salar Taranger and Hansen (1993), Whal and Loffler (2009), Pankhurst and King (2010)	Reductions in steroidogenesis and vitellogenesis. Proportion of ovulating females is reduced from 98% at 5–7°C to 57% at 13–14°C. Elevated temperatures impaired maturation, ovulation and fertility of females
Oncorhynchus mykiss Pankhurst et al. (1996); Pankhurst and Thomas (1998)	Reduction in steroidogenesis. Proportion of ovulating females is significantly decreased at > 15°C. Reduced fertility and survival at > 18°C
Pagrus major Lim et al. (2003)	Inhibition of aromatase and 11β-hydroxylase at higher temperatures
Gadus morhua Johansen et al. (1999); Van der Meeren and Ivannikov (2001)	Proportion of ovulating females is decreased from 89% at 8°C to 75% at 12°C; fertilization success is also decreased at 10°C
Hippoglossus hippoglossus Brown et al. (1995)	Fecundity decreases at temperatures above > 8°C
Dicentrarchus labrax Carrillo et al. (1993)	Does not spawn at 16°C and above
	Warm temperate
Odontesthes bonariensis Soria et al. (2008)	Reduced expression of LH βsub-unit FSH receptor gene, testosterone and estradiol levels and fertility at 24°C
Stegestes beebei Ruttenberg et al. (2005)	Relative fecundity rapidly declines from a peak at 25°C to a very low level at 27°C
	Tropical
Acanthochromis polyacanthus Donelson et al. (2010)	Reduced aromatase activity, estradiol level and fecundity at 30°C
Pomacentrus ambonensis Gagliano et al. (2007)	Pre-hatching mortality is higher at 31°C than at 29°C
Dascyllus aruanus Pankhurst et al. (in press)	Reduced estradiol production at 33°C

index and completely blocked spawning. There was also clear reduction in sperm count and diameter of spermatogenic tubules. Soria et al. also found that gonadotropic hormones exerted their action through FSHr and LHr receptors. Hence gonads were the primary targets for the elevated temperatures, which disrupted the endocrine process. These results suggest

Table 8.3. Comparative account on the effects elevated temperatures on reproductive performance of temperate food-fish *Pollachius pollachius* (source: Suquet et al., 2005) and tropical ornamental fish *Hippocampus kuda* (source: Lin et al., 2006) .

Parameter	*P. pollachius*		*H. kuda*	
	10°C	12°C	28°C	32°C
Days required to develop ovary	-	-	85	88
Spawning period (day)	42	47	-	-
GSI	-	-	16.6	10.8
Spawning frequency (times)†	17.3	3.0	-	-
Fecundity	3.3×10^{5}*	2.6×10^{5}*	608**	372**
Viable eggs	5.6×10^{4}*	4.2×10^{3}*	-	-
Fertilization success (%)	16.0	7.6	92	54
Hatching success (%)	9.8	-	94	51
Survival (%)	-	-	91	46

† Number of times nine females spawned, * Egg number/kg body weight,
**Egg number/female.

shortened spawning season and reduced fecundity in *O. bonariensis*. With increase in trend in global warming, which has become evident in the Argentinean pejerrey in less than 40 years, shortening of spawning season and reduction in annual fecundity may become not only recurrent in the pejerrey but also in other sensitive fishes (see Strussmann et al., 2010).

The fishes may suffer greatest impacts of increased temperature effect on breeding season, which may be advanced, postponed, reduced or extended or bimodelled. In salmonids, elevated temperatures truncate spring spawning and delay autumn spawning (Pankhurst and Munday, 2011). Annual mean Sea Surface Temperature (SST) in the south west Bay of Bengal increased from 28.42°C in 1980s to 28.74°C in 2005. For first time in Indian seas, the percentage occurrence of spawners of the two threadfin breams *Nemipterus japonicus* and *N. mesoprion* decreased from 40% during the warmer period from April to September during 1980s to just 10% in 2000s. In response to increase of 0.32°C in SST, the mean number of spawners of *N. japonicus* decreased from 4,917 during 1980s to 2355 during 2000s (Table 2.10). Vivekanandan (2013a,b,c) has shown that extended spawning season in this tropical fish is shifted to cooler months.

In response to climate change especially temperature, many fishes either shifted or extended the range of geographic distribution. Munday et al. (2008) lists selected examples for the expected range of shifts in response to increased Sea Surface Temperature (SST). Within the North Sea, plaice *Pleuronectes platessa* has shifted northward to deeper waters in response to climate change (Engelhard et al., 2011). Interestingly, the Indian pelagic fishes provide a dramatic extension of distribution along the Indian coast due to climate change. The oil sardine *Sardinella longiceps* and the Indian mackerel *Rastrelliger kanagurta* are tropical small pelagic fishes, forming massive fisheries along India. The length of the Indian West coast is 3,340 km, of which the Southwest coast (latitude: from 8° N to 14° N; longitude: from 75° E to 77° E, about 1,140 km) has been the 'home' of the oil sardine, where SST ranged from 27 to 29°C. Due to climate change, the SST began to climb at 0.04°C/decade and the warmer tongue (27.0–28.5°C) of the surface waters began to expand beyond 14°N, enabling the sardine to extend its range to the northern latitude (22° N, 90° E) both in the West and East coasts of India (Fig. 8.3). In the last 30 years, the distribution has expanded by about 2,200 km of the coastal length at the rate of 88 km per year (cf. Parmesan and Yohe, 2003, p 222). Considering the peak distribution by oil sardine up to 15 km from the shore, the fish has extended its range of distribution to more than five times, i.e., from 17,000 km^2 (along Southwest

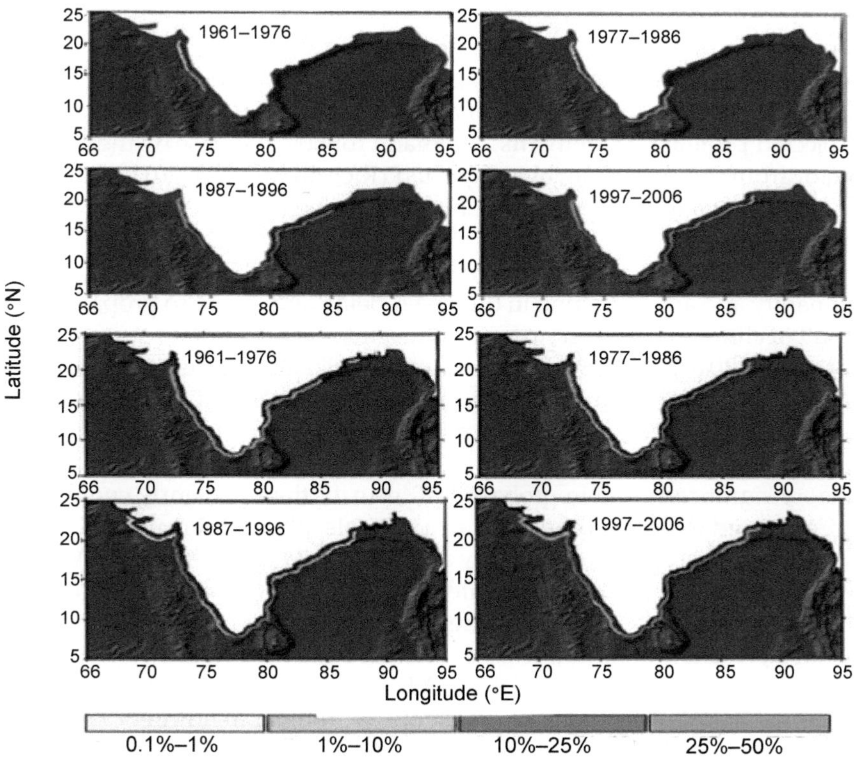

Fig. 8.3. Extension of distributional boundary of oil sardine *Sardinella longiceps* (upper 4 panels) and Indian mackerel *Rastrelliger kanagurta* (lower 4 panels) along the Indian west and east coasts during the period from 1961–1976 to 1997–2006. The colored lines indicate the percentage contribution of catch by different maritime states of India (from Vivekanandan, 2011, permission by Central Marine Fisheries Research Institute, Kochin).

Color image of this figure appears in the color plate section at the end of the book.

coast of India) until the 1980s to 90,000 km^2 along the entire Indian coast in about 30 years (Vivekanandan, 2011). Similarly, mackerel was also found to have expanded its range to the northern latitudes of both East and West coasts of India. In addition to the extension to the northern boundaries, the mackerel is found to descend to deeper water during the last two decades in response to the warming of sea surface water (Vivekanandan, 2011).

8.3 Global food security

As shown earlier, global fish production from fisheries is stabilized at around 90 mt. Aquaculture production is 60 mt and at the current rate of its expansion, it may reach 132 mt fishes and 43 mt seaweeds by 2020 (Gjedrem et al., 2012) and cumulative fish production may be more than 200 mt by 2020. As the aquatic products are rich in micronutrients, and contain high quality protein and omega-3-fatty acids, their nutritional value is higher than that of farm animals. Fishes may provide about 7.5 billion people with 20% of their intake of animal protein (FAO, 2012). To meet the projected protein requirements, a demand for 50% increase by the fisheries sector from the current level is expected (Rice and Garcia, 2011) and many scientists have expressed their concern over feeding the world (e.g., Frid and Paramor, 2012).

Fisheries constitute one of the most important natural resources for humanity and any reduction in their well-being and ability to propagate as a result of climate change may have significant socio-economic consequences (Brander, 2010). There is voluminous criticism against the policy of Maximum Sustainable Yield (MSY) from both scientists and economists, as the policy disguised as science was imposed by developed countries (Finley and Oreskes, 2013). For the following reasons, fisheries may play a more significant role in global food security than presently considered. 1. Compared to homoeothermic farm animals, poikilothermic features are more efficient converters of energy and protein (Pandian, 1987). 2. Whereas the number of farm animals is limited to a dozen species, there are hundreds of candidate species for aquaculture; however, the techniques developed for culturing them are limited to 54 species in South Korea and to less than a dozen in India (Pandian, 2001). 3. In 2005, the coastal aquaculture production of China averaged to 103 t/km coast line with the maximum of 172 t/km. The mean freshwater aquaculture in Bangladesh was 0.17 t/ha with the maximum of 6.0 t/ha. These ranges indicate that there is considerable scope to realize a 20-fold increase in global aquaculture production (Gjedrem et al., 2012). 4. Aquaculture lags far behind plant and animal farming in utilizing selective breeding as a tool to improve the biological efficiency of aquaculture production. For example, increased use of genetically improved strains and stocks in aquaculture is expected to produce 12.5% genetic gain per generation (Gjedrem and Thodesen, 2005). Clearly, fisheries and aquaculture have a very significant role to play in feeding the world. To develop genetically improved stocks using selective breeding and other techniques (e.g., transgenics, see Pandian and Marian, 1994), there is a need to conserve the available stocks and strains (e.g., a dozen strains in goldfish, see Pandian, 2011, p. 258).

References

Ainley, D.G., N. Nur, J.T. Eastman et al. 2013. Decadal trends in abundance, size and condition of Antarctic toothfish in McMurdo Sound, Antarctica, 1972–2011. Fish Fisher, 14: 343–363.

Allen, G.R. 2008. Conservation hotspots of biodiversity and endemism for Indo-Pacific coral reef fishes. Aquat Conserv Mar Freshwat Ecosyst, 18: 541–556.

Baroiller, J.F., D. Chourrout, A. Fostier and B. Jalabert. 1995. Temperature and sex chromosomes govern sex ratios of the mouthbrooding cichlid fish *Oreochromis niloticus*. J Exp Zool, 273: 216–223.

Brander, K. 2010. Impacts of climate change on fisheries. J Mar Systems, 79: 389–402.

Brown, N.P., N.R. Bromage and R.J. Shields. 1995. The effect of spawning temperature on egg viability in the Atlantic halibut (*Hippoglossus hippoglossus*). *In*: Proc 5th Internatl Symp Reprod Physiol Fish, pp. 95–97.

Campbell, P.M., T.G. Pottinger and J.P. Sumpter. 1994. Primary evidence that chronic confinement stress reduces quality of gametes produced by brown trout and rainbow trout. Aquaculture, 120: 151–169.

Carrillo, M., S. Zanuy, F. Prat et al. 1993. Environmental induction of spawning in sea bass. *In*: *Recent Advances in Aquaculture*. Blackwell, Oxford, 4: 43–54.

CMFRI. 2009. Annual Report 2008–2009. Central Marine Fisheries Research Institute, Kochi, India, p. 104.

Conover, D.O. and B.E. Kynard. 1981. Environmental sex determination: Interaction of temperature and genotype in a fish. Science, 213: 577–579.

Cziko, P.A., C.W. Evans, C.C. Cheng and A.L. DeVries. 2006. Freezing resistance of antifreeze-deficient larval Antarctic fish. Polar Sci, 4: 295–308.

Denman, K., J.R. Christian, N. Steiner et al. 2011. Potential impacts of future ocean acidification on marine ecosystems and fisheries: current knowledge and recommendations for future research. ICES J Mar Sci, 68: 1019–1029.

de Souza, R.C.R., M.C. Carvalho and A.C. Truzzi. 1998. *Cylindrospermopsis raciborskii* (Wolosz.) Seenaya and Subba Raju (Cyanophyceae) dominance and a contribution to the knowledge of Rio Pequeno Arm, Billings Reservoir, Brazil Environ Toxicol Water Quality, 13: 73–81.

Despatie, S.P., M. Casonguay, D. Chabot and C. Audet. 2001. Final thermal preferendum of Atlantic cod: effect of food ration. Trans Am Fish. Soc, 130: 263–275.

DeVries, A.L. and C.-H.C. Cheng. 2005. Anti-freeze proteins and organisamal freezing avoidance in polar fishes. *In*: A.P. Farrell and J.F. Steffensen (eds.). *The Physiology of Polar Fishes*, Fish Physiology. Elsevier Academic Press, San Diego, 22: 155–201.

Diaz, R.J. and R. Rosenberg. 2008. Spreading dead zones and consequences for marine ecosystems. Science, 321: 926–929.

Donelson, J.M., P.L. Munday, M.I. McCormick et al. 2010. Effects of elevated temperatures and food availability on the reproductive performance of a coral reef fish. Mar Ecol Prog Ser, 401: 233–243.

Eddy, F.B. and R.D. Handy. 2012. *Ecological and Environmental Physiology of Fishes*. Oxford University Press, Oxford.

Engelhard, G.H., J.K. Pinnegar, L.T. Kell and A.D. Rijnsdorp. 2011. Nine decades of North Sea sole and plaice distribution. ICES J Mar Sci, 68: 1090–1104.

FAO. 2012. The State of World Fisheries and Aquaculture in Asia Pacific. FAO, Rome, p. 230.
Ficke, A.D., C.A. Myrick and L.J. Hansen. 2007. Potential impacts of global climate change on freshwater fisheries. Rev Fish Biol Fish, 17: 581–613.
Finley, C. and N. Oreskes. 2013. Maximum sustainable yield: a policy disguised as science. ICES J Mar Sci, 17: 245–250.
Frid, C.L.J. and O.L. Paramor. 2012. Feeding the world: What role for fishes? ICES J Mar Sci, 69: 145–150.
Gagliano, M., M.I. McCormick and M.G. Meekan. 2007. Temperature-induced shifts in selective pressure at a critical developmental transition. Oecologia, 152: 219–225.
Gale, M.K., S.G. Hinch and M.R. Donaldson. 2013. The role of temperature in the capture and release of fish. Fish Fisher, 74: 1–33.
George, D.G., D.P. Hewitt, J.W.G. Lund and W.J.P. Smyly. 1990. The relative effects of enrichment and climate change on the long-term dynamics of *Daphnia* in Esthwaite Water, Cumbria. Freshwater Biol, 23: 55–70.
Gillet, C. 1991. Egg production in an Arctic charr (*Salvelinus alpinus* L.) brood stock: effects of temperature on the timing of spawning and the quality of eggs. Aquat Living Resour, 4: 109–116.
Gillet, C. and B. Breton. 2009. LH secretion and ovulation following exposure of Arctic charr to different temperature and photoperiod regimes: responsiveness of females to gonadotropin-releasing hormone analogue and dopamine antagonist. Gen Comp Endocronol, 162: 210–218.
Gjedrem, T. and J. Thodesen. 2005. Selection. *In*: T. Gjedrem (ed.). *Selection and Breeding Programs in Aquaculture*. Springer Verlag, Berlin, p. 364.
Gjedrem, T., N. Robinson and M. Rye. 2012. The importance of selective breeding in aquaculture to meet future demands for animal protein: A review. Aquaculture, 350: 117–129.
Hansson, T., E. Lindesjoo, L. Forlin et al. 2006. Long term monitoring of the health status of female perch (*Perca fluviatilias*) in the Baltic Sea shows decreased gonad weight and increased hepatic EROD activity. Aquatic Toxicol, 79: 341–355.
Hoegh-Guldberg, O.P.J., Mumby, A.J. Hooten et al. 1999. Climate change, coral bleaching and the future of the world's coral reefs. Mar Freshwat Res, 50: 839–866.
Hofmann, G.E., B.A. Buckley, S. Airaksinen et al. 2000. Heat-shock protein expression is absent in the Antarctic fish *Trematomus bernacchii* (Family Nototheniidae). J Exp Biol, 203: 2331–2339.
IPCC. 2013. Working Group. 1. Climate change 2013: The Physical Science Basis: Summary to Policy makers. Intergovernmental Panel on Climate Change, Contribution to the Fifth Assessment Report, p. 36.
Irion, G. and W.J. Junk. 1997. The large Central Amazonian River floodplains near Manaus. pp. 23–46. *In*: W.J. Junk (ed.). *The Central Amazon Floodplain: Ecology of a Pulsing System*. Springer-Verlag, Berlin, Heidelberg, Germany.
Johansen, T., G. Nyhammer, E. Otterlei and G. Naevdal. 1999. Sexual maturation in cod (*Gadus morhua* L.) in relation to strain, temperature, sex, size and hemoglobin type. Proc Sixth Internatl Symp Reprod Biol Fish. Bergen, Norway, p. 343.
Jones, G.P., M.I. McCormick, M. Srinivasan and J.V. Eagle. 2004. Coral decline threatens fish biodiversity in marine reserves. Proc Natl Acad of Sci USA, 101: 8251–8253.
Kangur, A., P. Kangur and E. Pihu. 2002. Long-term trends in the fish communities of Lakes Peipsi and Vortsjarv (Estonia). Aqua Ecosyst Health Manage, 5: 379–389.
King, H.R. and N.W. Pankhurst. 2000. Ovulation of Tasmanian Atlantic salmon maintained at elevated temperatures: Implications of climate change for sustainable industry development. Proc Sixth Internatl Symp Reprod Biol Fish. Bergen, pp. 396–398.
King, H.R., N.W. Pankhurst, M. Watts and P.M. Pankhurst. 2003. Effect of summer elevated temperatures on gonadal steroid, vitellogenesis and egg quality in female Atlantic salmon. J Fish Biol, 63: 153–167.

Kleypas, J.A., R.A. Feely, V.J. Fabry et al. 2006. Impacts of Ocean Acidification on Coral Reefs and other Marine Calcifiers. A Guide for Future Research NSF, NOAA and the US Geographical Survey St. Petersburg, FL, p. 88.

Koch, K.H. 1992. *Antarctic Fish and Fisheries*. Cambridge University Press, Cambridge.

Kokita, T. and A. Nakazona. 2000. Rapid response of an obligatory corallivorous filefish *Oxymonocanthus longirostris* (Monocanthidae) to a mass coral bleaching event. Coral Reefs, 20: 155–158.

Lim, B.S., H. Kagawa, K. Gen and K. Okuzawa. 2003. Effects of warm waters temperature on the gonadal development and expression of steroidogenic enzymes in the gonad of juvenile of red seabream *Pagrus major*. Fish Physiol Biochem, 28: 161–162.

Lin, Q., J. Lu, Y. Gao et al. 2006. The effect of temperature on gonad, embryonic development and survival rate of juvenile seahorses, *Hippocampus kuda* Bleeker. Aquaculture, 254: 701–713.

Magee, A., C.A. Myrick and J.J. Cech. 1999. Thermal preference of female three spine sticklebacks underfed and food-deprived conditions. Calif Fish Game, 85: 102–112.

Meekan, M.G., G.M. Wellington and I. Axe. 1999. El Nino-Southern oscillation events produce checks in the otoliths of coral fishes in the Galapagos Archipelago. Bull Mar Sci, 64: 383–390.

Mesa, M.L. and J.T. Eastman. 2012. Antarctic silverfish: life strategies of a key species in the high-Antarctic ecosystem. Fish and Fisher, 13: 241–266.

Mora, C. and A.F. Ospina. 2001. Tolerance to high temperatures and potential impact of sea warming on reef fishes of Gorgona Island (tropical Eastern Pacific). Mar Biol, 139: 765–769.

Munday, P.L. 2004. Habitat loss, resource specialization, and extinction on coral reefs. Global Change Biol, 10: 1642–1647.

Munday, P.L., G.P. Jones, M.S. Pratchett and A.J. Williams. 2008. Climate change and the future for coral reef fishes. Fish Fisher, 9: 261–285.

Ospina-Alvarez, N. and F. Piferrer. 2008. Temperature-dependent sex determination in fish revisited. Prevalence, a single sex ratio response pattern and possible effects of climate change. PLoS One, 3(7): e2837. doi: 10.1371/journal.pone 0002837.

Pandian, T.J. 1987. Fish. *In*: T.J. Pandian and F.J. Vernberg (eds.). *Animal Energetics*. Academic Press, San Diego, 2: 357–464.

Pandian, T.J. 2001. *Sustainable Indian Fisheries*. National Academy of Agricultural Sciences, New Delhi, p. 327.

Pandian, T.J. 2011. *Sex Determination in Fish*. Science Publishers, Enfield/CRC Press, Boca Raton, USA, p. 282.

Pandian, T.J. 2013. Endocrine Sex Differentiation in Fish. CRC Press, Baca Raton, USA, p. 303.

Pandian, T.J. and L.A. Marian. 1994. Problems and prospects of transgenic fish production. Curr Sci, 66: 635–648.

Pankhurst, N.W. and P.M. Thomas. 1998. Maintanance at elevated temperature delays the steroidogenic and ovulatory responsiveness of rainbow trout *Oncorhynchus mykiss* to lutenizing hormones releasing hormone analogue. Aquaculture, 166: 163–177.

Pankhurst, N.W. and H.R. King. 2010. Temperature and reproduction: Implications for aquaculture. J Fish Biol, 76: 69–85.

Pankhurst, N.W. and P.L. Munday. 2011. Effect of climate change on fish production and early life history stages. Mar Freshwater Res, 62: 1015–1026.

Pankhurst, N.W., G.J. Purser, G. Van Der Kraak et al. 1996. Effect of holding temperature on ovulation, egg fertility, plasma levels of reproductive hormones and *in vitro* ovarian steroidogenesis in the rainbow trout *Oncorhynchus mykiss*. Aquaculture, 146: 277–290.

Parmesan, C. and G. Yohe. 2003. A globally coherent fingerprint of climate change impacts across natural systems. Nature, 421: 37–42.

Perry, A.L, P.J. Low, J.R. Ellis and J.D. Reynolds. 2005. Climate change and distribution shifts in marine fishes. Science, 308: 1912–1915.

Pfeiffer, L. and A.C. Haynie. 2012. The effect of decreasing seasonal sea-ice cover in the winter Berring Sea pollock fishery. ICES J Mar Sci, 69: 1148–1159.

Portner, H.O. 2002. Physiological basis of temperature-dependent biogeography: trade-offs in muscle design and performance in polar ectotherms. J Exp Biol, 205: 2217–2230.
Pratchett, M.S., S.K.Wilson and A.H. Baird. 2006. Declines in the abundance of chaetodon butterflyfishes following extensive coral depletion. J Fish Biol, 69: 1269–1280.
Pratchett, M.S., P.L. Munday and S.K. Wilson. 2008. Effects of climate-induced coral bleaching on coral-reef fishes. Ecological and economic consequences. Oceanogr Mar Biol: Annu Rev, 46: 251–296.
Pulgar, J.M., M. Aldana, F. Bozinovic and F.P. Ojeda. 2003. Does food quality influence thermoregulatory behavior in the intertidal fish *Girella larveifrons* J Therm Biol, 28: 539–544.
Reid, P.C., M. de Fatima Borges and E. Svendsen. 2001. A regime shift in North Sea circa 1988 linked to changes in the North Sea horse mackerel fishery. Fish Res, 50: 163–171.
Rice, J.C. and S.M. Garcia. 2011. Fisheries, food security, climate change, and biodiversity: characteristics of the sector and perspectives on emerging issues. ICES J Mar Sci, 68: 1343–1353.
Roessig, J.M., C.M. Woodley, J.J. Cech et al. 2004. Effects of global climate change on marine and estuarine fishes and fisheries. Rev Fish Biol Fisher, 14: 251–275.
Rose, K.A. 2000. Why are quantitative relationships between environmental quality and fish populations so elusive. Ecol Appl, 10: 367–385.
Ruttenberg, B.I., A.J. Haupt, A.I. Chiriboga and R R. Warner. 2005. Patterns, causes and consequences of regional variation in the ecology and life history of a reef fish. Oecologia, 145: 394–403.
Sogard, S.M. and B.L. Olla. 2002. Contrasts in the capacity and underlying mechanisms for compensatory growth in two pelagic marine fishes. Mar Ecol Prog Ser, 243: 165–177.
Sogard, S.M. and M.L. Spencer. 2004. Energy allocation in juvenile sablefish: effects of temperature ration and body size. J Fish Biol, 64: 726–738.
Somero, G.N. and A.L. DeVries. 1967. Temperature tolerance of some Antarctic fishes. Sci, 156: 257–258.
Soria, F.N., C.A. Strussmann and L.A. Miranda. 2008. High water temperatures impair the reproductive ability of the pejerrey fish *Odontesthes bonariensis*. Physiol Biochem Zool, 81: 898–905.
Strussmann, C.A., D.O. Conover, G.M. Somoza and L.A. Miranda. 2010. Implications of climate change for the reproductive capacity and survival of New World silversides (family Atherinopsidae). J Fish Biol, 77: 1818–1834.
Strussmann, C.A. 1989. Basic studies on seed production of pejerrey *Odontesthes bonariensis.* Ph.D. Thesis, Tokyo University of Fisheries, Tokyo.
Suquet, M., Y. Normant, J.L. Gaignon et al. 2005. Effect of water temperature on individual reproductive activity of pollack (*Pollachius pollachius*). Aquaculture, 243: 113–120.
Taranger, G.L. and T. Hansen. 1993. Ovulation and egg survival following exposure of Atlantic salmon, *Salmo salar* L. broodstock to different water temperatures. Aquacult Res, 24: 151–156.
Tveiten, H. and H.K. Johnsen. 1999. Temperature experienced during vitellogenesis influences maturation and timing of ovulation in common wolffish. J Fish Biol, 55: 809–819.
Tveiten, H., S.E. Soleveg and H.K. Johnsen. 2001. Holding temperature during the breeding season influences final maturation and egg quality in common wolffish. J Fish Biol, 58: 374–385.
Van der Meeren, T. and V. Ivannikov. 2001. Seasonal shift in spawning of cod broodstocks by light manipulation: egg quality and larval rearing. *In*: C.I. Hendry, G.V. Van Stappen, M. Wile and P. Sorgeloos (eds.). Larvi 2001, Proc Third Fish Shellfish Larvicult Symp, EAS, Oostende, Sp Publ, 30: 616–617.
Vivekanandan, E. 2011. Climate change and Indian Marine Fisheries. Central Marine Fisheries Research Institute, Kochi, Spl Publ, 105: 1–97.
Vivekanandan, E. 2013a. Climate change: Challenging the sustainability of marine fisheries and ecosystems. J Aquat Biol Fish, 1: 58–71.

Vivekanandan, E. 2013b. Seawater warming coincides with shift in the spawning season of a tropical bream in the southwest Bay of Bengal. Rev Fish Biol Fish (in press).
Vivekanandan, E. 2013c. Precocious maturity and variations in reproduction of threadfin bream *Nemipterus japonicus* southwest Bay of Bengal. Sexuality, Early Development of Aquatic Organisms (communicated).
Vivekanandan, E., H. Ali, B. Jasper and M. Rajagopalan. 2009. Vulnerability of corals to warming of the Indian seas: A projections for the 21st century. Curr Sci, 97: 1654–1658.
Whal, B. and H. Loffler. 2009. Influences on the natural reproduction on whitefish (*Coregonus larvetus*) in Lake Constance. Can J Fish Aquat Sci, 66: 547–556.
Wallace, J. 1993. Environmental considerations. pp. 127–143. *In*: K. Heen, R. Monahan and F. Utter (eds.). *Salmon aquaculture*. Fishing News Books. Oxford.
Yamashira, K. and D.O. Conover. 2003. Interpopulation variability in temperature-dependent sex determination of tidewater silverside *Menidia peninsulae* (Pisces: Athernidae). Copeia, 2003: 155–159.

9

Conservation

The previous account on the anthropogenic activities, that disrupt sex differentiation cascade of fish in one way or other, has evidently shown the need for conservation of fishes. The need for conservation of the fish genome has become indeed urgent for many commercially and some biologically (e.g., *Tetradon fluviatilis*, Pandian, 2002a) important fishes. *In situ* conservation demands legislation and its implementation but is of local importance. Interestingly, the Indian Emperor Asoka (270–232 BC) is the first to ban inland fishing operations during spawning season. On the other hand, *ex situ* conservation of fishes requires scientific development and is of global importance. This chapter is limited to the latter.

9.1 Cryopreservation

During the checkered history of evolution, plants and microbes 'invented' seeds and spores, from which life can be recovered after thousands and millions years of dormancy. Animals too have 'invented' cysts and the like but the dormancy in them does not last longer than a few decades (Pandian, 2002b). Fishes do not produce cysts. Necessarily, offspring of fishes have to be recovered from their gametes and/or embryos.

Cryopreservation of gametes and embryos involves a series of complex and dynamic physico-chemical processes of temperature and water transportation between the 'cryopreserved' and surrounding medium. The techniques developed for mammalian eggs and embryos are not applicable to the yolk-filled eggs and embryos of fishes. Due to lecithality, fish eggs/embryos are not amenable to cryopreservation (Hagedorn and Kleinnans, 2000), despite the claims of new approaches to cryopreserve fish embryos (e.g., Bart, 2000). The earliest technique of obtaining viable eggs/embryos protected with 8–14% dimethyl sulfoxide (DMSO) frozen at –20°C (for eggs) and –55°C (for embryos) may at the best represent supercooling but not cryopreservation (Tiersch, 2001; Diwan et al., 2010). Hence the scope for cryopreservation of eggs and embryos of fishes is at present limited and the technique may not be useful to conserve fish.

Cryopreservation of fish sperm has been a hot topic of research for the last 60 years (Diwan et al., 2010). Cryopreserved sperm is mostly used in breeding programs and maintenance of genetic diversity. Protocols for successful cryopreservation of sperm are available for about 200 species of mostly commercial importance (Tiersch, 2000). Live, fertile sperm can also be drawn from specimens post-mortem preserved at –20°C for periods of 30 days in *Puntius conchonius* (Kirankumar and Pandian, 2004a), 40 days in *Hemigrammus caudovittatus* (David and Pandian, 2006b) and 240 days in *Heteropneustes fossilis* (Koteeswaran and Pandian, 2002). In *H. caudovittatus,* the sperm count is 8.6 x 10^3/ml with 19% fertilizability of the cadarveric sperm drawn from the 30-day post-mortem preserved specimen. It increased to 6.8 x 10^4/ml with 24% fertilizability, when the specimens were glycerol packed prior to preservation at –20°C (David and Pandian, 2006b). There is considerable scope to improve the technique and to increase sperm count and fertilizability as well as effective duration of preservation of cadarveric sperm. These cadarveric sperm are also capable of activation of homospecific and heterospecific eggs (Kirankumar and Pandian, 2004a; David and Pandian, 2006b). Nevertheless, the recovery of the desired strain/species using preserved sperm requires homospecific or compatible heterospecific eggs, and the appropriate induction protocol to diploidize the haploid androgenics. Even then, maternally inherited genes and mitochondrial DNA are not recovered in the androgenics (Yasui et al., 2011).

9.2 Vitrification

Although claims have been for successful attempts to cryopreserve embryos (e.g., zebrafish, Chao and Liau, 2001, flounder, Chen and Tian, 2005; Robles et al., 2005), a reproducible method per se has not been developed. The large size, high sensitivity to cooling and low membrane permeability of fish eggs are the major impediments in the process of cryopreservation. Slow cooling (see Table 9.1), vitrification (Higaki et al., 2009, 2010a) and yolk removal (Higaki et al., 2010b) are a few methods that improved survival of the cryopreserved embryos, especially the PGCs in the embryos.

The syncytial layer surrounding the embryo is known to prevent permeation of cryoprotectant into embryos; consequently, the Primordial Germ Cells (PGCs) suffer co-injuries during the process of cryopreservation (Hagedorn et al., 1998). To reduce the loss of PGCs during cooling-thawing and handling, cryopreservation by vitrification was attempted. Survival of

Table 9.1. Viable durations of fish blastomeres that were cryopreserved by slow-cooling.

Species	Viability (%)	References
Sillago japonica	19.9	Strussmann et al. (1999)
Oryzias latipes	34.1	Strussmann et al. (1999)
Carassius auratus	54	Kusuda et al. (2004)
Oncorhynchus keta	59.3	Kusuda et al. (2002)
O. mykiss	36–93	Kusuda et al. (2004); Calvi and Maisse (1998)
Misgurnus anguillicaudatus	62	Yasui et al. (2011)
Odontesthes bonariensis	67.4	Strussmann et al. (1999)
Danio rerio	70–85	see Lin et al. (2009)
Cyprinus carpio	96	Calvi and Maisse (1999)

the zebrafish from its dechorionated embryos at 14–20 somite stage was 20 and 40% in solutions containing DMSO and ethylene glycol, respectively (Higaki et al., 2009, 2010a). Preventing the ice formation during cooling and yolk removal may further improve the viability of PGCs in embryos. Expectedly, the viablility of PGCs in yolk-removed vitrified embryos, that were exposed to the optimal exposure to equilibration solution and vitrification solution, was 90% (10.6 cells/11.8 cells); of them 50% (5 cells/embryo) alone displayed pseudopodial movement, a feature required for PGC migration to genital ridge (Higaki et al., 2010b).

To their credit, Higaki et al. (2010b) also demonstrated the generation of fertile male and female allogenics recovered from vitrified embryos. From the *Gfp*-labeled PGCs recoverd from vitrified-warmed embryos of the dark-spotted and dark-striped zebrafish strain, a single PGC was transplanted into the dechorionated blastula of golden strain that was earlier sterilized by injecting *dnd* morpholino (see Ciruna et al., 2002). Following 2-days (post transplantation), the *Gfp*-labeled PGC was detected in the genital ridge of about 7.5% recipients and finally five males and one female mature, gold-colored chimeras were produced. The presence of one female is interesting, as it may question the earlier report by Slanchev et al. (2005), who found that all the *dnd*-merpholino-treated zebrafish embryos automatically developed into males. When the gold colored chimeras were crossed with normal gold colored zebrafish, all the F_2 progenies displayed the donor-derived dark-spots and dark-stripes clearly indicating that a single PGC following cryopreservation by vitrification of yolk-removed embryos is adequate to generate the donor-derived male and female allogenic F_2 progenies.

9.3 Androgenesis

Whereas cloning involves the transfer of nucleus from differentiated cell to an enucleated egg, the androgenic technique includes the transfer of sperm of donor to genome-inactivated/eliminated recipient egg. Hence androgenesis is a developmental process that facilitates exclusive inheritance of the paternal genome. Essentially, induction of androgenesis comprises of three steps: 1. Inactivation of maternal genome in smaller eggs of < 2 mm (diameter) by UV-irradiation, and fragmentation and elimination of maternal genome in larger eggs of > 2 mm by radiation at ^{60}Co (25kR–88kR) or X-ray (Komen and Thorgaard, 2007). 2. Activation of the genome inactivated/eliminated eggs by homospecific or heterospecific sperm to generate intra-specific or inter-specific haploid androgenics and 3. Activiation of a genome-inactivated egg by a haploid sperm and suppression of the first mitotic cleavage to diploidize the androgenic progeny.

Since survival of the androgenics at hatching is less than 15% and in most cases < 5% (Table 9.2), it has become necessary to critically analyze the available information to assess the loss of androgenics during the processes

Table 9.2. Selected examples of intraspecific androgenesis in fish to show the effect of egg irradiation and diploidization. (D) = Donor, (R) = Recipient.

Experimental design	Remarks
Oreochromis niloticus UV-irradiated gray egg (R) x genetically marked sperm (D) Marengoni and Onoue (1998)	Survival at hatching 1.6%
Puntius tetrazona UV- irradiated gray egg (R) x blond sperm (D) Kirankumar and Pandian (2003)	Survival at hatching 15% and at sexual maturity 7% F_1 Sex ratio = 1 ♀ : 1 ♂
Hemigrammus caudovittatus UV-irradiated gray egg (R) x golden sperm (D) David and Pandian (2006a)	Survival at hatching 10% and at maturity 6% F_1 Sex ratio = 1 ♀ : 1 ♂
Misgurnus anguillicaudatus UV-irradiated (1500 J/m^2) egg x n sperm	Survival at hatching = 7.1%, of which 46% alone were normal fry
UV-irradiated (1500 J/m^2) egg x 2n sperm Arai et al. (1995)	Survival at hatching = 8%, of which > 70% were normal fry
Cyprinus carpio, 25–30 kR X-ray irradiated egg Grunina et al. (1990)	Survival of n androgenic 12% No bi-parental hatchling
C. carpio, UV-irradiated (1750 J/m^2) gray egg (R) x n blond sperm (D) Bongers et al. (1994)	Survival of n androgenic 54%, with 3% bi-parental hatchling. Survival of 2n androgenic 13%
Oncorhynchus mykiss ^{60}Co-irradiated (36 kR) egg x n sperm pressure shocked for diploidization	Survival at hatching 1.6%
^{60}Co irradiated (36 kR) egg x cryopreserved 2n sperm Thorgaard et al. (1990)	Survival at hatching 48.5%

of (i) diploidization by heat or pressure shock, (ii) irradiation by UV to genome inactivation or X-ray to eliminate the maternal genome in surrogate eggs, and (iii) due to homozygosity. Using diploid sperm, the diploidization procedure from the protocol of androgenic induction can be avoided. Arai et al. (1995) used haploid as well as diploid sperm to generate androgenics in *Misgurnus anguillicaudatus* eggs that were previously UV-irradiated at 1500 J/m^2 (see also Fujimoto et al., 2007). With the use of diploid sperm, the androgenic embryos were not subjected to diploidization by heat shock. Survival of the diploid androgenics in both cases were almost equal (7–8%). However, among the surviving fry, there were 54 and 30% abnormal fry, when haploid and diploid sperm were used, respectively. As a result, the net survival of normal fry was 3.2% in the former, but 5.6% in the latter. Hence avoiding diploidization by heat shock almost doubles survival of normal androgenic at hatching. Yasui et al. (2010) have recovered the loach using cryopreserved diploid sperm. To eliminate maternal genome from eggs of *Oncorhynchus mykiss*, the relatively large eggs have to be irradiated by X-ray (36 kR). Thorgaard et al. (1990) generated androgenic *O. mykiss* using cryopreserved haploid and diploid sperm and reported 1.6% survival, when induced with haploid sperm followed by diploidization by pressure shock, against 48.5% survival, when diploid sperm was used. Hence diploidization by pressure shock is more injurious than heat shock.

Hatching percentages of haploid androgenics provide an opportunity to assess the injurious effect of irradiation of fish eggs by UV or X-rays. X-irradiating (25–30 kR) *Cyprinus carpio* eggs, Grunina et al. (1990) reported 12% survival of carp haploid androgenics. Bongers et al. (1994) used UV-irradiation at 1750 J/m^2 to genome-inactivate the carp eggs, and reported as high as 54% survival of haploid androgenics, albeit there were 3% with maternal contamination. To generate androgenics in the mud loach *M. mizolepis*, Nam et al. (2002) also UV-irradiated (1750 J/m^2) the surrogate carp eggs and reported 57% survival of haploid mud loach androgenics. Hence X-ray irradiation is more injurious than UV-irradiation.

For sustenance, it is important to know survival and sex ratio among successive generations of androgenics. Of 16 one-year-old F_2 androgenic loach *M. anguillicaudatus* that were generated crossing tetraploid androgenics, Fujimoto et al. (2010) reported all male progenies. Crosses between these F_2 diploid males and normal diploid females yielded a fewer fertile females and large number of partially sterile males. Contrastingly, survival of diploid androgenic progenies of *Puntius conchonius* that arose from the crosses between androgenic F_1, F_2 and F_3 males with corresponding androgenic females remained constant at 14% at hatching and 7% at sexual maturity (see Table 9.2). Their sex ratio also remained constant at 0.42♀ : 0.58♂ (Kirankumar and Pandian, 2004b). Hence the androgenics generated by diploid parents are desirable.

To introduce heterozygosity, dispermic androgenics were induced. The dispermic induction required the complete fusion between two haploid sperm. The fusion was ensured by prior incubation of semen in polyehylene glycol (PEG). Incidentally, the dispermic induction avoids the damage caused by application of heat or pressure shock for diploidization. Kirankumar and Pandian (2004c) successfully induced dispermic interspecific androgenics of the golden rosy barb *P. conchonius* using surrogate eggs of the gray tiger barb *P. tetrazona*. They reported just 1.7% survival for the interspecific dispermic androgenics, against 7 and 14% survival for the inter- and intra-specific monospermic androgenics induced in heterologous and homologous eggs, respectively (Tables 9.2, 9.3). Inducing dispermic interspecific androgenics of *H. caudovittatus* using surrogate eggs of *Gymnocorymbus ternetzi*, David and Pandian (2008) also reported equally low survival values. Tracing the causes, they found that the PEG-incubated semen contained only 78% motile sperm, of which 37% remained single and induced haploids alone and others suffered incomplete fusion or wrong fusion like the head to head fusion (Fig. 9.1). As a result, 14% of sperm alone were completely fused from head to tail and they activated only 61% of the UV-irradiated surrogate eggs. The causes for the low survival was thus traced to 49% of the dispermic androgenic embryos suffering haploid syndrome and another 10% suffering other developmental incompatibities, and 39% inactivated eggs. Consequently, 1.7% dispermic interspecific androgenics alone survived at hatching. However, the survival could be increased to 4.3% with availability of 19% completely fused sperm, when sperm concentration was quantified. Nevertheless, the technique of dispermic interspecific androgenesis may not be useful to conserve fish.

Interestingly, an attempt was made by Pandian and Kirankumar (2003) to understand the role played by haploid and diploid genome in regulation

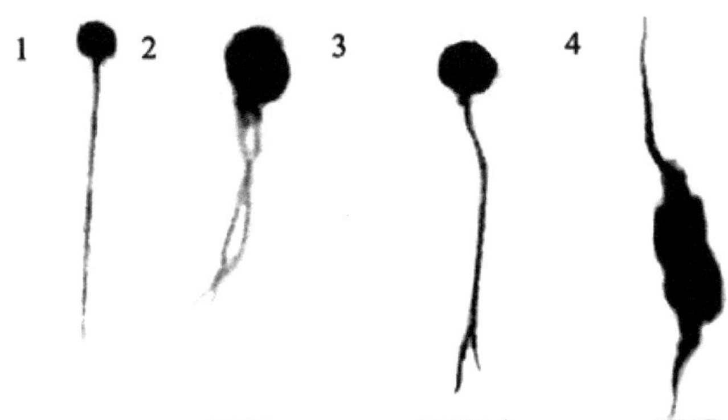

Fig. 9.1. Representative magnifications of *Hemigrammus caudovittatus* sperm (1) a single sperm, (2) attached/fused sperm each with two tails, (3) head to tail fully attached/fused sperm, and (4) head to head fused sperm (from David and Pandian, 2008).

of temporal sequence of development and DNA code from the nucleus of donor species being translated by cytoplasmic RNA of the recipient belonging to alien species. From the trends obtained for the progression of embryonic development, as a function of embryonic survival of *P. conchonius* (egg size 882 µm) and *P. tetrazona* (egg size 1230 µm), characterized by differences in egg size (Fig. 9.2), they arrived at following conculsions: 1. The alien haploid genome regulated the chronological sequence of development as effectively as that of diploid genome; this observation has implication to the haploid embryonic stem cells supporting embryonic development to hatching (Araki et al., 2001). 2. The haploid and diploid genome drawn either from fresh or cadraveric sperm of rosy barb regulated developmental sequence in alien surrogate egg of tiger barb and 3. The time sequence characteristic of the rosy barb is maintained in tiger barb eggs with about 25% excess yolk.

Hitherto, the technique of monospermic interspecific androgenesis, which requires cadarveric or cryopreserved sperm of the desired species, has been achieved only in four species of fishes (Table 9.3). Surprisingly, successful induction of interspecific androgenesis and recovery of androgenics derived from preserved sperm through surrogate eggs is possible crossing taxonomic borders of species, as in *P. conchonius* x *P. tetrazona*, genus, as in *Carassius auratus* x *Cyprinus carpio*, *Hemigrammus caudovittatus* x *Gymnocorymbus ternetzi* and family *M. mizolepis* (Cobitidae) x *C. carpio* (Cyprinidae). Hence the technique of monospermic interspecific androgenesis using cadarveric or cryopreserved sperm of desired species and eggs of compatible surrogate species may prove useful to conserve fish.

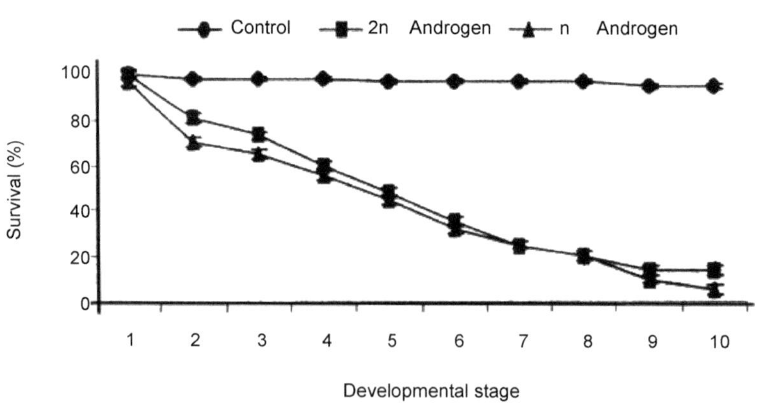

Fig. 9.2. Survival of androgenic clone of golden rosy barb *Puntius conchonius* as a function of developmental stage. Numbers in the X-axis indicate selected embryonic stages (from Pandian and Kirankumar, 2003).

Table 9.3. Survival and sex ratio of interspecific androgenic fishes. D = Donor, R = Recipient.

Experimental design	Remarks
Carassius auratus sperm (D) x *Cyprinus carpio* egg (R) Bercsenyi et al. (1998)	Survival at hatching with fresh sperm 19% but with cryopreserved sperm 23%
Misgurnus mizolepis sperm (D) x *C. carpio* egg (R) Nam et al. (2002)	Survival 19% at hatching and 5.7% at 4 weeks after hatching for homozygous androgenics but it is 7.9% for heterozygous androgenics
Puntius conchonius sperm (D) x *P. tetrazona* egg (R) Kirankumar and Pandian (2003, 2004a,b,c) (Pandian and Kirankumar, 2003)	Survival 14 and 7% at hatching with fresh and cadarveric sperm; 7 and 3% at sexual maturity with fresh and cadarveric sperm. Survival of F_1,F_2 and F_3 androgenic progenies 15% at hatching and 7% at maturity. Survival 1.7% at hatching with dispermic induction. Sex ratio o.42♀ : 0.58♂
Hemigrammus caudovittatus sperm (D) x *Gymnocorymbus ternetzi* eggs (R) David and Pandian (2006a,b, 2008)	Survival 15 and 7% at hatching with fresh and cadarveric sperm, 7 and % at maturity with fresh and cadarveric sperm. Survival 1.8% at hatching and 1.3% at maturity with dispermic induction. Sex ratio = 0.5♀ : 0.5♂

9.4 Cloning

Since the birth of the celebrated Dolly, the first cloned sheep (Wilmut et al., 1997), attempts are being made to clone fish. However, most of these attempts to clone fish are limited to the homospecific and heterospecific transplantation of nucleus from egg or at the best embryonic cell but not from differentiated somatic cell to the enucleated or non-enucleated egg. The very first successful attempt was made by Anon (1980) under the guidance of Prof TC Tung in China. Nucleus from just fertilized carp egg was sucked out into a fine glass needle and transferred into an enucleated crucian carp egg. The 'cloned'* offspring possessed all characteristics of the carp such as the barbel pairs, one pair on each side of mouth, 36 scales on the lateral line and so on, against the absence of barbels and presence of 28 scales on the lateral line. Since then, this type of 'cloning' has been achieved by transfer of nucleus of carp *Cyprinus carpio* into the enucleated egg of crucian carp *Carassius auratus* (Sun et al., 2005), and nucleus of zebrafish *Danio rerio* (Rasborinae) into the enucleated eggs of rare minnow *Gobiocypris rarus* (Gobioninae) (Pei et al., 2007). Remarkably, these transfers were achieved crossing the borders of genus and sub-family. Interestingly, Bubenshchikova et al. (2005) furthered the series of cloning experiments transferring nucleus

*By oversight, it was earlier regarded as androgenesis (Pandian, 2011, p. 70).

from blastula and four-somite-stage embryonic cell into non-enucleated medaka eggs that were activated but not fertilized. Following that, Hattori et al. (2011) transferred nucleus of the blastula cell of pigmented medaka into a non-enucleated, and activated but not fertilized egg of golden medaka. As a result, triploid and tetraploid, if the egg nucleus underwent mitosis, and diploid, if the egg nucleus was eliminated, hatchlings were present. Only 2.9% of hatchlings survived and 1.7% survived to sexual maturity. Of 31 nuclear transplants, just three (1.6%) were golden medaka. They neither displayed male mating behavior nor produced eggs. Apparently, they were all triploid, sterile males. Briefly, the technique of cloning fish is at every nascent status and may not serve now as a technique to conserve fish genome. In mammals, it is now known that enucleated metaphase II oocytes are more suited as surrogates than mature oocytes and non-enucleated eggs (Thuan et al., 2010); the nucleus of embryonic cell at G1 stage is better suited for cloning (Kasinathan et al., 2001).

9.5 Reprogramming

Another possibility of recovering fish from its somatic cells is to reprogram the entire sequence of differentiation process in a unipotential somatic cell to totipotency. In vertebrates, maternal genes direct fertilization, zygotic activation, the first few cell division(s) and initiation of zygotic transcripition (Dorsch et al., 2004; see also Pandian, 2012, p. 145). The zygotic genome activation (ZGA) is a point of transition from maternal to embryonic control of development. In embryos, the totipotency is lost at 2-cell stage in mice, 4–8-cell stage in human and at about 1,000-cell stage (≈3 hour post-fertilization, hpf) in zebrafish. Hence, as an ideal model, the zebrafish provides more time to study the process of ZGA. Potok et al. (2013) analyzed the DNA methylation patterns of sperm cells, egg cells and three stages prior to ZGA. When methyl groups attached to certain areas of DNA, gene activity is 'switched off'. Hence studies on methylation pattern of totipotent cell and comparison of it with that of pluripotent and unipotent cells may open the vistas of reprogramming the unipotent cell to totipotency.

9.6 Tissue culture

Besides cryopreservation, development of techniques for long term culture of embryonic cells may prove rewarding. However, Lee et al. (2002) alone seem to be engaged in this frontier area of research. Interestingly, the remarkable achievement by Lee et al. (2002) raises a ray of hope for cloning fish. Lee et al. transferred the *Gfp*-marked nucleus from embryonic fibroblast G1 cell cultured at 28–29°C over long (> 12 weeks) period into an enucleated eggs of zebrafish. Following surgical transplantation, 80% of the eggs failed to develop but at least 2%, i.e., eight females and three males of the nuclear transplants survived to sexual maturity.

Instead of cultured embryonic cells, cryopreserved blastomeres can also be used for cloning. Viability of fish blastomeres that were cryopreserved by slow cooling ranges from just 20 days in *Sillago japonica* to 96 days in *Cyprinus carpio* (Table 9.1). Yasui et al. (2011) injected cryopreserved-thawed blastula cells drawn from 30% epiboly stage of the loach *M. anguillicaudatus* into the surrogate embryos of *D. rerio*. Hatching success and proportion of germ-line chimera among the hatchling were just 3 and 1.1%, respectively. These values were 90 and 16.5% for the xenogenics, whose embryos received the loach cryopreserved blastomeres following incubation for 16 hours at 20ºC after thawing. Hence an incubation period after thawing significantly increased not only hatchability but also the cloning success. Understandably, the experiments of Yasui et al. may lead to the development of a successful technique.

9.7 Xenogenesis

The PGCs are the progenitors of germ cell lineage and possess bisexual potency to differentiate into either oogonia or spermatogonia (Wylie, 2000). Hence they carry the heritable information to the next generation and are considered 'immortal' (Yan et al., 2010). Being the germ line precursor, a presumptive PGC undergoes asymmetric cell division to produce one daughter cell committed to be somatic lineage and the other to the germ line. During the process of asymmetric divisions, at a point of time, when both daughter cells are committed to the germ line, the parent cell is the PGC. The PGCs arise from maternally supplied mRNA (Hashimoto et al., 2004). Origin of the PGCs can histologically be traced to the late blastula or somitogenic stage at the best (Nagai et al., 2001). Recently, powerful molecular and cellular techniques have been developed, facilitating identification, marking, isolation and transplantation of the PGCs: for example, 1. *Vasa* can be employed as a molecular marker to identify PGCs (Olsen et al., 1997; Yoon et al., 1997). 2. To visualize PGCs and their migration to the genital ridge *in vivo,* a transcript with *Gfp* driven by *vasa* regulatory regions has also been developed (Yoshizaki et al., 2000a). The RT-PCR analysis with *vasa* specific primers has shown that *vasa* is expressed only in *Gfp*-positive cells, confirming that they are the PGCs (Yoshizaki et al., 2000b). 3. To mass isolation of PGCs carrying *Gfp* driven by *vasa,* flow cytometric technique has been developed (Takeuchi et al., 2002). To reduce contamination by somatic cells from the isolated PGCs, a flow cytometric separation technique has been developed to sort out *Gfp*-positive from *Gfp*-negative somatic cells (Kobayashi et al., 2004; Yano et al., 2008). Alternatively, germ cell suspension can be subjected to percoll gradient by centrifugation (800 g) for 30 minutes at 25°C to isolate the germ cells (Lacerda et al., 2006). However, this may not be as effective and precise as that developed by Yoshizaki's team (see Pandian, 2011, p. 147). Further, by labeling the donor's PGCs with *Gfp* and endogenous PGCs with DsRed fluorescence, the frequency of these PGCs successfully arriving in the recipient's gonadal primordium can also be estimated (Yasui et al., 2011). 4. Two molecular factors, the chemokine SDF-1 secreted by the presumptive gonadal region and its receptor *Cxcr4* expressed in the PGCs are mainly responsible for guiding the migration of PGCs to the genital ridge (Doitsidou et al., 2002; Sassado et al., 2008). The proportion of transplanted PGCs successfully reaching the genital ridge of the host zebrafish *Danio rerio* (Rasborinae, Cyprinidae) ranges from 40% for the transplanted single PGC of pearl danio *D. albolineatus* and 63% for goldfish *Carassius auratus* (Cyprinae, Cyprinidae) and 43% for the loach *Misgurnus anguillicaudatus* (Cobitidae) (Saito et al., 2010; Yamaha et al., 2010). Hence the migration success does not reflect the phylogenetic distance between the donor and recipient species, clearly indicating the scope for successful transplantation of PGCs and possibly their derivatives SSCs and OSCs. 5. Notably, spermatogonium asymmetrically divides into undifferentiated type-A stem cells and differentiated spermatogonia, which share the same morphological characteristics of type-A but with greatly reduced potential for self renewal (see Pandian, 2013, p. 71). The donor germ cells, especially type-A spermatogonia can be stained with the fluorescent membrane dye PKH26, which excites at a wave length of 551 nm and emits at 567 nm. The incorporation of PKH26-labeled donor cells into the gonad can be observed under a fluorescent microscope (B x 51N-34 FL) (Takeuchi et al., 2009). It is known that only type-A spermatogonia (Yano et al., 2008) and oogonia and small oocytes (Yoshizaki et al., 2010) are capable of migration and incorporation into the gonad of recipient.

For transplantation of germ cells in fishes, a graft containing the posterior layer of blastoderm (Shimada and Takeda, 2008), its equivalent called sandwich presumably containing PGCs (Yamaha et al., 2001, 2003) was used: transplantion of a suspension containing a single PGC (Saito et al., 2008) or 5–30 PGCs (Takeuchi et al., 2003, 2004) or SSCs (Okutsu et al., 2006a) or OSCs (Yoshizaki et al., 2010) was also used. The *Gfp* reporter gene is expressed specifically in type-A spermatogonia and spermatocysts (Okutsu et al., 2006b; Yano et al., 2008) but not in meiotic products namely spermatids and sperm; as against the continued expression of *Gfp* throughout oogenesis from oogonia to maturing oocytes (see Pandian, 2012, p. 92). This may explain why a suspension containing only about 1,800 OSCs (and others namely oocytes, etc.) was adequate to successfully generate xenogeneic rainbow trout of the masu alevin (Yoshizaki et al., 2010), whereas suspension containing over 10,000 SSCs (and others namely spermatids, spermatocytes, etc.) was required to produce allogeneics/

xenogeneica (Okutsu et al., 2007) (Table 9.4). With the restricted accessibility of PGCs during the short period of embryogenesis and the limited number of PGCs (16 in loach – 43 in goldfish, see Pandian, 2011), the technique of transplantation of PGCs suffers inherent limitations. The same is true of OSCs. Thanks to their availability in large numbers almost throughout the year and life-time of a male, spermatogonia of fish may serve as an eternal source of SSCs (see Pandian, 2011, p. 147).

The developmental stage of the recipients also ranges from blastula of hybrid carp (Yamaha et al., 2003) to fertile alevin (Takeuchi et al., 2004), sterile alevin (Okutsu et al., 2007)—all of them possessing an undifferentiated gonad and to mature artificially sterilized testis (Lacerda et al., 2006, 2010) or alevin, whose gonad is committed to differentiate into testis (Wong et al., 2010) (Pandian, 2013, p. 83). Using a cleverly designed experiment in nibea croaker *Nibea mitsukurii*, Takeuchi et al. (2009) identified the prerequisites to increase the proportion of incorporation of the transplanted SSCs into recipient's gonad and survival of the recipient. They found that (i) 3 months-old immature male is exclusively comprised of type-A spermatogonia, and (iia) the recipient larva of 6 mm length has the lowest number (≈25) of endogenous PGCs, against ≈180 in larger one (≈12 mm) and (iib) the maximum of 70% survival of the 3 weeks pt (post-transplantation) recipient larva of 6 mm size.

Table 9.4. Generation of PGCs, OSCs- or SSCs-derived allogenic fishes.

Experimental design	Results
Oncorhynchus mykiss, 5–10 *p-vasa-Gfp* labeled PGCs from 35 d-old dominant orange colored alevin to blastula of recessive gray trout Takeuchi et al. (2003)	26% survival of F_1 chimeric allogenics; of them, 17% were males and 14% females. 4% males and 2.3% females at F_2
O. mykiss, 20–30 nl suspension containing 15,000 ovarian cells comprising of *Gfp*-labeled 1,800 oogonia from 6–9 months old albino immature female to 2.5 days old gray alevin Yoshizaki et al. (2010)	3.3 *Gfp*-postive cells in each of 14% recipients by 20-day pt. The cell number increased to 133 in each of 47% recipient in 5 months old recipient. No progeny testing was undertaken
O. mykiss, suspension containing 18,000 tesicular cells comprising of 10,000 *Gfp* labeled spermatogonia from 9 months-old immature dominant alevin to 32–35 days old gray alevin, Okutsu et al. (2006a,b, 2008)	SSCs differentiated into spematazoa in recipient males and fertile eggs in recipient females. At age 1+, 50% F_1 chimeric recipient males generated 5.5% F_2 pure allogenics. At age 2+, 37–40% F_1 chimeric recipient females generated about 2% F_2 pure allogenics
Seriola quinqueradiata, 15 nl suspension containing PKH 26-labeled 20,000 cells from 10 months-old immature male to 8 days-old yellowtail progeny larva of = 6.0 mm size Morita et al. (2012)	Survival decreased from 30 at pt to 5% at sexual maturity. 86% PKH26–labeled cells detected in 28 days-old larva. At the age of 1.5–2.5 years, tests showed that 67% males and 63% females were F_1 chimeric allogenics

Table 9.4 summarizes the experimental designs to generate allogenics in *Oncorhynchus mykiss* and *Seriola quinqueradiata*. In F_1 receipients of these fishes, the transplanted PGCs or OSCs or SSCs differentiated into both male and female chimeras, clearly indicating (i) The bisexual potency of all the three germ cells to differentrate into males and females in F_1 recipient offspring. (ii) The increase in pure F_1 allogenics from 2–3% in PGCs transplanted *O. mykiss* (Takeuchi et al., 2003) to 63–67% in the SSCs transplanted allogenic F_1 progenies of *S. quinqueradiata* (Morita et al., 2012) shows that the SSCs are more efficiently incorporated into the recipients gonad than the endogenous PGCs. This may be due to the choice of donor namely the germ cell suspension 'abundant' with type–A spermatogonia and recipient larva containing less number of endogenous PGCs (see Takeuchi et al., 2009) and (iii) With transplantation of PGCs, both poloidy and fertility status of the donor are also inherited, as in *C. auratus* (see Table 9.5).

The transplantation of *O. mykiss* PGCs into *O. masou* generated chimeric male and female xenogenic F_1 progenies. The SSCs colonized hetrospecific recipients more efficiently than the PGCs; the colonizing efficiency of the PGCs increased from 13.5% in F_1 chimeric xenogenic trout to 70–88% in *Nibea mitsukurii* (Table 9.5), when its SSCs were transplanted into heterospecific recipients. However, the PGCs transplantation did generate both male

Table 9.5. Generation of PGCs, or SSCs-derived xenogenic fishes. D = Donor, R = Recipient.

Experimental design	Results
2n *O. mykiss* (D) x 2n *O. masou* (R) 10–20 *Gfp*-labeled PGCs from newly hatched trout alevin into newly hatched masu alevin Takeuchi et al. (2004)	On 30-days pt, *Gfp*-labeled PGCs detected in each of 10 recipients. Donor-derived testis in one-year-old recipient male and ovary in 2-years-old recipient female detected. 13.5% F_1 chimeric recipient males produced trout sperm. *Gfp* F_1 male was crossed with trout female, 0.4% trout progenies were alone produced
3n ♀ fertile but ♂ sterile Crucian carp (D). Sandwich containing donor PGCs was inserted into blastula containing endogenous PGCs of 2n fertile goldfish (R) Yamaha et al. (2001)	F_1 chimeric female produced donor-derived 3n large clonal eggs and recipient-derived n smaller eggs. F_1 male chimeras produced recipient-derived haploid sperm but not donor's sperm
Nibea mitsukurii (D) x *Scomber japonicus* (R) 15 nl containing PKH26-labeled spermatogonia of 3 months-old croaker into 5 mm size larva of mackerel Yazawa et al. (2010)	On 21 days pt, 21% survival. Seventy per cent of F_1 chimera recipients exhibited donor-derived germ cells. Recipient gonad colonized by 76% transplanted SSCs and the remaining 24% by endogenous PGCs
N. mitsukurii (D) x *Seriola quinqueradiata* (R) Suspension containing PKH 26-labeled testicular cells transplanted into 4 mm nibea croaker larva, Higuchi et al. (2011)	The detected PKH26-labeled SSCs in genital ridge of recipient decreased from 88% in 3 week-old recipient to 10% in 11 months-old recipient. Neither the donor-derived sperm nor eggs were differentiated in the recipients

and female chimeric xenogenics but that of SSCs could only colonize the recipient gonad but failed to produce the donor-derived sperm and eggs in recipients that belonged to different family and order. The causes for the failure of SSCs to express their bisexual potency in surrogates that are alien are to be investigated. As may be seen, the use of partially sterile triploid recipients *O. masou* generated pure allogenic and xenogenic F_1 male and female progenies. Induction of triploidy involves only a simple procedure (Pandian, 2011, pp. 90–94). It is likely that partially sterile triploid nibea croaker recipient generates fertile F_1 male and female, if not, at least F_1 male progenies in mackerel and yellowtail.

To minimize or eliminate the interference of endogenous PGCs of the recipient, the following recipients may be selected: (i) dead end antisense *morpholino oliogonucleotide*-treated embryos eliminating the expression/activity of endogenous PGCs, (ii) hybrid embryos/alevins possessing undifferentiated gonad but failing to undergo oogenesis/spermatogenesis due to interspecific incompatibility between parental PGCs and (iii) partially sterile triploid embryos/alevins. In the absence endogenous PGCs, the *dnd* morpholino-treated recipient embryos differentiate into all male progenies (Slanchev et al., 2005; Saito et al., 2008). In the absence or with reduced number of endogenous PGCs/SSCs, the artificially sterilized testis of Nile tilapia (Lacerda et al., 2006, 2010) and pejerries (Majhi et al., 2009), on transplantation of SSCs, generated allogenic male tilapia and xenogenic male pejerry alone, respectively. However, no experiment has been made so far to transplant OSCs into the sterilized testis, or SSCs or OSCs into sterilized ovary of fishes. The use of partially sterile hybrid embryos, as in *Oryzias latipes* ♀ x *O. curvinotus* ♂ (Shimade and Takeda, 2008) and fry, as in *Danio rerio* ♀ x *D. albolineatus* ♂ (Wong et al., 2010) generates only sterile male and partially sterile female, respectively. The cross between hybrid goldfish XX egg, as donor and 3n crucian carp (carrying fertile XXX eggs + sterile XXY/XYY sperm) as recipient is complicated (Yamaha et al., 2003). Hence these three methods of elimination or minimization of endogenous PGCs/SSCs may not be useful to generate allogenics or xenogenics of desired fishes.

In many fishes, triploid females are hormonally and cytogenetically sterile, whereas triploid males are hormonally fertile but cytogenetically partially or completely sterile (Pandian, 2011, p. 99, 106). In some fishes like *Carassius auratus langsdorfi*, triploid females carrying XXX or XXY genotype are fertile and produce clonal triploid eggs, but the triploid males are sterile (Yamaha et al., 2001). Partial fertility is restored in 3-years old triploid brook trout *Salvelinus fontinalis* (see Pandian, 2011, pp. 107–108). Clearly, both male and female triploids of many fishes do possess PGCs and/or SSCs/OSCs, although in reduced numbers. With the presence of reduced number of endogenous germ cells, the triploid recipient fishes may provide greater chances for the transplanted germs cells to colonize the recipient's gonad.

To generate xenogenics using sterile triploid recipient, the team of Yoshizaki selected North American iteroparous dominant orange colored diploid rainbow trout *O. mykiss* as a donor and North-east Asian semelparous triploid masu salmon *O. masou* as a recipient (Table 9.6). Interestingly, the transplantation of OSCs generated 5% 2n females and 27% 2n male pure xenogenics among F_1 progenies and SSCs 10% 2n female and 34% 2n male pure xenogenics among F_1 progenies. Secondly, the presence of masu progenies among F_1 progenies clearly indicated that the triploid sterile masu also possess reduced number of PGCs. Thirdly, a cross between recipient F_1 male carrying XX genotype and normal XX female generated all female progenies, clearly indicating that despite carrying XX genotype, the F_1 male xenogenics were in fact male phenotypes. Hence the ploidy, gamety and fertility status of the donor is also inherited by F_1 xenogenics.

The achievements and limitation of the described *ex situ* conservation techniques are briefly summarized in Table 9.7. Firstly, there are 577 viviparous fish species belonging to 135 genera in 14 families; they constitute nearly 2% of teleost fishes (see Pandian, 2011, p. 18). Cloning may prove to be the only method to recover them using long-term cultured embryonic cells. But no attempt has yet been made. Regarding androgenesis, dispermic induction and pressure shock for diploidization can readily be avoided, as there are alternatives like monospermic induction and diploidization by heat shock. An alternate to irradiation by X-ray or ^{60}Co (kR30–80) may have to be discovered to eliminate/inactivate the maternal genome in eggs of > 2.0 mm; incidentally, many commercially important fishes like salmonids do

Table 9.6. Generation of xenogenic rainbow trout transplanting OSCs or SSCs into newly hatched alevin of partially sterile triploid masu.

Experimental design	Results
2n *O. mykiss* (D) x 3n sterile *O. masou* (R) 20–3-nl suspension containing 15,000 ovarian cells comprising of *Gfp*-labeled 1,800 oogonia from 6.9 months-old mature orange female to 30–35 days-old 3n-masu Yoshizaki et al. (2010)	5% of 2 years-old F_1 recipient masu female crossed with gray trout produced orange-colored *Gfp*-positive and gray-colored *Gfp*-negative F_2 progenies at 0.5♀ : 0.5♂ ratio indicating that F_1 recipient female is a pure xenogenic. Twenty Seven per cent of 2 years-old F_1 recipient XX male, on pairing with trout XX female, produced all female progenies
2n *O. mykiss* (D) x 3n sterile *O. masou* (R) Suspension containing *pvasa-Gfp* labeled spermatogonia of the trout into newly hatched alevin of 3n masu Okutsu et al. (2007)	At age 2+, 34.4% F_1 recipient males produced *Gfp*-labeled milt. At age 7 mo, 10% F_1 recipient females contained vitellogenic *Gfp*-labeled oocytes. Their eggs, on fertilization by F_1 male recipients, produced F_2 progenies at the ratio of 3 *Gfp* + : 1 *Gfp* –. Hence the 34% and 10% F_1 males and females are pure xenogenics

Table 9.7. Achievements and limitations of *ex-situ* conservation techniques to recover of fish genome using preserved gametes and germ cells.

Technique	Achievements	Limitations/Recommendations
Cryopreservation	Protocols for sperm cryopreservation of 2000 species. Simple protocol for cadarveric sperm developed	Eggs and embryos are not amenable. 577 viviparous fishes are not amenable even for sperm preservation
Androgenesis	Fresh, cryopreserved or cadarveric sperm induce using surrogate eggs crossing the borders of species, genus and family	Viviparous fishes are not amenable. Avoid dispermic induction, pressure shock and X-ray irradiation
Cloning	Using nucleus from cultured fibroblast, male and female zebrafish cloned. Nucleus of cryopreserved blastomeres of loach, on transplantation after thawing and incubation, into enucleated egg of zebrafish produced 16.5% xenogenics	The use of nucleus of somatic cells has not even attempted. Avoid using non-enucleated recipient eggs
Xenogenesis	Successfully induced in half dozen fishes. Male and female xenogenics produced from cryopreserved PGCs and SSCs	Viviparous fishes may not be amenable. No attempt to produce in sequential hermaphrodites. Use of type-A spermatogonia as donor and triploid sterile or early larva, with no PGC, as recipient

spawn larger eggs. Partially sterile triploid recipient may have to be used to generate the allogenics and xenogenics in many more fishes like the nibea croaker. Fishes are ideal animals that are readily amenable to endocrine sex reversal (Pandian, 2013, Chapter 7) and ploidy manipulations (Pandian, 2011, Chapters 4 to 7). In species, in which male or female allogenics and xenogenics (as it is likely in xenogenic nibea croaker, Table 9.5) are generated, endocrine sex reversal (cf. Saito et al., 2008) can readily produce the opposite sex and facilitate sustainable recovery of desired strain/species.

Table 9.8 summarizes the land-mark achievements in this area of *ex situ* conservation of fish. Clearly, it is possible to preserve sperm cells, blastomeres, PGCs and SSCs to successfully induce interspecific androgenesis, allogenesis and xenogensis in some fishes. A look at the available techniques suggests that the techniques are at their nascent status. Successful recovery of the desired species from its preserved sperm, as in inter-specific androgenesis, or blastomeres, as in cloning or through xenogenesis is possible crossing the borders of species (e.g., rainbow trout),

Table 9.8. Land-mark achievements in *ex-situ* conservation of fish.

Authors	Achievements
Bercsenyi et al. (1998)	Recovered androgenic F_1 male and female goldfish out of carp egg using cryopreserved sperm of goldfish
Kirankumar and Pandian (2004a)	Recovered androgenic F_1 male and female rosy barb out of tiger barb egg using cadarveric sperm of rosy barb
Yasui et al. (2010)	Recovered androgenic loach using its natural diploid Cryopreserved sperm
Lee et al. (2002)	Recovered clones of F_1 male and female zebrafish transferring nucleues from embryonic fibroblast G1 cell cultured over long period into an enucleated egg of zebrafish
Kobayashi et al. (2003, 2007)	Recovered allogenic chimeric F_1 male and female rainbow trout, using cryopreserved *Gfp*-labeled PGCs of trout
Yasui et al. (2011)	Recovered xenogenic loach offspring using cryo-preserved loach blastomere that was thawed and incubated prior to transplantation into surrogate zebrafish embryo
Higaki et al. (2010b)	Recovered dark-spotted, dark-striped F_2 males and a pure allogenic female zebrafish following the transplantation of a single PGC into yolk-removed embryo of gold colored zebrafish. The PGC was drawn removed embryo that was cryopreserved by vitrification
Okutsu et al. (2007)	Recovered diploid chimeric allogenic F_1 male and female rainbow trout using cryopreserved sperm of trout and surrogate triploid alevin masu

genus (e.g., goldfish out of carp) and family (loach out of zebrafish). This possibility provides hope for the future of *ex-situ* conservation techniques. The author hopes that the new techniques for *ex-situ* conservation shall emerge in the next few years.

References

Anon, 1980. Nuclear transplantation in teleosts. 1. Hybrid fish from the nucleus of carp and the cytoplasm of crucian. Sci Sinica, 23: 517–523.

Arai, K., M. Ikeno and R. Suzuki. 1995. Production of androgenetic diploid loach *Misgurnus anguillicaudatus* using spermatozoa of natural tetraploids. Aquaculture, 137: 131–138.

Araki, K., H. Okamoto, A.C. Graveson et al. 2001. Analysis of haploid development based on expression patterns of development genes in the medaka *Oryzias latipes*. Dev Growth Differ, 47: 591–599.

Bart, A. 2000. New approaches in cryopreservation of fish embryos. pp. 179–187. *In*: T.R.Tiersch and P.M. Mazik (eds). *Cryopreservaion in Aquatic Species*. World Aquacult Soc, Baton Rouge, Lousiana, USA.

Bercsenyi, M., I. Magyary, B. Urbanyi et al. 1998. Hatching out goldfish from common carp eggs. Interspecific androgenesis between two cyprinid species. Genome, 41: 573–579.

Bongers, A.B.J., E..P.C. in't Veld, K. Abo-Hashema et al. 1994. Androgenesis in common carp (*Cyprinus carpio* L.) using UV-irradiation in a synthetic ovarian fluid and heat shocks. Aquaculture, 122: 119–132.

Bubenshchikova, E., B. Ju, I. Pristyazhnyuk et al. 2005. Generation of fertile diploid fish medaka (*Oryzias latipes*) from nuclear transplantation of blastula and four-somite-stage embryonic cells into non-enucleated unfertilized eggs. Cloning Stem Cells, 7: 255–264.

Calvi, S.L. and G. Maisse. 1998. Cryopreservation of rainbow trout (*Oncorhynchus mykiss*) blastomeres: Influence of embryo stage on post-thaw survival rate. Cryobiology, 36: 255–262.

Calvi, S.L. and G. Maisse. 1999. Cryopreservation of carp (*Cyprinus carpio*) blastomeres. Aquat Living Resour, 12: 71–74.

Chao, N. and I. Liau. 2001. Cryopreservaion of kiwifish and shellfish gametes and embryos. Aquaculture, 197: 161–189.

Chen, S. and Y. Tian. 2005. Cryopreservaion of flounder (*Paralichthys olivaceus*) embryos by vitrificaion. Theriogenology, 63: 1207–1219.

Ciruna, B., G. Weidinger, H. Knaut et al. 2002. Production of maternal zygotic mutant zebrafish by germ line replacement. Proc Natl Acad Sci USA, 99: 14919–14924.

David, C.J. and T.J. Pandian. 2006a. *GFP* reporter gene confirms paternity in the androgenote Buenos Aires tetra *Hemigrammus caudovittatus*. J Exp Zool, 305A: 83–95.

David, C.J. and T.J. Pandian. 2006b. Cadaveric sperm induces intergeneric androgenesis in the fis*h Hemigrammus caudovittatus*. Theriogenology, 65: 1048–1070.

David, C.J. and T.J. Pandian. 2008. Dispermic induction of interspecific androgenesis in the fish Buenos Aires tetra using surrogate eggs of widow tetra. Curr Sci, 95: 63–74.

Diwan, A.D., S. Ayyappan, K.K. Lal and W.S. Lakra. 2010. Cryopreservation of fish gametes and embroyos. Ind J Anim Sci, 80: S109–124.

Doidsidou, M., M. Reichman-Fried, J. Stebler et al. 2002. Guidance of primordial germ cell migration by the chemokine SDF-1. Cell, 111: 647–659.

Dorsch, R., D.S. Wagner and K.A. Mintzer. 2004. Maternal control of vertebrate development before the midblastula transition: Mutants from the zebrafish. T Dev Cell, 6: 771–780.

Fujimoto, T., S. Saito, K. Sakau et al. 2006. Development stages and germ-line lineage of the loach (*Misgurnus anguillicaudatus*). Zool Sci, 21: 747–755.

Fujimoto, T., S. Sakao, E. Yamaha and K. Arai. 2007. Evaluation of different doses of UV-irradiation to loach eggs for genetic inactivation of the maternal genome. J Exp Zool, 307A: 449–462.
Fujimoto, T., G.S. Yasui, M. Hayakawa et al. 2010. Reproductive capacity of neotetraploid loaches produced using diploid spermatozoa from a natural tetraploid male. Aquaculture, 308: 5133–5139.
Grunina, A.S., B.I. Gomelsky and A.A. Neifakh. 1990. Diploid androgenesis in pond carp. Genetika, 26: 2037–2043.
Hagedorn, M. and F.W. Kleinnans. 2000. Problems and prospects of cryopreservation of fish embryos. pp. 161–178. *In*: T.R. Tiersch and P.M. Mazik (eds.). *Cryopreservation in Aquatic Species*. World Aquacult Soc, Baton Rouge, Lousiana, USA.
Hagedorn, M., F. Kleinnans, D. Artemov and U. Pilatus. 1998. Characterization of a major permeability barrier in the zebrafish embryo. Biol Reprod, 59: 1240–1250.
Hashimoto, Y., S. Maegwa, T. Nagai et al. 2004. Requirement of localized maternal factors of zebrafish germ cell formation. Dev Biol, 268: 152–161.
Hattori, M., H. Hashimoto, E. Bubenshchikova and Y. Wakamatsu. 2011. Nuclear transplantation of embryonic cell nuclei to non-enucleated eggs in zebrafish *Danio rerio.* Internatl J Biol Sci, 7: 460–468.
Higaki, S., K. Mochizuki, H. Baba et al. 2009. Feasibility of cryopreservation of zebrafish (*Danio rerio*) primordial germ cells by whole embryo freezing. Jap J Vetiner Res, 57: 119–128.
Higaki, S., K. Mochizuki, Y. Akashi et al. 2010a. Cryopreservation of primordial cells by rapid cooling of whole zebrafish (*Danio rerio*) embryos. J Reprod Dev, 56: 212–218.
Higaki, S., Y. Eto, Y. Kawakami et al. 2010b. Production of fertile zebrafish (*Danio rerio*) possessing germ cells (gametes) originated from primordial germ cells recovered from vitrified embryos. Reproduction, 139: 733–740.
Higuchi, K., Y. Takeuchi, M. Miwa et al. 2011. Colonization, proliferation and survival of intraperitoneally transplanted yellowtail *Seriola quinqueradiata* spermatogonia in nibe croaker *Nibea mitsukurii*. Fish Sci, 77: 69–77.
Kasinathan, P., J.G. Knott, Z. Wang et al. 2001. Production of calves from G1 fibroblasts. Nat Biotechnol, 19: 1176–1178.
Kawakami, Y., R. Goto-Kazeto, T. Saito et al. 2010. Generation of germ-line chimera zebrafish using primordial germ cells isolated from cultured blastomeres and cryopreserved embryoids. Int J Dev Biol, 51: 1493–1501.
Kirankumar, S. and T.J. Pandian. 2003. Production of androgenetic tiger barb *Puntius tetrazoma.* Aquaculture, 228: 37–51.
Kirankumar, S. and T.J. Pandian. 2004a. Interspecific androgenetic restoration of rosy barb using cadaveric sperm. Genome, 47: 66–73.
Kirankumar, S. and T.J. Pandian. 2004b. Production and progeny testing of androgenetic rosy barb *Puntius conchonius*. J Exp Zool, 301A: 477–490.
Kirankumar, S. and T.J. Pandian. 2004c. Use of heterolozous sperm for the disperonic induction of androgenesis in barbs. J Fish Biol, 64: 1485–1497.
Kobayashi, T., Y. Takeuchi, G. Yoshizaki and T. Takeuchi. 2003. Cryopreservation of trout primordial germ cells. Fish Physiol Biochem, 28: 479–480.
Kobayashi, T., Y. Takeuchi and T. Takeuchi. 2004. Isolation of highly pure and viable primordial germ cells from rainbow trout by *Gfp* dependent flow cytometry. Mol Reprod Dev, 67: 91–100.
Kobayashi, T., Y. Takeuchi, T. Takeuchi and G. Yoshizaki. 2007. Generation of viable fish from cryo-preserved primordial germ cells, Mol Reprod Dev, 74: 207–213.
Komen, J. and G.H. Thorgaard. 2007. Androgenesis, gynogenesis and the production of clones in fishes: A review. Aquaculture, 269: 150–173.
Koteeswaran, R. and T.J. Pandian. 2002. Live sperm from post-mortem preserved Indian catfish. Curr Sci, 82: 447–450.
Kusuda, S.T., T. Teranishi and N. Koide. 2002. Cryopreservation of chum salmon blastomeres by the straw method. Cryobiology, 45: 60–67.

Kusuda, S.T., T. Teranishi, N. Koide et al. 2004. Pluripotency of cryopreserved blastomeres of the goldfish. J Exp Zool, 301A: 131–138.

Lacerda, S.M.S.N., S.R. Batlouni, S.B.G. Silva et al. 2006. Germ cells transplantation in fish: the Nile tilapia model. Anim Reprod, 3: 146–159.

Lacerda, S.M.S.N., S.R. Batlouni, G.M.J. Costa et al. 2010. A new and fast technique to generate offspring after germ cells transplantation in adult fish. The Nile tilapia (*Oreochromis nilotica*) model. PLoS ONE: e10740. doi: 10-1371/journal.pone.0010740.

Lee, K.Y., H. Huang, B. Ju et al. 2002. Cloned zebrafish by nuclear transfer from long-term-cultured cells. Nat Biotech, 20: 795–799.

Lin, C., T. Zhang and D.M. Ranson. 2009. Cryopreservation of zebrafish (*Danio rerio*) blastomeres. Cryo-Letters, 30: 132–141.

Majhi, S.K., R.S. Hattori, M. Yokota et al. 2009. Germ cell transplantation using sexually competent fish: An approach for rapid propagation of endangered and valuable germ lines. PLoS ONE, 4: e6132. Doi: 10/1371/journal. Pone, 0006132.

Marengoni, N.G. and Y. Onoue. 1998. Ultraviolet-induced androgenesis in Nile tilapia *Oreochromis niloticus* (L.) and hybrid Nile x blue tilapia *O. aureus* (Steindachner). Aquacult Res, 29: 359–366.

Meisheng, Y.I., H. Ni, Z.D. Li et al. 2010. Medaka fish stem cells and their applications. Sci China Life Sci, 53: 426–434.

Morita, T., N. Kumakura, K. Morishma et al. 2012. Production of donor derived offspring by allogeneic transplantation of spermatogonia in the yellowtail (*Seriola quinqueradiata*). Biol Reprod, 86: 176–176.

Nagai, T., E. Yamaha and K. Arai. 2001. Histological differentiation of primordial germ cells in zebrafish. Zool Sci, 18: 215–223.

Nam, Y.K., Y.S. Cho, H.J. Cho and D.S. Kim. 2002. Accelerated growth performance and stable germ-line transmission in androgenetically derived homozygous transgenic mud loach *Misgurnus mizolepis*. Aquaculture, 209: 257–270.

Okutsu, T., A. Yano, K. Nagasawa et al. 2006a. Manipulation of fish germ cell: Visulaization, cryopreservation and transplantation. J Reprod Dev, 52: 686–693.

Okutsu, T., K. Suzuki, Y. Takeuchi et al. 2006b. Testicular germ cells can colonize sexually undifferentiated embryonic gonad and produce functional egg in fish. Proc Natl Acad Sci USA, 103: 2725–2729.

Okutsu, T., Y. Takeuchi, G. Yoshizaki et al. 2007. Production of trout offspring from triploid salmon parents. Science, 317: 1517.

Okutsu, T., Y. Takeuchi and G. Yoshizaki. 2008. Spermatogonial transplantation in fish: Production of trout offspring from salmon parents. pp. 209–2019. *In*: K. Tsukamoto, T. Kunamura, T. Takeuchi et al. (eds.). Fisheries for Global Welfare and Environment. Fifth World Fish Cong.

Olsen, L.C., R. Aasland and A.A. Fjose. 1997. *Vasa*-like gene in zebrafish identifies putative primordial germs cells. Mech Dev, 66: 95–105.

Pandian, T.J. 2002a. Cloning the fish. Curr Sci, 83: 1063–1064.

Pandian, T.J. 2002b. Biodiversity : Status and endeavours of India. ANJAC J Sci, 1: 21–32.

Pandian, T.J. 2011. *Sex Determination in Fish*. Science Publishers, Enfield/CRC Press, Boca Raton, USA, p. 282.

Pandian, T.J. 2012. *Genetic Sex Differentiation in Fish*. CRC Press, Boca Raton, USA, p. 214.

Pandian, T.J. 2013. *Endocrine Sex Differentiation in Fish*. CRC Press, Boca Raton, USA, p. 240.

Pandian, T.J. and S. Kirankumar. 2003. Androgenesis and conservation of fishes. Curr Sci, 85: 917–931.

Pei, D.S., Y.H. Sun, S.P. Chen et al. 2007. Identification of differentially expressed genes from the cross sub-family cloned embryos derived from zebrafish nuclei and rare minnow enucleated eggs. Theriogenology, 68: 1282–1291.

Potok, E., D.A. Nix, T.J. Parnell and B.R. Cairns. 2013. Reprogramming the maternal zebrafish genome after fertilization to match the paternal methylation pattern. Cell, 153: 759–772.

Robles, V., E. Cabrita, G.L. Fletcher et al. 2005. Vitrification assays with embryos from a cold tolerant sub-arctic fish species. Theriogenology, 64: 1633–1646.

Robles, V., E. Cabrita, J.P. Acker and P. Herraez. 2008. Embryo cryopreservation: What we know until now. pp. 265–286. *In*: E. Cabrita, V. Robles and P. Herraez (eds.). *Methods in Reproductive Aquaculture: Marine and Freshwater Species*. CRC Press, Boca Raton.

Saito, T., R. Goto-Kazeto, K. Arai et al. 2008. Xenogenesis in teleost fish through generation of germ line chimeras by single primordial germ cell transplantation. Biol Reprod, 78: 159–166.

Saito, T., R. Goto-Kazeto, T. Fujimoto et al. 2010. Interspecies transplantation and migration of primordial germ cells in cyprinid fishes. Int J Dev Biol, 54: 1481–1486.

Sassado, T., A. Yasuka, K. Abe et al. 2008. Distinct contributions of CXCR4b and CXCR7/RDC1 receptor systems in regulation of PGC migration revealed by medaka mutants kazura and yanazi: Dev Biol, 320–329.

Shimada, A. and H. Takeda. 2008. Production of a maternal zygotic medaka using hybrid sterility. Dev Growth Diff, 50: 421–426.

Slanchev, K., J. Stebler, G. de la Cueva-Mendez and E. Raz. 2005. Development without germ cells: The role of germ line in zebrafish sex differentiation. Proc Natl Acad Sci USA, 102: 4074–4079.

Strussmann, C.A., H. Nakatsugawa, F. Takashima et al. 1999. Cryopreservation of isolated fish blastomeres. Effects of cell stage, cryoprotectant concentration and cooling rate on post-thawing survival. Cryobiology, 39: 252–261.

Sun, Y.H., S.P. Chen, Y.P. Wang et al. 2005. Cytoplasmic impact on cross-genus cloned fish derived from transgenic common carp (*Cyprinus carpio*) nuclei and goldfish *(Carassius auratus)* enucleated eggs. Biol Reprod, 72: 510–515.

Takeuchi, Y., G. Yoshizaki, T. Kobayashi et al. 2002. Mass isolation of primordial germ cells from rainbow trout carrying the green fluorescent protein gene driven by the *vasa* gene promoter. Biol Reprod, 67: 1087–1092.

Takeuchi, Y., G. Yoshizaki, T. Kobayashi et al. 2003. Generation of live fry from inteperitoneally transplanted primordial germ cells in rainbow trout. Biol Reprod, 69: 1142–1149.

Takeuchi, Y., G. Yoshizaki, T. Kobayashi et al. 2004. Surrogate broodstock produces salmonids. Nature, 430: 629–630.

Takeuchi, Y., K. Higuchi, T. Yatabe et al. 2009. Development of spermatogonial cell transplantation in nibe croaker *Nibea mitsukurii* (Perciformes, Sciaenidae). Biol Reprod, 81: 1055–1063.

Tiersch, T.R. 2000. Introduction. pp. 19–26. *In*: T.R. Tiersch and P.M. Mazik (eds.). *Cryopreservation in Aquatic Species*. World Aquacult Soc, Baton Rouge, Lousiana,USA.

Tiersch, T.R. 2001. Cryopreservation in Aquarium Fishes. Mar Biotechnol, 3: s212–223.

Thorgaard, G., P. Scheerer, W.K. Hershberger and J.M. Myers. 1990. Androgenetic rainbow trout produced using sperm from tetraploid males show improved survival. Aquaculture, 85: 215–221.

Thuan, N.V., S. Kishigami, T. Wakayama et al. 2010. How to improve the success rate of mouse cloning technology. J Reprod Dev, 56: 20–30.

Wilmut, I., A.E. Schnieke, J. McWhir et al. 1997. Viable offspring derived from fetal and adult mammalian cells. Nature, 385: 810–813.

Wong, T.T., T. Saito, J. Crodian et al. 2010. Zebrafish germ line chimeras produced by transplantation of ovarian germ cells into sterile host larvae. Biol Reprod, 84: 1190–1197.

Wylie, C. 2000. Germ cells. Cell, 96: 165–174.

Yamaha, E., M. Kazama-Wakabayashi, S. Otani et al. 2001. Germ line chimera by lower-part of blastoderm transplantation between goldfish and crucian carp. Genetica, 111: 227–236.

Yamaha, E., M. Murakami, K. Hada et al. 2003. Recovery of fertility in a male hybrid of a cross between goldfish and common carp by transplantation of PGC (Primordial germ cell) containing graft. Genetica, 119: 121–131.

Yamaha, E., R. Goto-Kazeto, T. Saito et al. 2010. Primordial germ cell in teleost fish with special reference to its specification and migration. J Appl Ichthyol, 26: 816–822.

Yan, H.U.H., Li, J.F. Gui and Y.H. Hong. 2010. Fish germ cells. Sci China Life Sci, 53: 435–446.

Yano, A., K. Suzuki and G. Yoshizaki. 2008. Flow cytometric isolation of testicular germ cells from rainbow trout (*Oncorhynchus mykiss*) carrying the green fluorescent protein gene driven by trout vasa regulatory regions. Biol Reprod, 78: 151–158.

Yasui, G.S., T. Fujimoto and K. Arai. 2010. Restoration of the loach *Misgurnus anguillicaudatus* from cryopreserved diploid sperm and induced androgenesis. Aquaculture, 308: s140–s144.

Yasui, G.S., T. Fujimoto, S. Sakao et al. 2011. Production of loach (*Misgurnus anguillicaudatus*) germ-line chimera using transplantation of primordial germ cells isolated from cryopreserved blastomeres. J Anim Sci, 89: 2380–2388.

Yazawa, R., Y. Takeuchi, K. Higuchi et al. 2010. Chub mackerel gonad support colonization, and proliferation of intraperitoneally transplanted xenogeneic germ cells. Biol Reprod, 82: 896–904.

Yoon, C., K. Kawakami and N. Hopkins. 1997. Zebrafish *vasa* homologue RNA is localized to the cleavage planes of 2- and 4-cell stage embryos and expressed in primordial germ cells. Development, 124: 3157–3166.

Yoshizaki, G., S. Sakatani and I. Tominaga. 2000a. Cloning and characterization of a *vasa*-like gene in rainbow trout and its expression in the germ cell lineage. Mol Reprod Dev, 55: 364–371.

Yoshizaki, G., Y. Takeuchi, S. Sakatani and T. Takeuchi. 2000b. Germ cell specific expression of green fluorescent protein in transgenic trout under control of the rainbow trout. *Vasa*-like promoter. Int J Dev Biol, 44: 323–326.

Yoshizaki, G., M. Ichikama, M. Hayashi et al. 2010. Sexual plasticity of ovarian germ cells in rainbow trout. Development, 137: 1227–1230.

Yoshizaki, G., K. Fujinuma, Y. Iwasaki et al. 2011. Spermatogonial transplantation in fish: A novel method for the preservation of genetic resources. Comp Biochem Physiol Part D, 6: 55–61.

10

Differentiation and Conclusions

The description and analysis of the sex differentiation process of fishes in this book series has led to proposing a new conclusions. Sex is more or less decisively determined by *Dmy* prior to hatching in two species of medaka and *Dmrt1* after hatching in Nile tilapia. In all other fishes, it may indecisively be determined by a host of genes during the post-hatching stages; for example, *sox11b*, *sox21a* and *sox3l* are overexpressed in the ovary but *amh* in the testis of zebrafish, a secondary gonochore (Santos et al., 2007). Only 11% of fish species have cytologically and/or cytogenetically distinguishable male (XX/XY) or female (ZW/ZZ) specific heteromorphic chromosomes (Pandian, 2011, p. 23). In majority of fishes, the processes of sex determination and differentiation are continuous and accomplished by a cascade of genes (e.g., Zhang et al., 2011), each with small additive effects, that are randomly distributed in many or all chromosomes. The expression of most of these genes is under environmental control.

10.1 Labile periods and plasticity

Regarding sex differentiation in fishes, an obvious feature is the extension of the plastic labile period to later ontogenetic stages. Four such features are apparent:

1. The labile period is postponed till the end of adult stage in almost all hermaphrodites (Pandian, 2013, Fig. 56), whose gonads are commenced as differentiated ovaries or ovotestes (Pandian, 2012, Fig. 4). As adults,

all hermaphrodites (see Fig. 1.4) undergo natural sex change and are amenable to sex reversal by hormonal or social induction (Pandian, 2013, Chapter 8). Their sexuality remains very plastic that they undergo sex change or can be induced to undergo reversal more than once toward female or male direction (Table 10.1 H–O).

2. In secondary gonochores, the plastic labile period is also retained in adults but the sex reversal is limited from female to male only (e.g., zebrafish, Table 10.1F). In them too, the labile period can be postponed as late as possible but prior to attaining adult stage (e.g., *Dascyllus flavicaudus,* Table 10.1G, see also Pandian, 2012, pp. 8–11). Interestingly, adult *Macropodus opercularis* is capable of bidirectional sex change. An individual is capable of reversing sex maximum of 4–6 times.
3. In primary gonochores too, the plastic labile period is retained upto puberty in mullet and golden rabbitfish (Table 10.1C). The labile period, during which the ovarian differentiation precedes that of testis but occurs continuously one after the other in quick succession, resulting in a single labile period. However, the period can also be temporally separated and there can be two distinct labile periods; during the first, temperature (e.g., sea bass: Koumoundouros et al., 2000, medaka: Hattori et al., 2007) or hormone (e.g., medaka: Edmunds et al., 2000; Hano et al., 2005) can disrupt the ovarian differentiation and during the second, exogenous hormones including EDCs disrupt the process. For example, in the male heterogametic medaka *Oryzias latipes*, in which sex is determined by a single gene *Dmy* before hatching, the exposure of undifferentiated embryos to 34°C induces sex reversal from genetic (XX) female to phenotypic male (Table 10.1A); however, undifferentiated genetic (XY) male embryo is not amenable to sex reversal on exposure to 34°C (Hattori et al., 2007). But the labile period of medaka for hormonal induction of sex reversal is located during the post-hatching stage (see Pandian, 2013, p. 178). In *Dicentrarchus labrax*, the labile periods are temporally widely separated. In this sea bass, the ovarian differentiation lasts from half epibolic stage to mid-metamorphosis, i.e., from 0 ≈ 60 dpf (Pavlidis et al., 2000; Koumoundouros et al., 2002). During this period, exposure to 20°C increases the male ratio. However, the labile period for hormonal induction of sex reversal is located between 86 and 126 dpf (Blazquez et al., 2001). In general, the primary gonochores are amenable to hormonal induction of sex reversal only once during the labile period (Kuhl and Brouwer, 2006). However, experiments are to be undertaken in *O. latipes*, in which microinjection of estrogen into eggs induced the first sex reversal. But it is also to be known whether medaka is amenable to a second sex reversal by temperature elevation. Similarly, experiments are also desirable to know whether medaka, that underwent the first sex

Table 10.1. Ontogenetic pathways and locations of labile periods in fishes.

Groups/ Models	**Features**
	I Primary gonochores
A) Medaka	*Dmy* determines male sex prior to hatching. Two labile periods, one at embryonic and the other during post-hatching stages. In the first, elevated temperature induces in undifferentiated XX eggs sex change to males. During the second, hormones induce both masculinization and feminization
B) Nile tilapia	*Dmrt1* determines male sex after hatching. Labile period for hormonal manipulation at fry stage
C) Mullet	Labile period is postponed to juvenile-puberty stage. Sparids like panga and bronze bream with delimited gonads differentiate into male and female at sub-adult stage
D) Sea bass	Two labile periods: the first one at embryonic stage amenable to temperature manipulation and the second at juvenile stage amenable to hormonal manipulation
	II Tertiary gonochores
E) Eel	Postponed differentiation to silver eel stage. Two labile periods; at the first one, ovary and intersex gonads are differentiated. During the second, the intersex gonad differentiates into testis or ovary, depending on temperature and/or food supply
	III Secondary gonochores
F) Zebrafish	All hatchlings possess ovary. Transformation from the ovary to testis in about 50% individuals occurs during juvenile stage. Amenable to hormonal masculinization and feminization. However, on depuration, adult sex can be reversed from female to male but not from male to female. *Macropodus opercularis* is capable of bidirectional sex change
G) *Dascyllus*	The transformation is postponed to puberty/sub-adult stage
	IV Hermaphrodites: Labile period till the end of adult stage
H) Rivulus	Commences as female embryo but becomes hermaphrodite by adding testicular component after hatching. Two labile periods, Primary and Secondary males are induced by temperature at embryonic and hatching stages, respectively. Amenable to feminization as hatchlings
I) *Serranus baldwini*	At the late adult stage, large hermaphrodite loses the ovary and the sex changed male holds a harem of hermaphrodites
J) Monandric protogynics	Juvenile differentiates into adult female, which changes sex as adult to secondary male
K) Groupers	Amenable to hormonal induction to directly develop as male
L) Diandric protogynics	Juvenile differentiates into Primary male and adult female, which changes sex as adult to Secondary male
M) Monogynic protandrics	Juvenile differentiates into adult male, which changes sex to Secondary female
N) Digynic protandrics	Juvenile differentiates into Primary female and adult male, which changes sex as adult to Secondary female
O) Serials and hermaphrodites	Extended adult plastic labile period amenable for sex change/reversal more than once in either female or male direction

change from female to male induced by elevated temperature (Hattori et al., 2007) is amenable to endocrine sex reversal during post-hatching stage. Suggested experiments are also wanted for sea bass, in which two labile periods are more widely temporally separated. In *O. latipes* (Kuhl and Brouwer, 2006) and *Ictalurus punctatus* (Patino et al., 1996), the existence of temporally separated two labile periods is known. In *I. punctatus* (Goudie et al., 1983) and *O. latipes* (Kuhl and Brouwer, 2006), once feminization by exogenous estrogens is completed, a second hormonal induction of masculinization is not possible. For example, catfish is amenable for feminization but not masculinization.

4. In tertiary gonochores like *Anguilla anguilla* too, two distinct temporally separated labile periods are recognizable. Having remained as undifferentiated gonads for many years, the gonads of about 40% yellow eels undergo differentiation into the ovary and intersexual Syrski stage (Table 10.1E). Long after this differentiation, the intersexual Syrski differentiates either into testis or ovary (Geoffroy et al., 2013), depending upon temperature and food availability, when grown under different levels of crowding (Table 10.1E).

Remarkably, not all hormonal inductions produce 100% males or females. Even after treatment at optimal dose of a potent steroid and duration, a small portion of the treated individuals remain not amenable to hormonal sex reversal (e.g., Pandian, 2013, Table 46). Similarly, not all individuals/families subjected to elevated temperatures reverse sex to males (e.g., Nile tilapia, Table 7.1, Atlantic silverside, Table 7.4). These fish species, in which the sex of a portion of the treated individuals per se belong to the group GSD + A (see Fig. 1.1). On the other hand, fishes like *Hippoglossus hippoglossus*, sex remains insensitive to changes in temperature (Hughes et al., 2008). Similarly, *Ctenopharyngodon idella* is not amenable to hormonal sex reversal even after an induction period of 500 days (Shelton and Jensen, 1979). All the six protocols employed by different authors including the one releasing a high dose of 16.6 μg MT/day for a period of longer than 300 days have failed to induce masculinization (Hunter and Donaldson, 1983). In these fishes, a combination of autosomal genes decisively determine sex and the genes in their cascade of sex differentiation are neither sensitive nor responsive to changes in environmental factors including EDCs. Hence, the possibility does exist for the operation of GSD within ESD group (Fig. 1.1). Figure 1.1 also shows the ESD, in which sex differentiation genes virtually realize the sex, as determined by autosomal genes. This justifies the following statements: 1. In all fishes, sex is determined by genes and 2. Sex determination and differentiation processes are temporally continuous in many fishes but they are still distinctly recognizable as almost independent processes.

10.2 Morphotypes and plasticity

In 140 gonochoric species belonging to 28 families (Taborsky, 1998), the non-labile adult males that display Alternative Mating Strategy (AMS) remain flexible with regard to testicular differentiation. To familiarize different AMS types (Oliveira, 2006), a brief description is provided. Notably, flexibility in testicular differentiation is progressively reduced from Reversible Transformants to Fixed Morphotypes (Table 10.2). Type 1: Plastic morphotypes, in which a 'submissive' may reversibly change into 'dominant', as in *Astatotilapia burtoni* and from 'helper' to 'breeder', as in *Neolamprologus pulcher*. Evidently, the adults of plastic reversible have retained the potency to undergo the entire sequence of testicular differentiation from non-functional to functional testis—a situation that recalls the OP-treated sexually ambiguous medaka, on depuration, reversing from testis-ova to testis and from female specific to male specific secondary sexual characteristics (Seki et al., 2003). Social cues seem to switch on or off the synthesis and release of GnRH in the plastic reversibles (White et al., 2002). Another example, for social cues is the black goby. Exclusive possession of a female transforms the parasitic 'sneaker' to bourgeois morphotypes in the black goby (Immler et al., 2004). Hence, the testicular differentiation remains flexible throughout adult stage.

Type 2: The transformation of the non-reversible transformants involves from 2 morphotypes, i.e., sneaker territorial, as in *Symphodus*

Table 10.2. Some characteristics of different types of morphotypes in fishes that display AMS.

Species, Reference	Reported observations
Plastic Reversible Morphotype	
Astatotilapia burtoni White et al. (2002)	Depending on the social context and hormone levels, a 'submissive' male, with smaller testes containing no mature sperm, may reversibly become 'dominant' with large testes full of mature sperm
Non-reversibly Transforming Morphotype	
Symphodus ocellatus Bentivegna and Benedetto (1989)	Brown color sneaker (3.5–6.0 cm) transforms into greenish satellite (6.1–8.0 cm) and subsequently to reddish blue territorial (> 8.0 cm) but the reverse transformation is not possible
Leptoscarus vaigiensis Robertson et al. (1982)	Initial phase (IP) male (6–8 cm) non-reversibly transforms into Terminal Phase (TP) male (8.1–18 cm)
Fixed Morphotype	
Lepomis macrochir Gross (1982, 1984)	Cuckolders have a relatively shorter life span than parental
Oncorhynchus kisutch Gross (1991)	Precociously maturing smaller (34 cm) jack and large (52 cm) hooknose. Jacks generate a higher proportion of F_1 jacks and hooknoses

roissali to 4 morphotypes, i.e., sneaker (2.3–4.5 cm) satellite (4.6–5.0 cm) territorial (5.1 to 6.1 cm) pirate (> 6.2 cm), as in *Telematochromis vittatus*. Uniquely, in the scarid *Leptoscarus vaigiensis*, a gonochore reflecting its past dichromatic, monandric protogyny, the males switch from Initial Phase (IP) (4.2 to 7.1 cm) to Terminal Phase (TP) on attaining a body size of 7.2 cm (Robertson et al., 1982). Though sex is known to determine the size of fish (p. 212), size, i.e., body length seems to determine the differentiation in transformants from one phase to the next. By number, the IP males of *L. vaigiensis* may fertilize more number of eggs than the TP males (see p. 54). In this type of AMS, the testicular differentiation remains also flexible but on in a single direction only.

Type 3: In the fixed morphotypes, the traits responsible for the morphotypic differences are fixed for life time and are linked to the respective morphotypes. In this AMS type, the testicular differentiation remains fixed for life time in the adults either as bourgeois or parasitic male. The cuckolder of the bluegill sunfish *Lepomis macrochir* and jacks of the salmonids parasitically compete to fertilize eggs against parental and hooknose, respectively. These parasitic morphotypes grow fast and precociously attain sexual maturity. Consequently, they have shorter life span. For example, the non-migratory jacks of *Salmo salar* sexually mature as parr at the age of 1–3 years (Horth et al., 2005); the jacks of *Oncorhynchus kisutch* are smaller, fast growing and mature at the size of 34 cm, against the large migratory hooknose attaining sexual maturity at the size of 52 cm (Gross, 1991). In the sunfish, the cuckolder grows fast and attains sexual maturity at the age of 2–3 years, against the slow growing, large parental, which attains sexual maturity at the age of 7 years (Gross, 1982, 1984). Surprisingly, the females of sunfish release 3 times more number of eggs in the presence of competing cuckolders than when the spawn in the presence of parental only (Fu et al., 2000). Arguably, this observation brings a strong evidence to the increased reproductive output by fishes that display AMS though at the cost of altered sex ratio.

In a Nigerian cichlid, both flexible transformant and fixed morphotype are combined. Relevant publications (Heiligenberg, 1965; Martin and Taborsky, 1997; Reddon and Hurd, 2013) on the reproductive biology of fixed morphotypes seem to throw light on the protracted sex differentiation during the non-labile adult stage of fishes that display AMS. For example, the Nigerian cichlid *P. pulcher* are small, sexually dimorphic with color dimorphism (Fig. 10.1). Its sex ratio ranges from 0.75 ♀ : 0.25 ♂ in the upper part of the Sombreiro River to 0.5 ♀ : 0.5 ♂ in other locations. The females have intensely colored belly. In the males, there are yellow and red colored morphs. About 50% of the red morphic males are pair-forming, while the other 50% males hold 2–3 females each in their respective harem; the only difference between the haremic and the pairing males is that the former is more aggressive. The yellow morph develops into pair-forming male or becomes submissive parasitic satellite, when his partner or territory is lost. A few dominant satellites may, however, regain mate and territory. Briefly, the yellow morphic male, as plastic reversible, can change its status from pair-forming to satellite and satellite to pair-forming male. However, the adult yellow morph cannot change into the fixed red morph. Similarly, an adult red morph cannot change into the fixed yellow morph, although an adult red morph can change its status from paired male to haremic and vice versa. The fact that the haremic males produce 3 times more number of progenies than the non-haremic paired male and 7 times more number of progenies than a satellite clearly evidences that harem-formation significantly increases the reproductive output, albeit at the cost of altered (0.75 ♀ : 0.25 ♂) sex ratio.

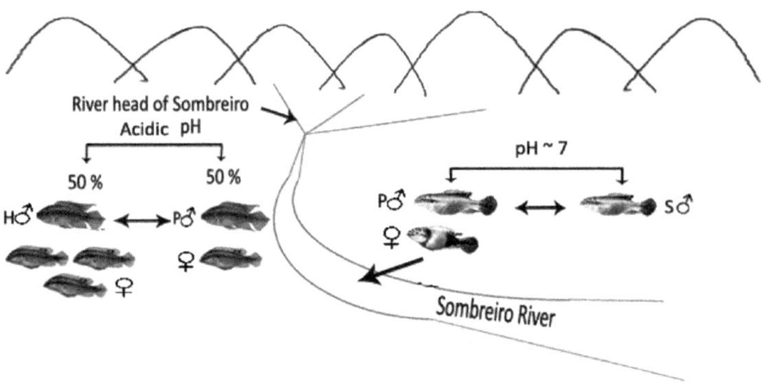

Fig. 10.1. The plastic reversible morphotypes within fixed red and yellow morphs of the Nigerian cichilid *Pelvicachromis pulcher*. The locations of occurrence of the red and yellow morphs in the Sombreiro River are indicated. Note the double headed arrows indicating the directions of plastic reversibility within red and yellow morphs. H♂ = Haremic male, P ♂ = Paired male, S♂ = Satellite male, ♀ = Female.

Color image of this figure appears in the color plate section at the end of the book.

Rearing the Nigerian cichlid in waters of selected pH during the labile period from 0 to 30 dph, Reddon and Hurd (2013) reported in their seminal publication that not only male ratio increased from 0.4 at pH 6.5 to 0.6 at pH 5.5 but also the red morph from 18% at pH 6.5 to 67% at pH 5.5. Understandably, the cichlid possesses a combination of 1. The plastic reversible morphotypes during the non-labile adult stage, i.e., haremic vs. pairing male morphotypes within red morphs and pairing vs. satellite morphotypes within yellow morphs as well as 2. The fixed red and yellow morphotypes. During the labile period from 0 to 30 dph, the cichlid has an option to differentiate into a male or female and also the male to differentiate into red or yellow morph. During the non-labile adult stage, neither the sex can be reversed nor the morph can cross into the other, i.e., red cannot cross into yellow morph and yellow cannot cross into red morph. Apparently, the pH of the water is the environmental factor that not only directs sex differentiation process but also switches on the expression of the genes responsible for differentiation into red or yellow morph within males. It is likely that during the respective labile periods, the option is made not only differentiate into a male or female but also the males to differentiate into cuckolder or parental in the sunfish and jack or hooknose in many salmonids that display AMS. Interestingly, Iwamoto et al. (1984) reported that the jacks sired both jacks and hooknoses, albeit significantly a higher proportion of jacks. Hence, the morphotypic trait is not linked to sex determining genes. Not surprisingly, the percentage of jacks in *O. tshawystcha* ranges from 10 to > 90% among populations (Heath et al., 1991). However, it is to be known which environmental factor is responsible to switch on the expression of the genes responsible for differentiation of jack and hooknose in salmonids and cuckolder and parental in the sunfish. But it must be noted that the changes in morphotypes are intra-sexual in gonochores but are intersexual in hermaphrodites. Hence, sex differentiation process is several times more flexible in hermaphrodites than in gonochores.

10.3 Reproduction vs. Genetic diversity

During favorable season and/or location, fishes seem to increase reproductive output in preference to ensuing genetic diversity. However, they ensure genetic diversity during unfavorable season and/or location. Considering the Nigerian cichlid, this may be exemplified. At the favorable head site, in which acidic water floods the River Sombreiro, 67% red morphs are generated. With male ratio of 0.25 only, a red morphic male holds 2–3 females in his harem to produce 3 and 7 times more progenies than paired and (satellite) parasitic males, respectively (see page 15, 59–61). In alkaline waters at the tail end of the River, 72% yellow morphs are generated. With male ratio of 0.75, the paired and parasitic males compete against each other to generate increased genetic diversity among the progenies, although at the cost of reproductive output (see p. 61). To encounter unfavorable situations like the oxygen depletion, altered pH and elevated temperature, male ratio is increased to generate increased genetic diversity among progenies. In fishes, selection seems to prefer reproductive output over genetic diversity and genetic diversity over balanced sex ratio. This is true of many aquatic invertebrate groups, e.g., crustaceans.

10.4 Sex determines size

In fixed morphotypes, the parasitic cuckolders and jacks are reported to grow fast and attain sexual maturity precociously at an early age and smaller body size. Whether the sex determines the size or the size determines the sex has been a theme of much discussion. Briefly, field observations (see p. 81) and experimental evidences described for the eel (p. 84–85), hornomoko (p. 181–190) and sea bass (p. 212) have confirmed that it is the sex that determines the size of fishes. Similarly, it seems it is the morphotype that determines the size and behavior of a male in species that display AMS.

References

Bentivegna, F. and F. Benedetto. 1989. Gonochorism and seasonal variations in the gonads of the labrid *Symphodus* (*Crenilabrus*) *ocellatus* (Forsskal). J Fish Biol, 34: 343–348.
Blazquez, M., A. Felip, S. Zanuy et al. 2001. Critical period of androgen-inducible sex differentiation in a teleost fish, the European sea bass (*Dicentrarchus labrax*). J Fish Biol, 58: 342–358.
Edmunds, J.S., R.A. McCarthy and J.S. Ramsdell. 2000. Permanent and functional male-to-female sex reversal in d-rR strain medaka (*Oryzias latipes*) following egg microinjection of o,p'-DDT. Environ Health Perspect, 108: 219–224.
Fu, P., B.D. Neff and M.R. Gross. 2001. Tactic-specific success in sperm competition. Proc R Soc Lond, 268B: 1105–1112.
Geoffroy, B., Y. Guiguen, A. Fostier and A. Bardonnet. 2013. New insights regarding gonad development in European eel: evidence for a direct ovarian differentiation. Fish Physiol Biochem, 39: 1129–1140.
Goudie, C.A., B.D. Redner, B.A. Simco and K.B. Davis. 1983. Feminization of channel catfish by oral administration of steroid sex hormones. Trans Am Fish Soc, 112: 670–672.
Gross, M.R. 1982. Sneakers, satellites and parentals: Polymorphic alternating strategies in North American sunfishes. Z Tierpsychol, 60: 1–26.
Gross, M.R. 1984. Sunfish, salmon and evolution of alternative reproductive strategies and tactics in fishes. *In*: G.W. Potts, R.J. Wooton (eds.). *Fish Reproduction: Strategies and Tactics*. Academic Press, London, pp. 55–75.
Gross, M.R. 1991. Evolution of alternative reproductive strategies, frequency-dependent sexual selection in male bluegill sunfish. Phil Trans R Soc Lond, B, 332: 59–66.
Hano, T., Y. Oshima, T. Oe et al. 2005. Quantitative bio-imaging analysis for evaluation of sexual differentiation in germ cells of olvas-GFP/ST-II YI medaka (*Oryzias latipes*) nanoinjected in sex change in ova with ethynylestradiol. Environ Toxicol Chem, 24: 70–71.
Hattori, R.S., R.J. Gould, T. Fujioka et al. 2007. Temperature-dependent sex determination in Hd-rR medaka *Oryzias latipes*: gender sensitivity, thermal threshold, critical period, and DMRT1 expression profile. Sex Dev, 1: 138–146.
Heath, D.D., L. Rankin, C.A. Bryden et al. 2002. Heritability and Y-chromosome influence in the jack male life history of Chinook salmon (*Oncorhynchus tshawytscha*). Heredity, 89: 311–317.
Heiligenberg, W. 1965. Colour polymorphism in males of an African cichlid fish. J Zool, 146: 95–97.
Herpin, A. and M. Schartl. 2011. *Dmrt1* gene at the cross roads: a widespread and central class of sexual development factors in fish. FEBS J, 278: 1010–1119.
Hughes, V., T.J. Benfey and D.J. Martin-Robichaud. 2008. Effect of temperature on sex ratio in juvenile Atlantic halibut *Hippoglossus hippoglossus*. Env Biol Fish, 81: 415–419.

Hunter, G.A. and E.M. Donaldson. 1983. Hormonal sex control and its application to fish culture. *In*: W.S. Hoar, D.J. Randall and E.M. Donaldson (eds.). *Fish Physiology: Reproduction, Part B. Behavior and Fertility Control*. Academic Press, New York, USA, 9: 223–303.

Immler, S., C. Mazzoldi and M.B. Rasotto. 2004. From sneaker to parental male: change of reproductive traits in the black goby *Gobius niger* (Teleostei: Gobiidae). J Exp Zool, 301A: 177–185.

Iwamoto, R.N., B.A. Alexander and W.K. Herschberger. 1984. Genotypic and environmental effects on the incidence of sexual precocity in coho salmon (*Oncorhynchus kisutch*). Aquaculture, 43: 105–121.

Koumoundouros, G., Pavlidis, L. Anezaki et al. 2002. Temperature sex determination in the European sea bass *Dicentrarchus labrax* (L. 1758) (Teleostei, Perciformes, Moronidae): critical sensitive ontogenetic phase. J Exp Zool, 292: 573–579.

Kuhl, A.J. and M. Brouwer. 2006. Antiestrogens inhibit xenoestrogen-induced brain aromatase activity but do not prevent xenoestrogen-induced feminization in Japanese medaka (*Oryzias latipes*). Environ Health Perspect, 114: 500–506.

Martin, E. and M. Taborsky. 1997. Alternative male mating tactic in a cichlid, *Pelvicachromis pulcher*: a comparison of reproductive effort and success. Behav Ecol Sociobiol, 41: 311–319.

Oliveira, R.F. 2006. Neuroendocrine mechanism of alternative reproductive tactics in fish. 24: 297–357. *In:* K. Sloman, S. Batshine and R. Wilson (eds.). *Fish Physiology, Behaviour and Physiology of Fish*. Elsevier, Amsterdam, The Netherlands.

Pandian, T.J. 2011. *Sex Determination in Fish*. Science Publishers, Enfield/CRC Press, Boca Raton, USA, p. 282.

Pandian, T.J. 2012. *Genetic Sex Differentiation in Fish*. CRC Press, Boca Raton, USA, p. 214.

Pandian, T.J. 2013. *Endocrine Sex Differentiation in Fish*. CRC Press, Boca Raton, USA, p. 303.

Patino, R., K.B. Davis, J.E. Schoore et al. 1996. Sex differentiation in channel catfish gonads. Normal development and effects of temperature. J Exp Zool, 276: 209–218.

Pavlidis, M., G. Koumoundouros, A. Sterioti et al. 2000. Evidence of temperature-dependent sex determination in the European sea bass (*Dicentrarchus labrax* L.). J Exp Zool, 287: 225–232.

Reddon, A.R. and P.L. Hurd. 2013. Water pH during early development influences sex ratio and male morph in a West African cichlid fish *Pelvicachromis pulcher*. Zoology, 116: 139–143.

Robertson, D.R., R. Reinboth and R.W. Bruce. 1982. Gonochorism, protogynous sex change and spawning in western Indian Ocean. Bull Mar Sci, 32: 868–879.

Santos, E.M., V.L. Wormann, G.C. Paull et al. 2007. Molecular basis of sex and reproductive status in breeding zebrafish. Physiol Genomics, 30: 111–122.

Seki, M., H. Yokota, M. Maeda et al. 2003. Effects of 4-nonylphenol and 4-tert-octylphenol on sex differentiation and vitellogenin induction in medaka (*Oryzias latipes*). Environ Toxicol Chem, 22: 1507–1516.

Shelton, W.L. and G.L. Jensen. 1979. Production of reproductively limited grass carp for biological control of aquatic weeds. WRRI Bull (Auburn Univ), 39: 1–74.

Taborsky, M. 1998. Sperm competition in fish bourgeois males and parasitic spawning. Trends Ecol Evol, 13: 222–227.

White, S.A., T. Nquyan and R.D. Fernald. 2002. Social regulation of gonadotropin releasing hormone. J Exp Biol, 205: 2567–2581.

Zhang, Z., Y. Wang, S. Wang et al. 2011. Transcriptome analysis of female and male *Xiphophorus maculatus* JP 163A. PLoS ONE, 6: e18379. doi: 10.1371/journal pone.0018379.

Author Index

A

B

I

J

K

L

M

N

O

P

R

S

Z

Species Index

D

E

F

G

H

Subject Index

Color Plate Section

Chapter 1

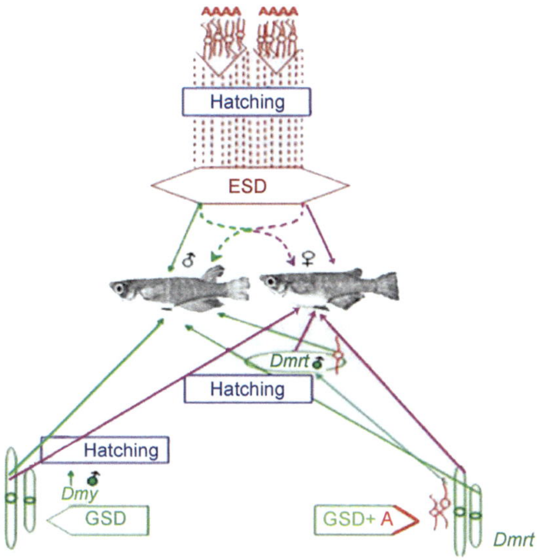

Fig. 1.1. Ontogenetic pathways of sex differentiation in fishes. Note the dimorphic sex differentiation is commenced before hatching in GSD but after hatching in GSD+A pathway. In GSD+A pathway, morphologically distinguishable sex chromosomes are identifiable and sex differentiation is altered by minor autosomal genes in some individuals alone. In the ESD pathway, one or other combinations of genes harbored on autosomes determine sex in all individuals but the genetic cascade of sex differentiation is mostly under the control of one or other environmental factor.

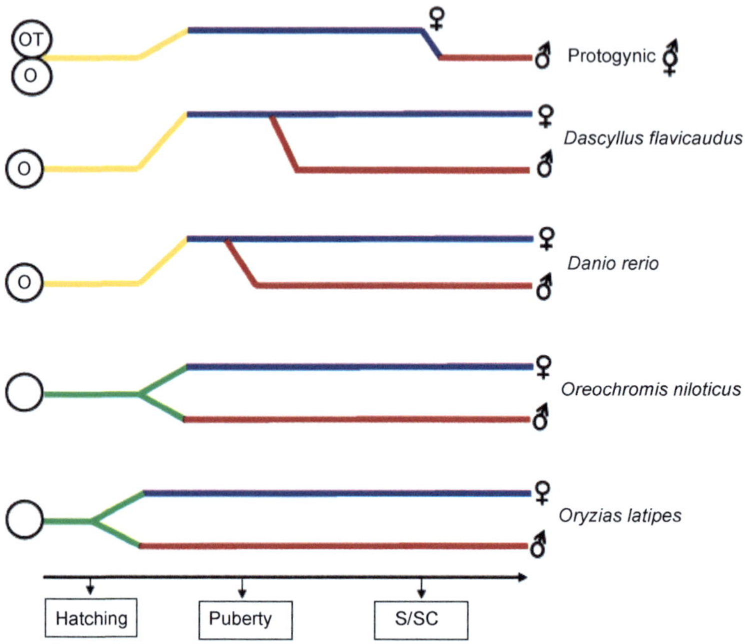

Fig. 1.2. Ontogenetic pathways of sex differentiation in selected fishes. O = Ovary, OT = Ovotestis, S/SC = Senesence/Sex change, Red = Female, Blue = Male, Green = Undifferentiated phase, Yellow = Differentiated but non-functional phase.

Chapter 2

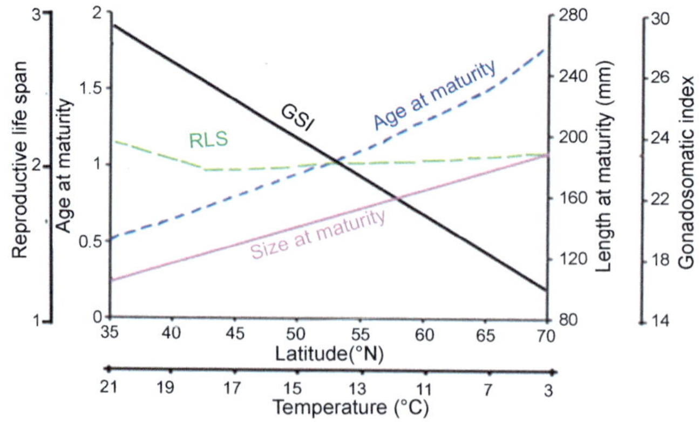

Fig. 2.2. Decreasing trends in body length and age at sexual maturity estimated for 75 populations of Eurasian perch *Perca fluviatilis* as function of latitude and corresponding temperature. Figure also shows the trends of GSI and RLS as functions of latitude and temperature (source: Heibo et al., 2005).

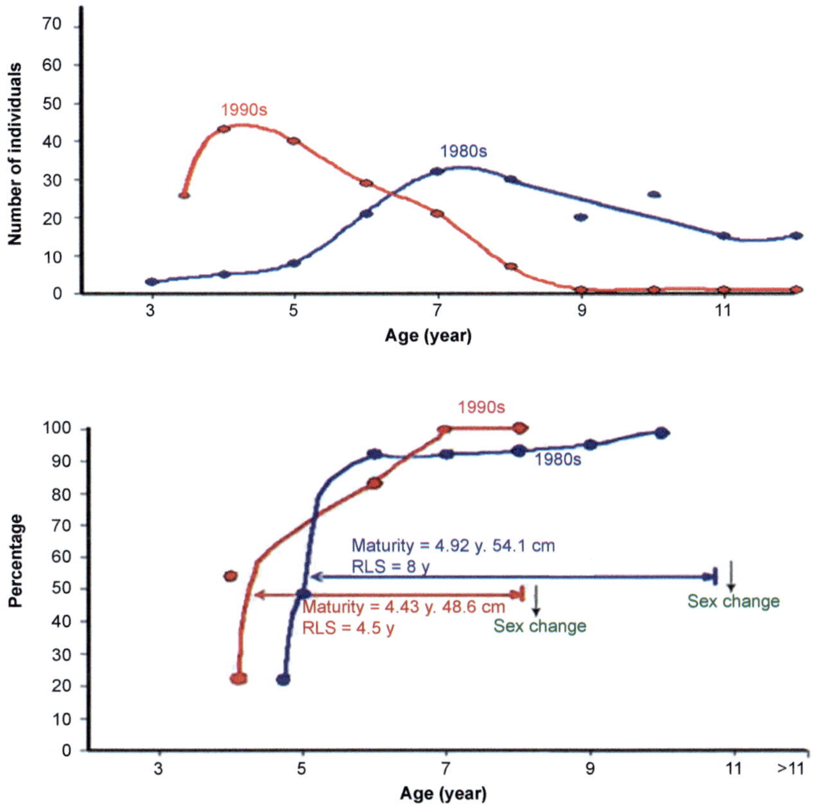

Fig. 2.17. Number (upper panel) and percentage (lower panel) of female snowy grouper *Epinephelus niveatus* collected off North Carolina during 1980s and 1990s. Note the precocious sexual maturity and sex change and decrease in Reproductive Life Span (RLS) in females of 1990s (compiled from Wyanski et al., 2000).

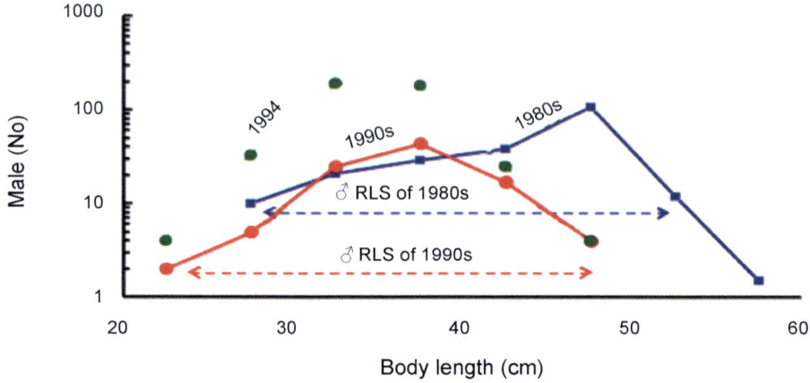

Fig. 2.18. Number of male red porgy *Pagrus pagrus* collected from North Carolina to Florida during 1980s and 1990s. Data are also shown for the number of males present after 1990s. Note the decrease in Reproductive Life Span (RLS) of males due to precocious sex change and capture of larger males (compiled from Harris and McGovern, 1997).

Chapter 4

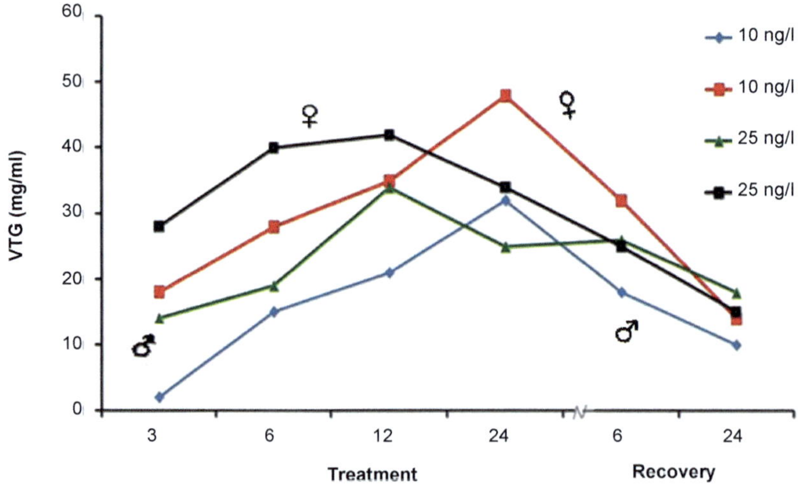

Fig. 4.6. Vitellogenesis following the exposure to 10 or 25 ng EE_2/1 and subsequent depuration as a function of time in *Danio rerio* (compiled from Van den Belt et al., 2002).

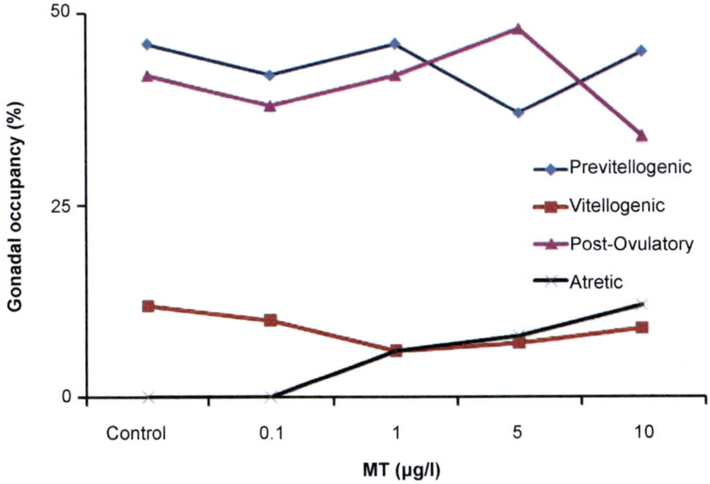

Fig. 4.7. Occupancy by oogenetic stages in the ovary of *Pimephales promelas* females exposed to different doses of MT for 21 days (compiled from Pawloski et al., 2004).

Chapter 7

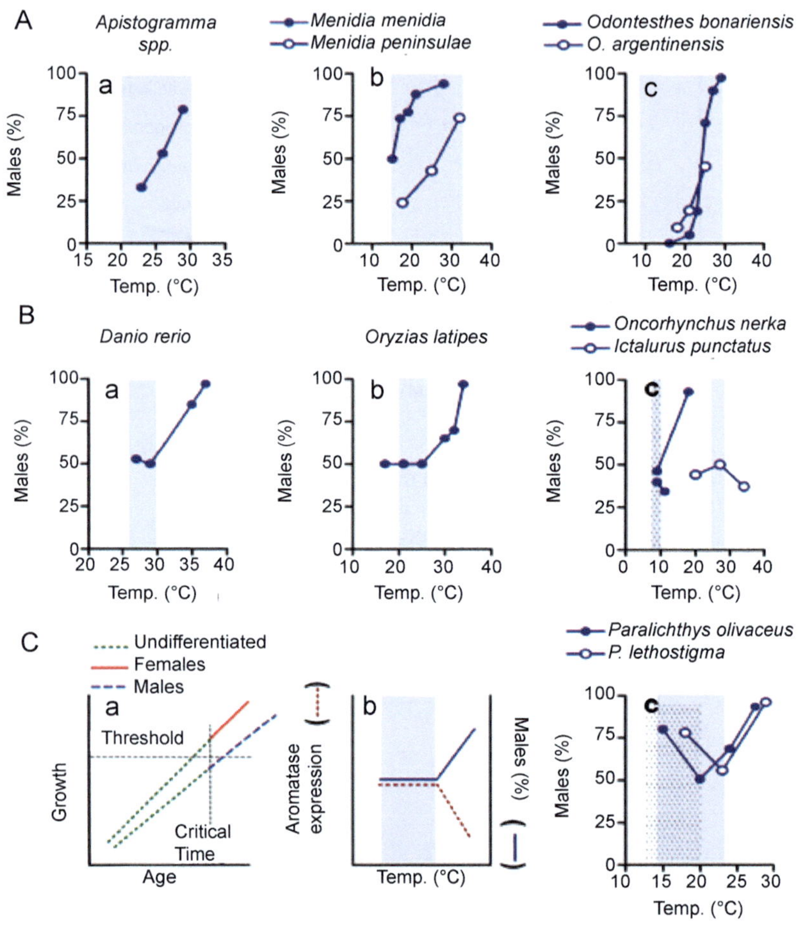

Fig. 7.2. *Patterns* of sex ratio response to temperature of fishes. A. Examples of *Pattern* 1 in which there are more males with increasing temperature. B. Examples for *TSD* in which sex ratio shifts only at extreme high temperatures and thus represent thermal effects on *GSD* (a, b). The fewer males at high temperatures, as in c is not supported by re-analysis of data. C. Formally proposed *Pattern* 3 in which more males occur at extreme low and high temperatures. However, these are proved to be wrong by subsequent publications (see Text) (permission by Dr. Piferrer, from Ospina-Alvarez, N. and F. Piferrer. 2008. Temperature-dependent sex determination in fish revisited. Prevalence, a single sex ratio response pattern and possible effects of climate change. PLoS One, 3(7): e2837. doi: 10.1371/journal.pone 0002837).

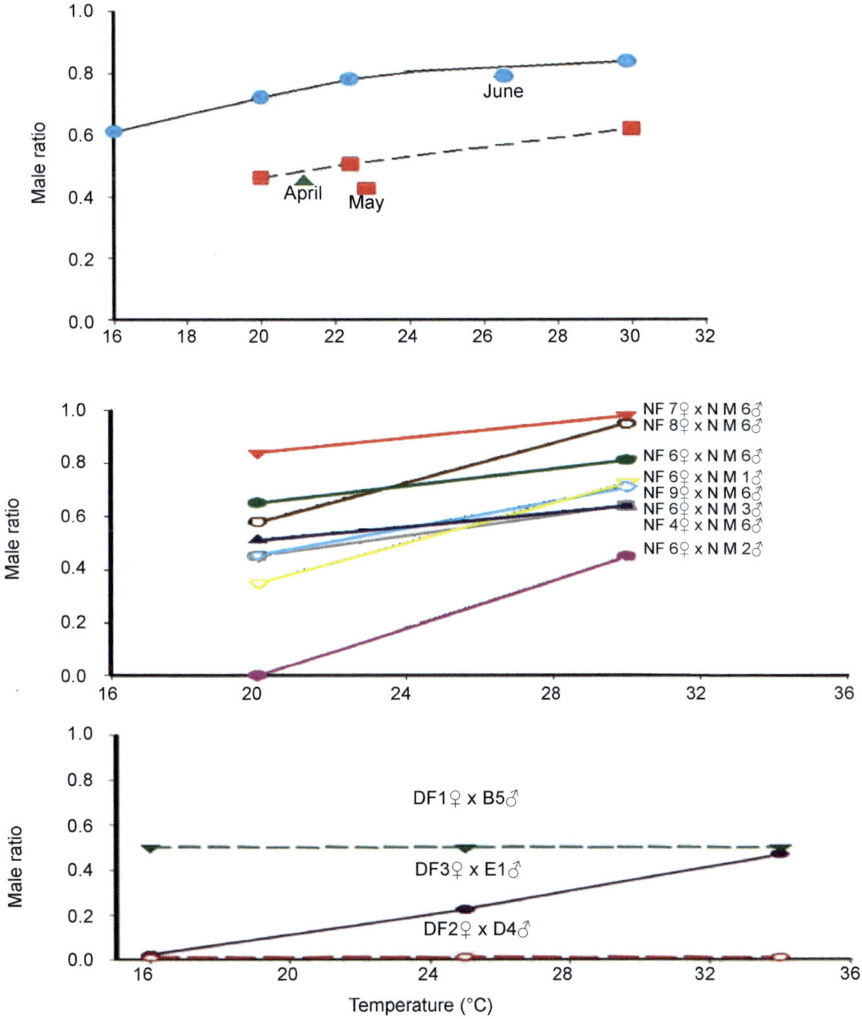

Fig. 7.5. Upper panel: male ratio of honmoroko *Gnathopogon caerulescens* in Lake Biwa during spring. Note the difference in male ratio response to increase in temperature by the progenies of early and late spring spawners. Middle panel: increasing trends in the male ratio of honmoroko with increase in temperature. Note the differences in the levels of responses by the progenies arising from different crosses. Lower panel: note that four of six crosses generated progenies that were insensitive to increase in temperature resulting in balanced sex ratio. One cross produced no male progenies, while the progenies of the sixth cross displayed increased male ratio with increasing temperature (compiled from Fujioka, 2006).

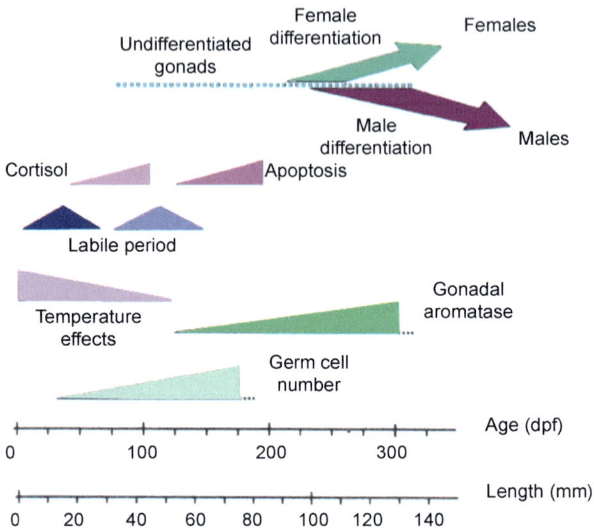

Fig. 7.10. The relationship between age and size on one hand and factors like germ cells gonadal aromatase and temperature on the other hand during the labile period of the European sea bass *Dicentrarchus labrax*. Note also the inclusion of locations of cortisol accumulation (Hattori et al., 2009, see also Fig. 7.7) and apoptosis (Uchida et al., 2004, see also Fig. 7.8) known for other fishes (source: Piferrer et al., 2005, modified).

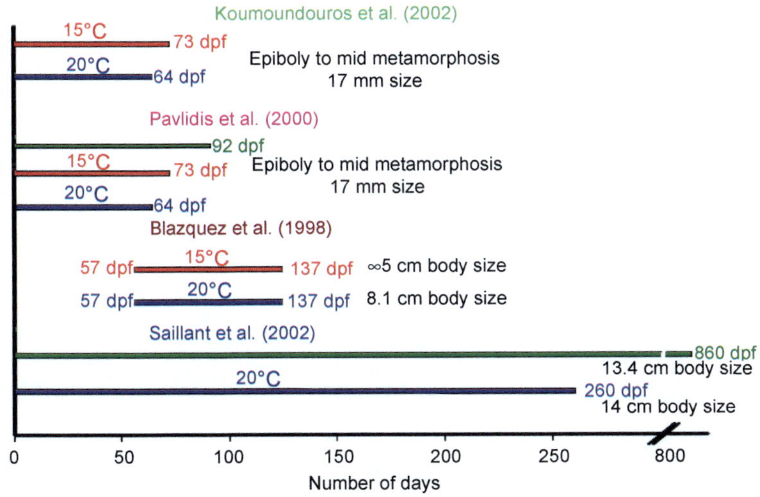

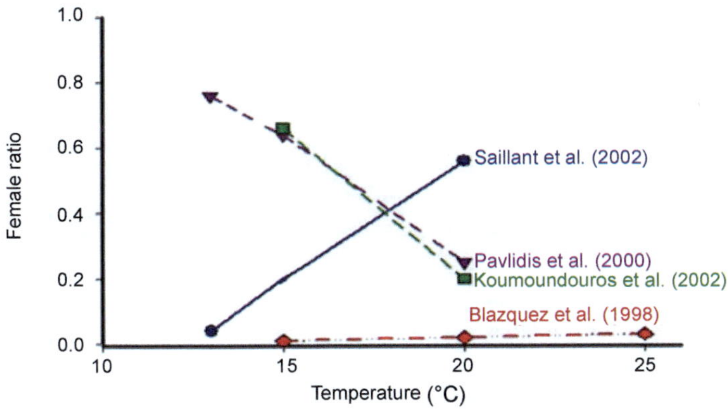

Fig. 7.11. Upper panel: labile period selected for temperature treatment of *Dicentrarchus labrax* by different European authors. Lower panel: Female ratio response to different temperatures. Note the completely different trends of female ratio response to temperature changes by the sea bass in relation to the difference in labile period shown above.

Chapter 8

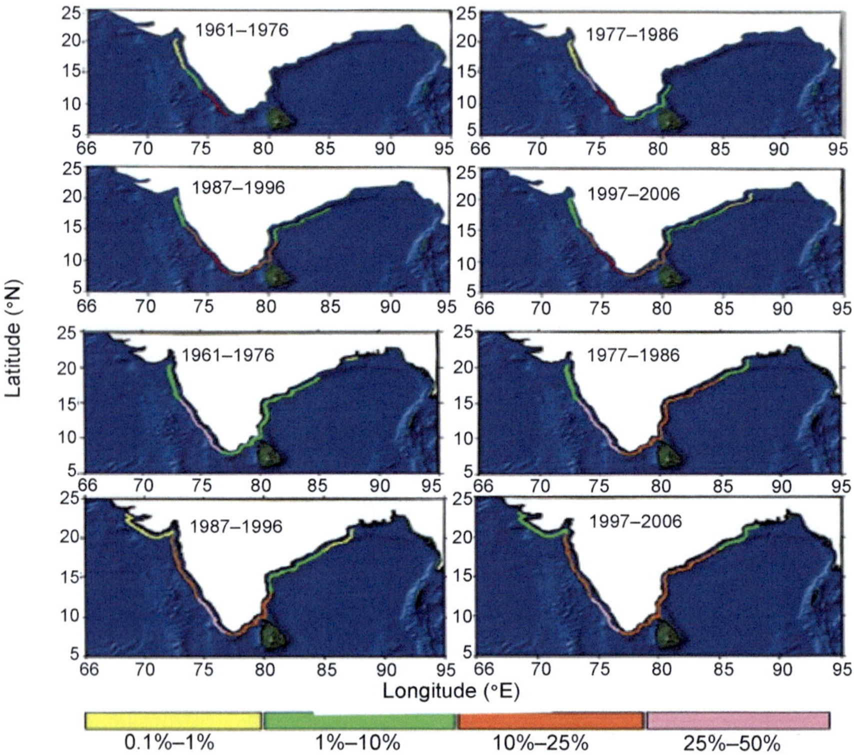

Fig. 8.3. Extension of distributional boundary of oil sardine *Sardinella longiceps* (upper 4 panels) and Indian mackerel *Rastrelliger kanagurta* (lower 4 panels) along the Indian west and east coasts during the period from 1961–1976 to 1997–2006. The colored lines indicate the percentage contribution of catch by different maritime states of India (from Vivekanandan, 2011, permission by Central Marine Fisheries Research Institute, Kochin).

Chapter 10

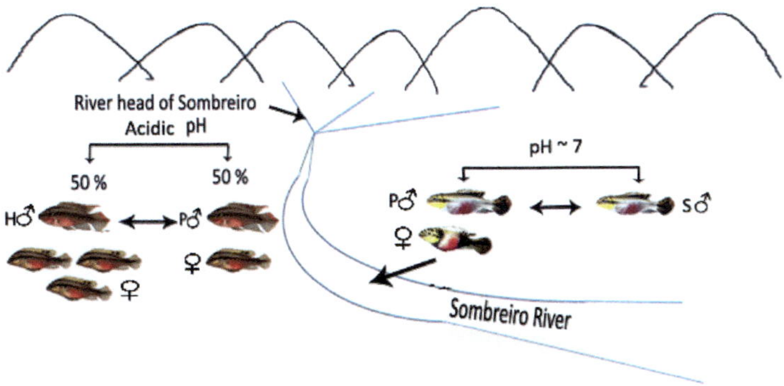

Fig. 10.1. The plastic reversible morphotypes within fixed red and yellow morphs of the Nigerian cichilid *Pelvicachromis pulcher*. The locations of occurrence of the red and yellow morphs in the Sombreiro River are indicated. Note the double headed arrows indicating the directions of plastic reversibility within red and yellow morphs. H♂ = Haremic male, P ♂ = Paired male, S♂ = Satellite male, ♀ = Female.

For Product Safety Concerns and Information please contact our EU representative GPSR@taylorandfrancis.com Taylor & Francis Verlag GmbH, Kaufingerstraße 24, 80331 München, Germany

Batch number: 10726532

Printed by Printforce, the Netherlands